ENERGY DELTA

Supply vs. Demand

AMERICAN ASTRONAUTICAL SOCIETY
Publications Office, Post Office Box 746—Tarzana, California
91356

AAS PUBLICATIONS

SCIENCE AND TECHNOLOGY

Vol. 1 MANNED SPACE RELIABILITY SYMPOSIUM
1965, 112 p. $10.00

Vol. 2 TOWARDS DEEPER SPACE PENETRATION
1965, 182 p. $10.00

Vol. 3 ORBITAL HODOGRAPH ANALYSIS
S. P. Altman, 1965, 150 p. $10.00

Vol. 4 SCIENTIFIC EXPERIMENTS FOR MANNED ORBITAL
FLIGHT 1965, 372 p. $16.00

Vol. 5 PHYSIOLOGICAL AND PERFORMANCE DETER-
MINANTS IN MANNED SPACE SYSTEMS 1965, 220 p. $12.00

Vol. 6 SPACE ELECTRONICS SYMPOSIUM
1965, 404 p. $16.00

Vol. 7 THEODORE VON KARMAN MEMORIAL SEMINAR
1966, 140 p. $10.00

Vol. 8 IMPACT OF SPACE EXPLORATION ON SOCIETY
1966 382 p. $16.00

Vol. 9 RECENT DEVELOPMENTS IN SPACE FLIGHT
MECHANICS 1966, 280 p. $14.00

Vol. 10 SPACE AGE IN FISCAL YEAR 2001
1967, 458 p. $18.00

Vol. 11 SPACE FLIGHT MECHANICS SYMPOSIUM
1967, 618 p. $20.00
Microfiche Suppl. $6.00

Vol. 12 THE MANAGEMENT OF AEROSPACE PROGRAMS
1967, 392 p. $16.00

Vol. 13 THE PHYSICS OF THE MOON
1967, 260 p. $14.00

Vol. 14 INTERPRETATION OF LUNAR PROBE DATA
1967, 270 p. $14.00

Vol. 15 FUTURE SPACE PROGRAM AND IMPACT ON RANGE
AND NETWORK DEVELOPMENT 1967, 583 p. $18.00

Vol. 16 THE VOYAGE TO THE PLANETS
1968, 184 p. $10.00

Vol. 17 USE OF SPACE SYSTEMS FOR PLANETARY
GEOLOGY AND GEOPHYSICS, 1968, 623 p. $20.00
Microfiche Suppl. $4.00

Vol. 18 TECHNOLOGY AND SOCIAL PROGRESS
1969, 170 p. $10.00

Vol. 19 EXOBIOLOGY—THE SEARCH FOR EXTRATER-
RESTRIAL LIFE 1969, 184 p. $10.00

Vol. 20 BIOENGINEERING AND CABIN ECOLOGY
1969, 162 p. $10.00

Vol. 21 REDUCING THE COST OF SPACE TRANSPORTATION
1969, 264 p. $14.00

Vol. 22 PLANNING CHALLENGES OF THE 70's IN THE
PUBLIC DOMAIN 1970, 504 p. $18.00
Microfiche Suppl. $12.00

Vol. 23 SPACE TECHNOLOGY AND EARTH PROBLEMS
1970, 418 p. $16.00
Microfiche Suppl. $12.00

Vol. 24 AEROSPACE RESEARCH AND DEVELOPMENT
1970, 500 p. $18.00

Vol. 25 GEOLOGICAL PROBLEMS IN LUNAR AND
PLANETARY RESEARCH 1971, 750 p. $25.00

Vol. 26 TECHNOLOGY UTILIZATION IDEAS FOR THE 70's
AND BEYOND 1971, 312 p. $20.00

Vol. 27 INTERNATIONAL COOPERATION IN SPACE
OPERATIONS AND EXPLORATION 1971, 194 p. . . $12.00

Vol. 28 ASTRONOMY FROM A SPACE PLATFORM
1972, 416 p. $20.00

Vol. 29 SPACE TECHNOLOGY TRANSFER TO COMMUNITY
AND INDUSTRY 1972, 196 p. (Microfiche only) $10.00

Vol. 30 SPACE SHUTTLE PAYLOADS
1973, 532 p. $25.00

Vol. 31 THE SECOND FIFTEEN YEARS IN SPACE
1973, 212 p. $15.00

Vol. 32 HEALTH CARE SYSTEMS
1974, 258 p. $15.00

Vol. 33 ORBITAL INTERNATIONAL LABORATORY
1974, 324 p. $20.00

ADVANCES IN THE ASTRONAUTICAL SCIENCES

Vol. 1 THIRD ANNUAL MEETING PROCEEDINGS
1957, 184 p. $15.00

Vol. 2 FOURTH ANNUAL MEETING PROCEEDINGS
1958, 440 p. $20.00

Vol. 3 WESTERN REGIONAL MEETING PROCEEDINGS
1958, 530 p. $20.00

Vol. 4 FIFTH ANNUAL MEETING PROCEEDINGS
1959, 462 p. $20.00

Vol. 5 SECOND WESTERN NATIONAL MEETING PRO-
CEEDINGS 1960, 364 p. $20.00

Vol. 6 SIXTH ANNUAL MEETING PROCEEDINGS
1961, 968 p. $25.00

Vol. 7 THIRD ANNUAL WEST COAST MEETING PRO-
CEEDINGS 1961, 464 p. $25.00

Vol. 8 SEVENTH ANNUAL MEETING PROCEEDINGS
1963, 602 p., (Microfiche only) $30.00

Vol. 9 FOURTH WESTERN MEETING PROCEEDINGS
1963, 910 p. $25.00

Vol. 10 MANNED LUNAR FLIGHT
1963, 310 p. $15.00

Vol. 11 EIGHTH ANNUAL MEETING PROCEEDINGS
1963, 808 p. $25.00

Vol. 12 SCIENTIFIC SATELLITES
1963, 262 p. $15.00

Vol. 13 NINTH ANNUAL MEETING: INTERPLANETARY
MISSIONS 1963, 690 p. $20.00

Vol. 14 PHYSICAL AND BIOLOGICAL PHENOMENA IN A
WEIGHTLESS STATE 1963, 382 p. $15.00

Vol. 15 EXPLORATION OF MARS
1963, 634 p. $20.00

Vol. 16 SPACE RENDEZVOUS, RESCUE, AND RECOVERY
1963; Part I, 1028 p. $25.00
Part II, 380 p. $15.00

Vol. 17 BIOASTRONAUTICS—FUNDAMENTAL AND
PRACTICAL PROBLEMS 1964, 128 p. $10.00

Vol. 18 LUNAR FLIGHT PROGRAMS
1964, 630 p. $20.00

Vol. 19 UNMANNED EXPLORATION OF THE SOLAR SYSTEM
1965, 1000 p. $25.00

Vol. 20 POST APOLLO SPACE EXPLORATION
1966; Part I, 572 p. $20.00
Part II, 648 p. $20.00

Vol. 21 PRACTICAL SPACE APPLICATIONS
1967, 508 p. $20.00

Vol. 22 THE SEARCH FOR EXTRATERRESTRIAL LIFE
1967, 388 p. $18.00
Microfiche Suppl. $3.00

Vol. 23 COMMERCIAL UTILIZATION OF SPACE
1968, 512 p., plus 24 microfiches $25.00

Vol. 24 EXPLORATION OF SPACE
1968, 363 p. $16.00

Vol. 25 ADVANCED SPACE EXPERIMENTS
1969, 530 p. $20.00

Vol. 26 PLANNING CHALLENGES OF THE 70's IN SPACE
1970, 470 p. $18.00
Microfiche Suppl. $10.00

Vol. 27 SPACE STATIONS
1970, 606 p. $20.00

Vol. 28 SPACE SHUTTLES AND INTERPLANETARY MISSIONS
1970, 488 p. $18.00

Vol. 29 THE OUTER SOLAR SYSTEM
1971; Part I, 618 p. $20.00
Part II, 740 p. $20.00

Vol. 30 INTERNATIONAL CONGRESS OF SPACE BENEFITS
1974, 528 p. $25.00

SPECIAL VOLUMES

1. WEIGHTLESSNESS—PHYSICAL PHENOMENA AND
BIOLOGICAL EFFECTS 1961, 182 p. $15.00

2. LUNAR EXPLORATION AND SPACECRAFT SYSTEMS
1962, 214 p. $15.00

AAS President
J. Ray Gilmer   Neuro Systems Incorporated

Vice President - Publications
Dr. Francis E. Fendell   TRW Systems

Series Editor
Dr. Horace Jacobs   Lockheed-California Company

Editors
Dr. George W. Morgenthaler
Dr. Aaron N. Silver   Martin-Marietta Corporation

Art Staff
J. C. Speas   Lockheed-California Company

Thanks are due Diana Will, Lockheed-California Company,
for final preparation of the manuscript for the printer.

French Solar Furnace in Pyrenees

ENERGY DELTA
SUPPLY vs. DEMAND

VOLUME 35, SCIENCE AND TECHNOLOGY

A Supplement to Advances in the Astronautical Sciences

Edited by
George W. Morgenthaler
Aaron N. Silver

Proceedings of Energy Symposium of the 140th
Annual Meeting of the American Association
for the Advancement of Science

Co-sponsored by:

- American Astronautical Society
- American Sociological Association
- Institute of Electrical and Electronic Engineers
- Operations Research Society of America

February 25 - 27, 1974, San Francisco, California

AAS PUBLICATIONS OFFICE, P.O. Box 746, Tarzana, California 91356

PREFACE

THE ENERGY DEFICIT

In an era of inexpensive and prevalent fossil energy sources, the world, and particularly the United States, developed a style of life that exponentially increased annual per capita energy consumption. This trend, coupled with the steady growth of population and the slower-than-anticipated growth of nuclear energy sources, has so taxed fossil fuel supplies that the "Delta" of supply minus demand appears headed for serious deficit during the next decades. A disruptive effect on world economic balance, military balance of power, and the standard of living is foreseeable. Developing nations, in particular, face a crisis situation, being unable to afford the higher prices for the ever more scarce but vital energy supplies.

Society must enter a new era based upon adequate and environmentally sound energy sources and wiser energy uses. For the long range future of mankind means of limiting population growth and curbing per capita human energy consumption must be considered. In the near term, however, active management of the energy "Delta" is absolutely <u>necessary</u> to prevent disruption of modern society. Accordingly, this important symposium heard from government, university, and industry experts in four areas: 1) forecasts of the magnitude of the energy "Delta"; 2) reducing the energy demand by more efficient means of transportation, more efficient industrial processes, and changes in personal life style; 3) increasing the energy supply by developing new energy sources and by expanding old sources; and 4) exploring alternative energy economic and R&D policies for near-term and long-term alleviation of the energy deficit.

A JOINT NATIONAL ENERGY SYMPOSIUM

The Symposium, held at the San Francisco Hilton Hotel, February 25-27, 1974 was sponsored by the American Association for the Advancement of Science, and co-sponsored by the Institute of Electrical and Electronic Engineers, the American Sociological Association, the Operations Research Society of America, and the American Astronautical Society. It was part of the 140th Annual

Meeting of the AAAS and was interdisciplinary and truly national in scope, speakers, and attendees.

The program, which cut across broad economic, technical and social issues, was arranged by Dr. George W. Morgenthaler, Corporate Director of Research and Development, Martin Marietta Corporation. Dr. Morgenthaler and Mr. Robert L. Gervais of McDonnell Douglas Astronautics were Symposium Co-Chairmen.

Specifically, Session I, chaired by Dr. S. William Gouse, Director, Energy Research and Development, U.S. Department of Interior, Washington, D.C., developed the overall perspective in terms of national energy policy alternatives and the structuring of viable energy strategies. The papers presented in this session included, "Energy Research and Development Alternatives for Future Supply" by Dr. Chauncey Starr, President, Electric Power Research Institute, Palo Alto, California; "Energy Supply and Demand Challenges and Some Possible Solutions" by Dr. Mr. Ray Thomasson; and "Reducing the Demand - A Viable Strategy" by Dr. Lester Lees, Director, Environmental Quality Laboratory, California Institute of Technology.

Session II, III, and IV focused upon <u>reducing energy demand</u> in the areas of transportation, industrial processes, and personal life styles. Some typical papers presented in these areas were "Evaluation and Effects of Energy Conservation in the Transportation Sector" by Dr. Richard L. Strombotne, Office of the Secretary, U.S. Department of Transportation, Washington, D.C.; "Energy Saving Developments in Metal Processing" by Dr. Alan S. Russell, Director of Research and Development, Alcoa; and "Energy Flow and Culture Change" by Prof. Wm. B. Kemp of McGill University, Montreal, Canada.

Session V through X emphasized <u>increasing the energy supply</u>. The areas covered were petroleum and natural gas, coal derived fuels, nuclear and fusion sources, solar sources, geothermal sources, and advanced conceptual techniques. Representative of the papers presented in this field were "The Outlook for Natural Gas Supplies" by Mr. Leonard W. Fish, American Gas Association; "The Environmental Impact of Large Scale Coal and Shale Extracted" by Mr. Joseph Brennan, Natural Coal Association; "The Development of the Nuclear Electric Energy Economy" by Mr. Philip N. Ross, Westinghouse Corporation; "The Solar Heating and Cooling of Buildings" by Dr. Roger D. Bourke, Jet Propulsion Laboratories, California Institute of Technology; "Current Worldwide Utilization and Ultimate Potential of Geothermal Energy Systems" by Mr. P. Muffler,

United States Geological Survey; and "The Hydrogen Economy" by Dr. A. P. Gregory, Director of Energy Systems, The Institute of Gas Technology.

Session XI dealt with the <u>economics of energy issues</u>, and included a paper by Dr. H. S. Houthakker and Mr. M. Kennedy entitled "Demand for Energy as a Function of Price"; "The Supply of Fuels as a Function of Price" by Mr. S. Schurr and Mr. M. Searl of Resources for the Future; and "The Energy Game: Fuels and Uncertainty" by Dr. Arnold Packer, Committee for Economic Development, Washington, D. C.

<u>MEETING ACCOMPLISHMENTS; ACKNOWLEDGMENTS</u>

The meeting was attended by over 400 scientists, engineers, industrial planners, students, and press and many oral and written comments were received stating that the perspective and exchanges of the Symposium were very valuable and would help attendees in planning effective personal contributions in the energy area.

The American Association for the Advancement of Science and the other sponsoring societies are very grateful to the Session Chairmen and to the speakers for their excellent work and for sharing their information.

The undersigned and the sponsoring societies also give special thanks to the Symposium Committee: Dr. Arnold Packer of Committee on Economic Development representing ORSA: Mr. Roger W. Sampson of McDonnell Douglas, representing IEEE; Prof. Samuel Z. Klausner, University of Pennsylvania, representing the American Sociological Association; Dr. Glen Werth of Lawrence Radiation Laboratory; and Martin Meyerson of Martin Marietta Corporation. Thanks are also given to Dr. Aaron Silver of Martin Marietta Corporation for his excellent staff assistance in interfacing members of the program, and to Mrs. Marie Heidbreder, secretary to Dr. Morgenthaler, for her unfailing efforts in following up on the many details attending the development of such a large Symposium Program.

The undersigned hope that the reader of this Proceedings will be stimulated by the challenge of the energy field herein portrayed and will be encouraged by the alternatives for increased supply and reduced demand suggested at the meeting.

Finally, Session XII consisted of a panel discussion entitled "Recommended Strategies for Minimizing the Energy 'Crunch': 1975 - 1990, and Beyond

1990" and included Dr. Edward Teller and other distinguished members of government, industry, and academia. This panel session was sparked by questions and statements by activist student groups who felt that the Establishment was not spending enough on R&D and not funding the most viable energy R&D alternatives. The panel seemed to arrive at a consensus regarding the best alternatives for present support.

George W. Morgenthaler
Corporate Director of Research &
  Development
Martin Marietta Corporation
Meeting Arranger and Co-Chairman

Robert L. Gervais
Project Manager
Douglas Missile & Space Systems Div.
McDonnell Douglas Astronautics Company
Huntington Beach, California
Meeting Co-Chairman

CONTENTS

SESSION I

THE PERSPECTIVE

Chairman: S. William Gouse
           Director, Energy, R&D, U.S. Dept.
           of Interior, Washington, D.C.

ENERGY SUPPLY AND DEMAND CHALLENGES

AND SOME POSSIBLE SOLUTIONS*

M. Ray Thomasson**

Shell's most probable estimate of U.S. domestic
energy supply during the years 1975 - 1990 falls
substantially below the demand we forecast last
year for this period. The gap can only be filled
by petroleum imports. However it now appears
unlikely that sufficient supply will be available
on international markets. Therefore, we have gone
back to the drawing board to estimate how the U.S.
might reduce energy demand without causing major
economic disruption. We also took an optimistic
look at energy supply. Our analysis shows that
with major commitment to energy efficiency and
conservation combined with a massive national
drive to achieve energy self sufficing the U.S.
could become essentially independent of energy
imports by 1990.

Thank you very much, Mr. Chairman. Most of us today are very

concerned about the energy situation in the United States. The energy

shortage is one of the most important and far-reaching problems facing

our nation today. It has already affected each of us and will certainly

continue to affect us in the future. The security of our jobs, our

comfort, our mobility, and our life-style in general are in jeopardy.

It is of utmost importance to us all to communicate with each other and

to strive together for understanding and help. We must work together to

assure our future and, to do that, we all need a better understanding of

our energy problems and the alternatives available to us.

* Text of presentation, AAAS Energy Symposium, San Francisco,
   February 25, 1974.
**Manager, Planning and Economics Forecasting Group, Shell Oil
   Company, Houston, Texas.

The materials I will show you today are those we use in
our own internal planning at Shell. In order to present you a realistic
appraisal of energy problems and possible solutions I would like to (1)
review the U.S. energy supply and demand forecasts that we in Shell Oil
Company saw last year as most probable for the period to 1990, (2) examine
the world supply and demand in this period of rapid changes in terms of
prices and supply in the international scene, and (3) present my views on
how the U.S. might achieve energy self-sufficiency.

DEMAND

Let me begin with the U.S. energy forecast which we prepared
early last year. Our forecast was based on the set of premises shown
in Fig. 1. We suggested that the population would grow at a rate of 1
percent per year, lower than at any time in the last 40 years. We also
assumed that there would be no major change in our life-styles, a premise
I suggest we must re-examine later in this talk. We also premised that a
national energy policy would evolve and would be directed by a central
agency which would (1) promote the development of our domestic energy
resources, (2) strive for a balance between our environmental and energy
needs, (3) develop a national land use policy which would cater to the
siting of energy facilities, (4) permit the deregulation of the price of
new natural gas, (5) promote energy conservation and encourage efficiency,
and (6) maintain imports at the lowest possible level.

Regarding economic activity, we premised continued economic
growth and 3 percent inflation versus our current view that it could be

much higher. We saw the gross national product growing at 5.7 percent per year through 1975, then growing at 4.2 percent and declining to 3.8 percent after 1980. We had also taken into account foreseeable technological innovation, mainly commercial stack gas scrubbing by 1977. Perhaps, the most important premise we made concerned the availability of imports. In order to first understand unconstrained demands, we assumed that the U.S. would have access to all the imported crude oil we required and to substantial imports of refined products. We did not predict the latest Middle East war. However, in my opinion, the long-range energy situation has not been changed materially by this event. We had a serious energy problem before the current embargo began, and we will have it after the embargo is lifted.

The post-World War II "baby boom" generation is now the 15-25 year age group. Between 1970 and 1990 they will increase the number of households from 63 to 90 million, and markedly add to presently increasing demand for the use of energy. Fig. 2 depicts our per capita income and energy consumption based upon 1970 data. It is but one way of showing our economic strength, as compared with other countries, and our high degree of dependence upon energy. If that economic strength is to be maintained during the working life of that post-war generation, total U.S. energy consumption must increase.

However, we will see that the increase is unlikely to be at the pace of past energy growth. Our energy consumption more than doubled in the last 20 years. We think, as shown in Fig. 3, that consumption without constraint would double again between 1970 and 1990, increasing from the

SUMMARY

1973 ENERGY FORECAST PREMISES

POPULATION
- SERIES E PROJECTION, 1% AAI, 205 MILLION IN 1970 TO 251 MILLION IN 1990.

POLITICAL AND SOCIAL
- NO MAJOR CHANGE IN LIFE STYLE, POLITICAL SYSTEM OR TAXATION POLICY.
- A NATIONAL ENERGY POLICY WILL EVOLVE DIRECTED BY A CENTRAL AGENCY WHICH WILL:
  - FAVOR DEVELOPMENT OF DOMESTIC RESOURCES
  - BALANCE ENVIRONMENTAL AND ENERGY NEEDS
  - DEVELOP NATIONAL LAND USE POLICY
  - ALLOW NEW GAS PRICE DEREGULATION
  - CONSERVE ENERGY AND ENCOURAGE EFFICIENCY
  - MAINTAIN IMPORTS AT LOWEST POSSIBLE LEVEL

ECONOMIC
- CONTINUED ECONOMIC GROWTH.
  - INFLATION HELD AT 3% AAI
  - GNP DECLINING FROM 5.7% ('71-'75) TO 3.8% ('80-'90) AAI.

TECHNOLOGICAL
- STACK GAS SCRUBBING WILL BE COMMERCIAL BY 1977.

IMPORTS
- ENERGY SUPPLY WILL RISE TO MEET THE DEMAND.

Figure 1

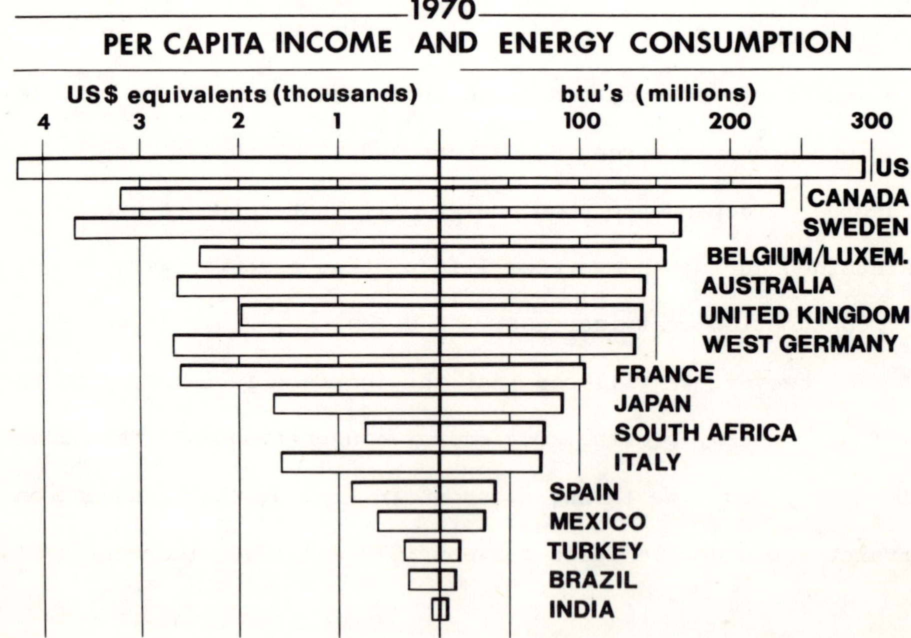

Figure 2

equivalent of 31.8 billion barrels of crude daily to 67 million barrels.
Even in this "unconstrained" forecast, the annual average growth rate
declines from 4.1 percent in the 1960's to 2.5 percent by the late
1980's.

The nation's energy is primarily used by five major markets:
transportation, industrial, residential, commercial, and electricity
generation. Of these markets, electricity generation, transportation,
commercial, and (within industry) chemical, grow faster than total energy
demand. Other markets grow more slowly.

In the transportation market, motor gasoline plays a dominant
role even though battery and fuel cell cars are estimated to be 2 million
by 1985 and 6 million by 1990. The prospect is that this will continue to
be a highly mobile society, and therefore, transportation will continue to
be a major energy consumer. Fuel consumption per mile is increasing
significantly each year. Emission control and safety devices fitted to
new automobiles will decrease average miles per gallon. This trend must
and can be reversed. The whole transportation sector is a market in
which major savings are possible. We will discuss this again later.

Several factors--particularly population and disposable income--
influence the demand for energy in the residential market. It is estimated
that this demand will increase from the equivalent of 5 million barrels in
1970 to 6.7 million barrels daily in 1990. We calculated a slower growth
in this market than in the past as the result of (1) better heat insulation
in new houses and (2) the trend toward smaller homes, mobile homes, and
multiple family dwellings--all of which will reduce requirements for space
heating and cooling.

U.S. ENERGY DEMAND

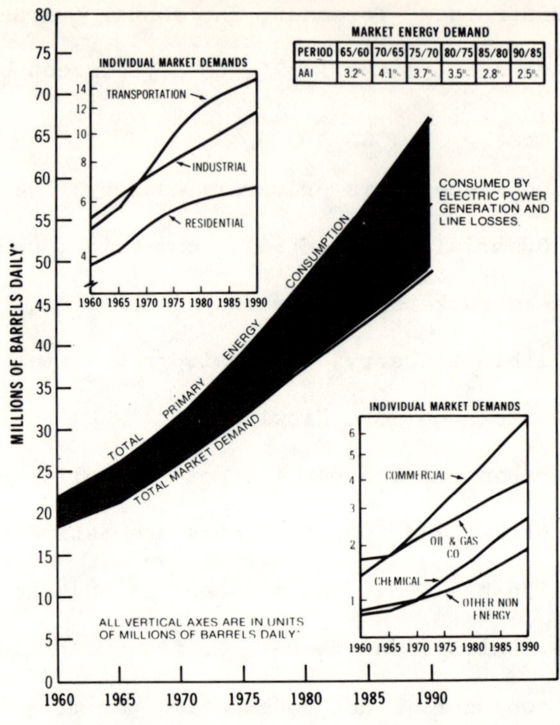

Figure 3

U.S. COAL SUPPLY

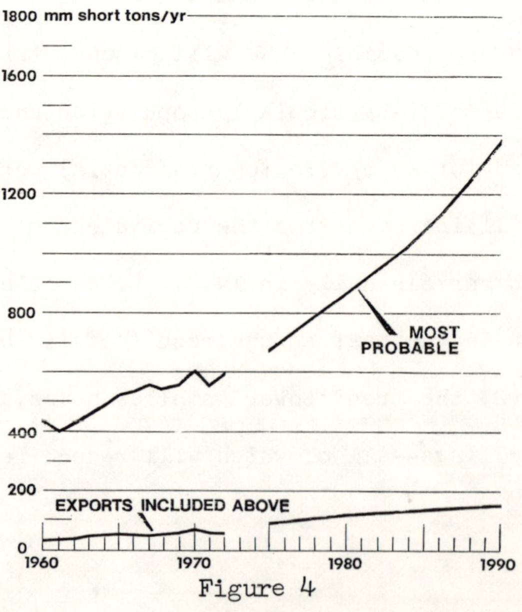

Figure 4

The commercial market includes stores, office buildings, schools, hospitals, and government buildings. Consumption of energy is directly affected by the level of business activity and the demand for public services. It is estimated to increase at more than 5 percent annually and in volume to about 7 million barrels daily by 1990. In both the residential and the commercial markets, gas will be the main supply source throughout the period but electricity will play a growing role.

If we break out chemical and allied products, other industry will show only a modest growth in energy consumption. The level of industry consumption will be moderated by improved efficiency in energy use. This trend is likely to be further stimulated by rising energy costs. For example, Shell Oil has committed itself to achieving a 10 percent reduction in energy use in refineries by the end of 1974.

The energy requirements of the electric utilities during the period to 1990 will be met from various sources: nuclear, coal, natural gas, oil and hydroelectric power. The use of natural gas is forecast to decline because of supply shortage, and hydroelectric power will show only modest growth because of lack of suitable sites. Short-term, oil will replace gas in the fuel supply, and coal will increase its role once stack gas scrubbing is commercial about 1977. But in the long term, nuclear power will be the fastest growing source of fuel for utilities and could be as much as 58 percent of total supply by 1990.

SUPPLY

Now, let us consider the domestic reserves we have available for meeting our energy demand. These are economically recoverable reserves at

1972 price levels and current technology. Coal is, by far, our most abundant fossil fuel. Proven reserves are estimated at 150 billion tons, which
represents 71 percent of our total energy reserves. Next, shale oil accounts
for 10 percent of our total reserves, followed by natural gas at 7 percent,
crude oil and natural gas liquids at 6 percent, and uranium at 6 percent.
Late in the forecast, the commercial development of the breeder reactor is
expected to make our entire supply of U_3O_8 fissionable and thus increase our
uranium reserves by a factor of 140.

Nuclear energy is the long-term key to the U.S. energy future.
However, because of the long lead times and environmental concern about
safety and pollution, nuclear energy will not develop as rapidly as
originally expected during this decade. We forecast an installed capacity
of 400,000 megawatts by 1990 which would represent about 19 percent of our
total energy needs. The problems of manpower and material resource requirements, and the economic and social problems associated with such a nuclear
energy program are enormous. It will require a massive effort to achieve
this goal.

Turning our attention now to coal, we see in Fig. 4 that although
coal is our most abundant fossil fuel, its growth has been inhibited because
of cheaper alternate fuels, increasingly severe air quality standards,
other environmental problems associated with strip mining, mine safety and
labor problems, as well as the prospect of nuclear energy rendering coal
obsolete. However, the trend must change. Our forecast projects coal
production more than doubling from the current level of 600 million tons
per year to 1400 million tons per year in 1990. Strip-mined western low

sulfur coal should be a major supply source in the future. Development
of this resource is contingent on reconciling environmental objections
to strip mining which today threaten its growth. Certainly, it is
imperative for the disturbed land area to be reclaimed. In a relative
sense, this will be quite inexpensive. It will amount to only between 1
and 2 percent of the cost of mining and transporting coal in 1980.
Furthermore, vast areas are not involved. If the coal strip mining
necessary to meet this forecast between now and 1990 were confined to
a single mine, an area about 10 miles by 20 miles would be necessary.

Now let us turn our attention to oil and gas. Fig. 5 shows a
dramatic moment in our exploration history. Starting in 1967, we produced
more gas than we found. This trend appears irreversible, and there was
little doubt that soon after this occurred production would peak in this
country. The prediction, made by Dr. M. King Hubbert in 1956, that U.S. production of conventional crude oil and natural gas would peak in the late 1960's,
finally occurred in the early 1970's. Because oil and gas are linked so
closely together Fig. 6 shows the same picture for oil.

There has been a downward trend in the total number of wells
drilled annually in the United States since 1956. In 1955 there were
40,000 small independents looking for, and developing oil and gas fields.
Today that group has dwindled to about 3,500, or less than 10 percent of
the figure 18 years ago. Historically, these independents have found over
60 percent of the new discoveries.

Natural gas currently provides about one-third of our energy
needs. There are two reasons for its importance: first, it is an

GAS

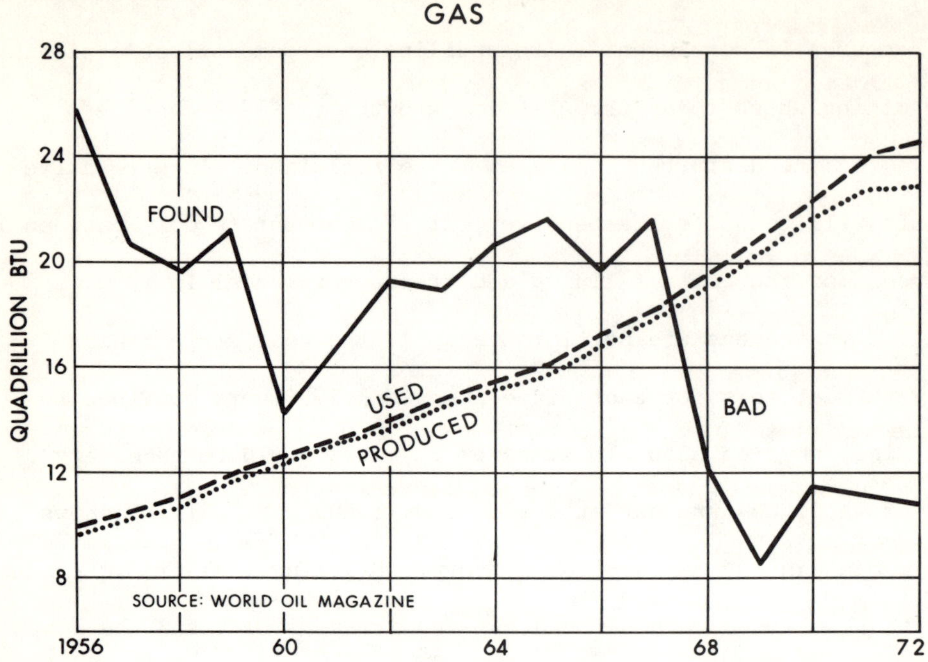

Figure 5

OIL

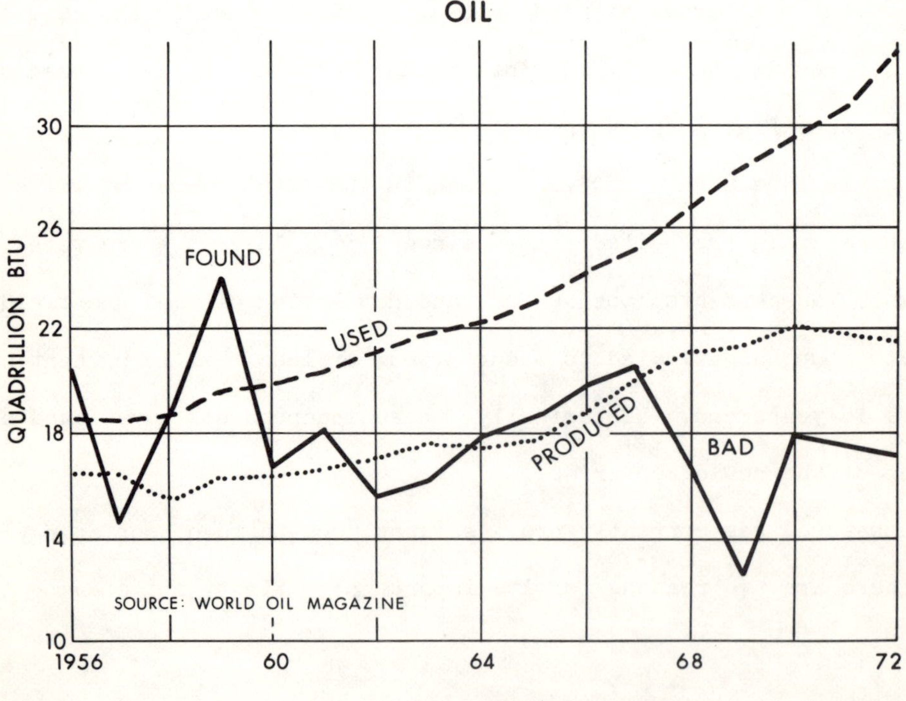

Figure 6

extremely clean fuel and secondly, it has been a relatively inexpensive
one. The Federal Power Commission has regulated the price of natural gas
since 1954 at an artificially low level which has served to both stimulate
demand and at the same time discourage exploration for new supplies. As
a result, demand grew rapidly during the 1960's at a rate of 7 percent per
year. Industrial users and electric utilities who use natural gas mainly
as a boiler fuel were chiefly responsible for this rapid growth. Fig. 7
shows that we believe gas production peaked in the 1972-1973 period and is
now on the way down, even though we assume substantial future exploration
success. The dotted line shows the projected decline in production without
future exploration success. This is, of course, theoretical as there will
almost certainly be some new discoveries. Nevertheless, about 1/2 of the
supply in our forecast will come from sources still to be found.

The decline in production will be offset somewhat in the late
1970's by gas from the U.S. Arctic, liquefied natural gas imports, and gas
derived from the gasification of coal. However, supply from these supplemental sources will be more costly, at least 4 times the current 25¢ per
MCF price of natural gas.

Last year we anticipated large amounts of Canada imports and
also large synthetic gas production from oil. These are not consistent
with the energy self-sufficiency policies evolving for both the U.S. and
Canada. The conversion of oil to synthetic natural gas is an inefficient
process resulting in considerable net energy loss and is only a short term
stop-gap measure to ease the gas supply situation. The cost of gasifying
naphtha is estimated at $1.50 to $2.00 per thousand cubic feet.

Imports of liquefied natural gas will be limited by the high
capital costs required for gasification plants, cryogenic tankers,
storage tanks, and re-gasification plants. The cost of LNG delivered
to the East Coast is estimated to be at least $1.00 per thousand cubic
feet.

The gasification of coal is expected to play an important role in
our future gas supply since it utilizes our most abundant fossil fuel.
However, as with other supplemental sources, gas derived from coal will
be expensive, costing at least $1.00 per thousand cubic feet. Taking into
account all the sources of natural gas, our most probable estimate is that
the current level of supply can barely be maintained.

Fig. 8 shows our forecast of U.S. conventional petroleum supply.
The demand for petroleum is currently about 17 million barrels per day and
is expected to grow rapidly in the future. This rapid growth results from
a reduction in natural gas supplies, environmental problems associated
with coal, and delays in nuclear power plant construction. These factors
have created an energy gap which can only be filled by oil. We have been
operating in this country under the assumption that foreign countries would
increase their oil production so that the United States could continue to
expand energy consumption at historic rates. Now that the danger of too
great a reliance on foreign sources of oil has been demonstrated dramatically,
the need for rapid development of our domestic energy resources should be
evident to everyone.

In our most probable case, we have assumed a reasonably significant
dedication to developing energy self-sufficiency with some constraints

USA DOMESTIC CONVENTIONAL NATURAL GAS PRODUCTION*

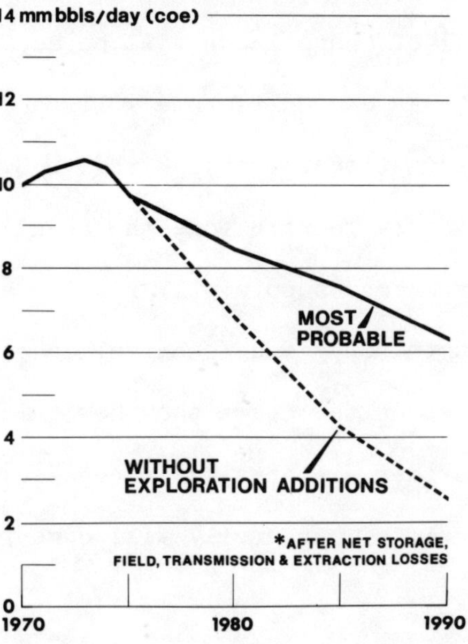

Figure 7

USA DOMESTIC CONVENTIONAL CRUDE & NATURAL GAS LIQUIDS PRODUCTION

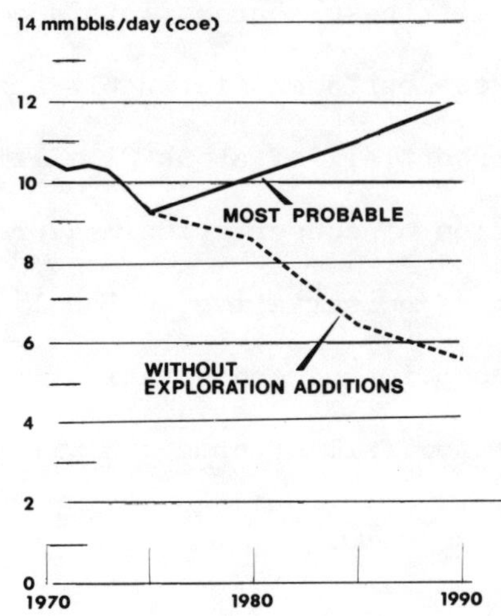

Figure 8

13

remaining in the supply of human and material resources available for the
job that might be removed by a stronger commitment. We have also assumed
that offerings of Federal offshore lands will be accelerated and that naval
petroleum reserves in Alaska and California will be developed and produced.

Under these assumptions, we think that U.S. oil production will
continue to decline for a few years before the trend is arrested and
reversed. Expanding our energy supply will require a long lead time and
massive investments of both money and technical manpower.

The dotted lines on the chart show how quickly our production
would decline without further exploration successes. We have estimated
that by 1990 more than half of our supply will come from sources still
to be found.

Fig. 9 represents what we consider to be the most probable energy
supply case, reflecting all energy sources. The top line, depicting demand
without conservation, rises from 39 million barrels per day crude oil
equivalent in 1973 to 48 million barrels per day in 1980 and 67 million
barrels in 1990. Taking all these sources into account, there will still
be a significant gap between our domestic supply and demand. Our most
probable case foresees a shortfall of about 21 million barrels per day
in 1990 without the adoption of conservation measures. The dotted line
shows the possible effect of conservation, which I will describe in more
detail later. With the adoption of conservation measures, our most
probable case foresees a shortfall of about 13 million barrels per day
in the period 1980 - 1990.

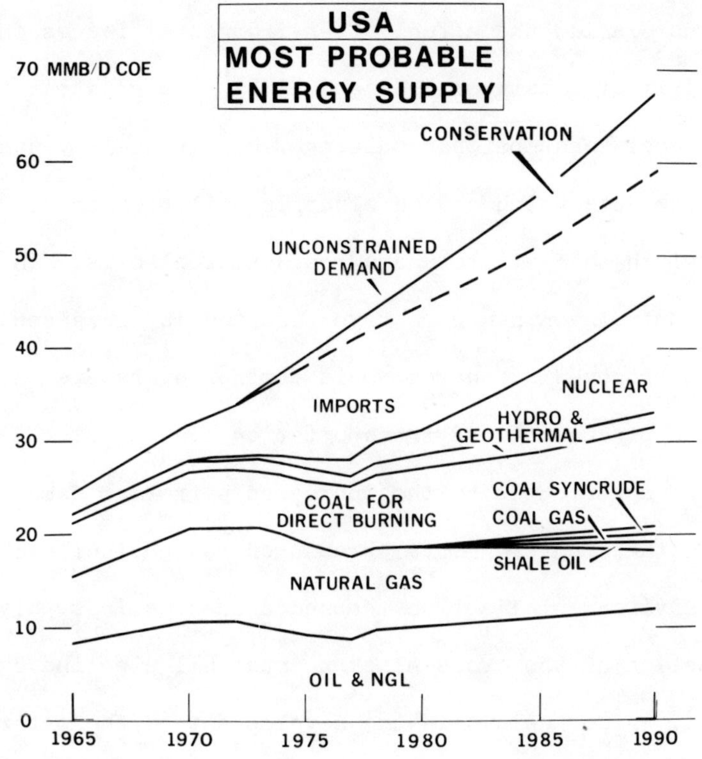

Figure 9

ENERGY DEMAND VS. DOMESTIC SUPPLY

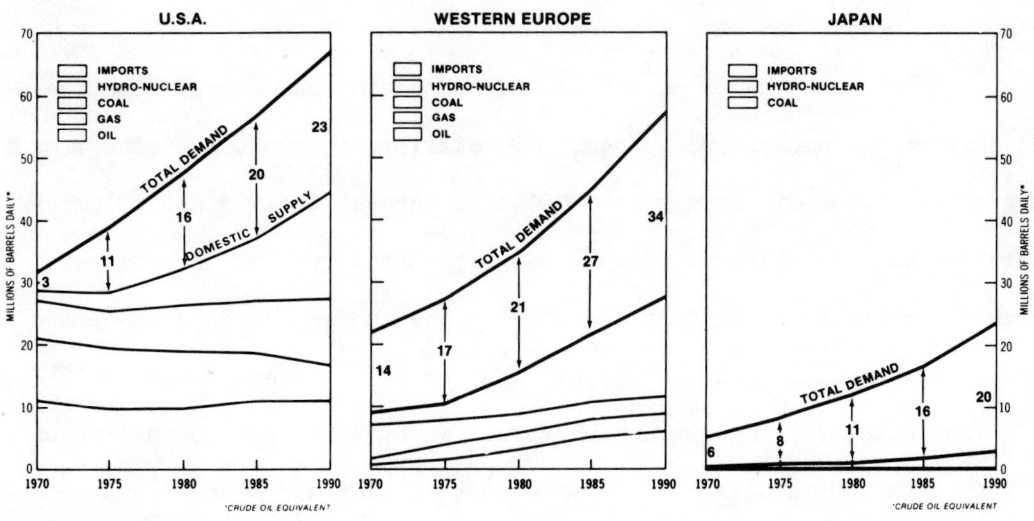

Figure 10

ENVIRONMENTAL IMPACT

Let us examine the effect of environmental issues on the rapid increase in oil imports between 1972 and 1975. The greatest impact is due to reduced coal consumption resulting from air quality standards and health and safety acts which caused electric utilities and industrial customers to switch to distillate and residual fuel oils. In addition, auto emission controls on new cars have resulted in increased fuel consumption and proposed standards would further aggravate this situation. The delay in the production and transportation of natural gas from Alaska's North Slope also contributes to the increased petroleum demand in 1975.

While these environmental pressures caused increases in petroleum demand, other environmental actions caused decreases in supply. The delay in the construction of the trans-Alaskan crude oil pipeline exerted the greatest impact. Also, delays in the development of the outer continental shelf resulted in curtailed petroleum supply. Most importantly to the shortages facing us just prior to the Arab oil embargo, we were unable to predict the exact coincidence of these environmental constraints with the peaking of production of oil and natural gas.

Without a major commitment to conserving our energy resources and developing domestic resources, U.S. oil imports could increase from 23 percent of our oil supply in 1970 to 50 percent by 1975 and 63 percent by 1990. This requirement includes significant increases in overseas product imports, primarily attributable to the sudden increase in demand for distillate and residual fuel oils, and to insufficient domestic refining capacity. The current shortfall in U.S. refining capacity can be traced to two factors: (1) environmental constraints and (2) uncertainty

over the supply of crude oil. Both these problems still exist and as a
result expansion plans announced by the oil industry last April have been for
the most part either delayed or dropped.

WORLD SITUATION

As Fig. 10 shows, the rapid and dramatic increase in oil imports
puts us in competition with the rest of the industrialized world. If the
trend of unconstrained demand continues to 1990 we in the United States
would be dependent on imports for about two-thirds of our requirements.
Western Europe is about 65 percent dependent and Japan about 90 percent
dependent on imported energy today, and they will both remain at high
levels.

In 1970 the Western Hemisphere was virtually independent of
supplies from the Eastern Hemisphere. Fig. 11 shows the logistics of the
world oil supply in 1985. The dependence of the world outside the Soviet block
and China on the oil reserves of the Middle East is clearly shown. The
proven crude oil reserves in the U.S.--39 billion barrels of crude oil,
including Alaska--comapres to 342 billion barrels--ten times as much--in
the Middle Eastern countries.

Since we will remain dependent on imports to a significant
extent for the foreseeable future, the question arises whether the oil
will be available in the desired quantities from the oil exporting
countries. Fig. 12 shows an estimate of the maximum ability of the Free
World outside North America to produce oil to 1980. The rate is subdivided between those producing countries which are not expected to have

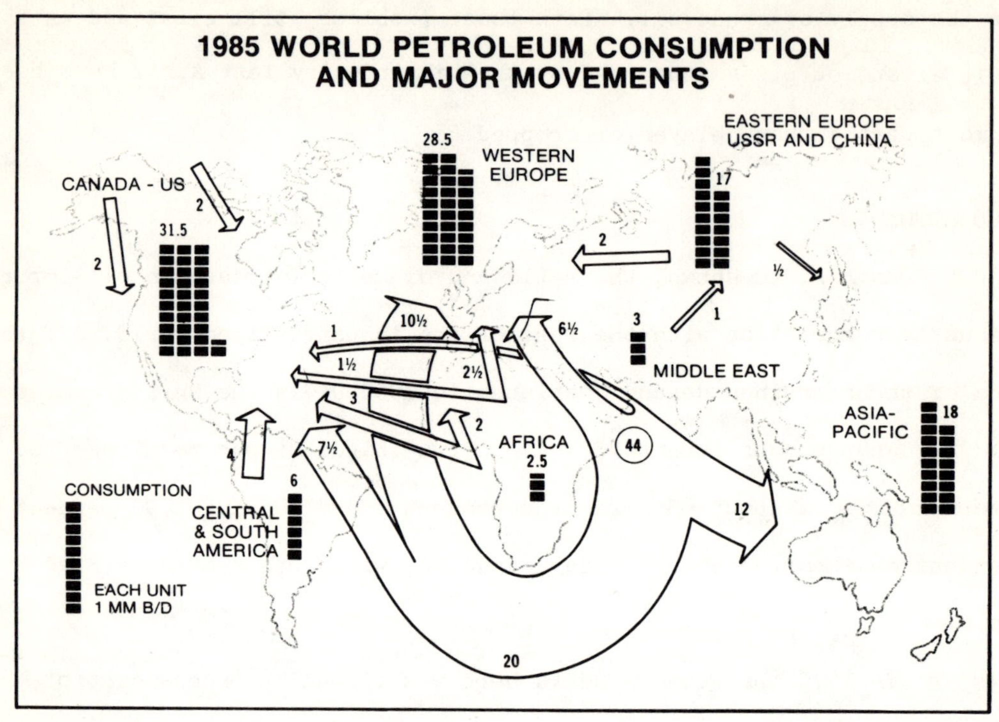

Figure 11

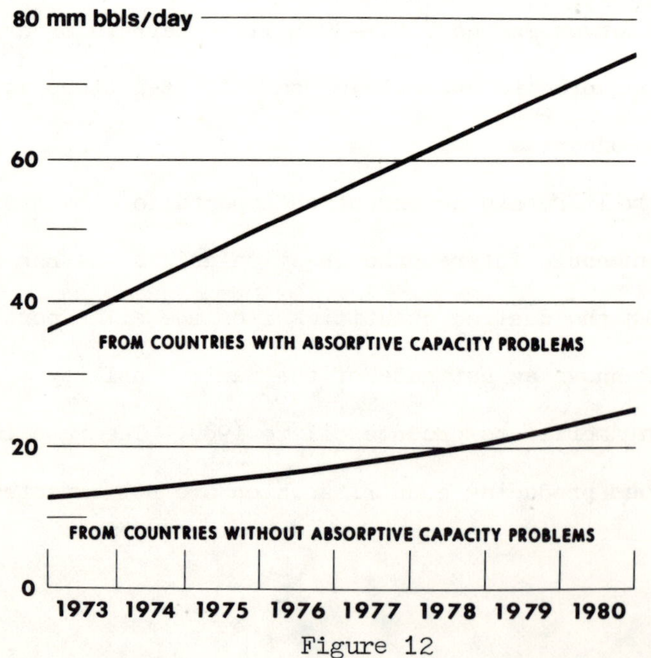

FREE WORLD CRUDE OIL AVAILABILITY OUTSIDE CANADA & USA

Figure 12

difficulty absorbing their oil revenues and those countries which may have problems. This factor undoubtedly will affect future levels of production in these countries.

However, we see a possible maximum production in 1980 of 72 million barrels per day. During the same year, U.S. demand for imports and unconstrained demand in the rest of the Free World outside North America is expected to be 58 million barrels per day. On this basis, there seems to be a reasonably good chance that oil could be physically available to meet the requirements of the Free World--at least until 1980.

Still, it appears that every major consuming country will run a large trade deficit in order to meet its oil needs. And most producing countries will be unable to absorb the huge volumes of money received for oil exports. Fig. 13 shows that by 1980, at $10 per barrel average price, surplus revenues in the oil exporting countries will, at a minimum, build up to a cumulative total of 450 billion dollars. Estimates for 1974 already range from 70 to 85 billion dollars revenue intake by these countries over what they can spend internally on their own domestic needs. These consid- erations certainly raise questions about the willingness of countries with low absorptive capacity to produce at maximum rates.

Furthermore, valuing the oil we think could be physically avail- able to be traded internationally between 1975 and 1990 at its current price of $10/barrel--gives a total value of about 3 trillion dollars. Since oil export revenues are expected to be large in relation to organized capital markets--and since much oil revenue is expected to be invested in these markets--the potential for serious disruption exists.

We are convinced that worldwide oil export transactions, during this period, must be at a lower and more manageable level.

Let me now show you our view of the world oil supply but projected in this particular forecast to the year 2000, as shown in Fig. 14. Using the largest estimate for world ultimate reserves from a recent paper by Meyerhoff, and subtracting the Communist reserves of over 850 million barrels, there would remain approximately 1.6 trillion barrels of conventional free world oil supply. We have used 1.8 trillion barrels and a reserves to production ratio of 15:1. This gives a peak in world production prior to 1990. The assumptions we have made are on the optimistic side. Fifteen years is about the minimum life of a refinery in order to justify the investment. After this point, there would be no need to build new refineries and no cause to increase the production rate. From the estimate of free world crude supply, we have subtracted estimates of both the U.S. production and the production in other countries that would be used internally as shown in the chart. This leaves the amount available for movement among the countries of the world, i.e., that available for import.

We have assumed that the U.S. share of imports (crude and products), in the constrained supply situation illustrated, would remain the same as in the total demand situation illustrated previously. This fraction rises to approximately 28 percent by 1975 and remains roughly the same throughout the forecast. The result is that after 1980 the volume of oil available for import by the U.S. falls below that required by the unconstrained demand case. The shortfall is 2.5 million B/D by 1985 and 6 million B/D by 1990.

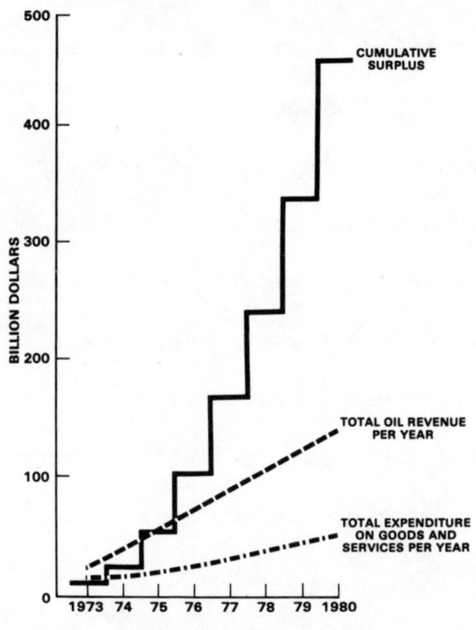

Figure 13

CONVENTIONAL
USA OIL IMPORT POTENTIAL

(ASSUMING USA HAS PRO RATA SHARE)

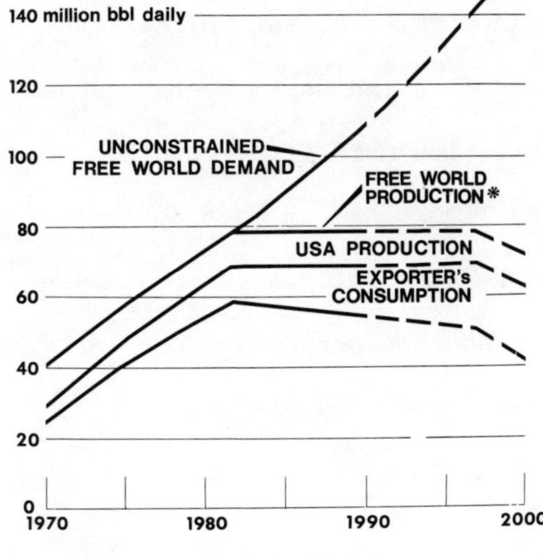

Figure 14

Quite obviously, without sufficient imports, our forecast
unconstrained demand cannot be met. So we at Shell accept the condition
that adjustments in our thinking and our lives are necessary. The United
States simply cannot practically buy the amount of oil we want. And to
complicate matters more, we don't even know how much will be available
at prices we can afford. These considerations make a tremendous impact
on the energy forecasts we have prepared in the past, so we decided to go
back to the drawing board to see what changes we must make to approach
self-sufficiency. We must not forget that we have in the United States
the coal, oil shale, and the future of nuclear energy which make possible
the setting of a goal of energy self-sufficiency.

CONSERVATION

We first looked again at what we might do to use less energy
without massive disruption to our economy. Fig. 15 shows that by 1980, we
think we could save annually more than 3 million barrels per day, and about
8.5 million barrels in 1990. In the long run, such savings will call for a
commitment to efficiency and conservation entirely new to this country, and
acceptance of the trade-offs involved. Higher fuel prices in themselves
will act as a considerable stimulus to energy conservation.

The transportation market represents the largest single
potential savings. By 1990 a faster move to smaller and more efficient
automobiles, use of more car pools, and more passengers per airplane could
save about 3.5 million barrels per day. More energy efficient industrial
plants could show a savings of 1.5 million barrels per day by 1990. In
the combined residential and commercial markets, increased insulation

standards, more efficient appliances, and lower levels of lighting can provide potential savings of over 3 million barrels per day by 1990.

OPTIMISTIC SUPPLY

The next thing we did was look at the supply side. We prepared an optimistic energy supply case, which we consider attainable only by a massive nationwide "crash program" type commitment to achieving energy self-sufficiency. Fig. 16 shows our optimistic picture for oil compared to our most probable forecast. Optimistically we might hope to increase domestic production to about 1.5 times today's levels. This assumes adding 90 billion barrels of oil to the reserves in the United States in the next 17 years. This is as much as we have added in the past 23 years. We still see a decline in domestic production but it is arrested sooner than in the most probable case and followed by more rapid increases. On this chart also we have shown the expected decline of production if no further reserves are added.

We are not going out of the petroleum-finding business in this country, on the contrary, we are maintaining our past levels of effort and we are prepared to increase them. However, achievement of our optimistic forecast will largely depend upon the government making available increased acreage for exploration--principally in the offshore and in Alaska. Our exploration success will have to be better than it has been in the past. Exploration, development and transportation in these frontier areas are increasingly difficult and expensive. A national commitment to provide equipment and manpower will be a necessity. The naval petroleum reserves must also be further explored and developed.

POTENTIAL ENERGY SAVINGS

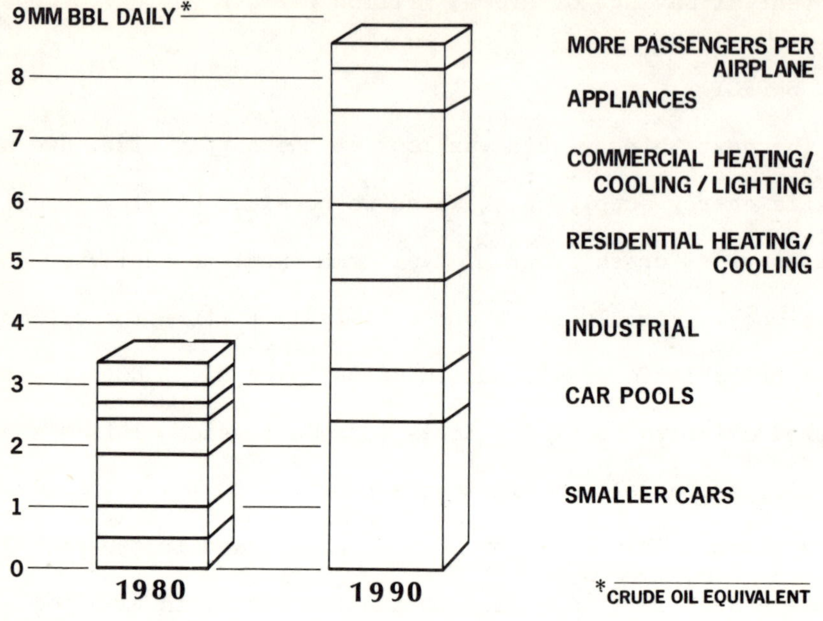

Figure 15

USA DOMESTIC CONVENTIONAL CRUDE & NATURAL GAS LIQUIDS PRODUCTION

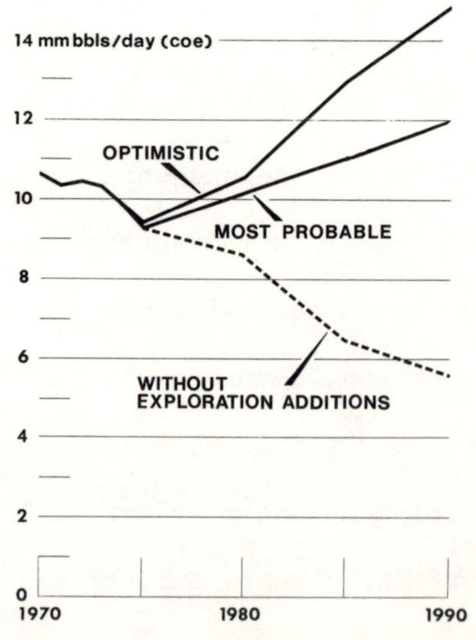

Figure 16

We then looked at U.S. shale oil production, as shown in Fig. 17 and, by extrapolating the National Petroleum Council projections, we estimated that shale oil production might be at a level of about one and one-quarter million barrels per day by 1990--over 500,000 barrels greater than our own best estimate.

With a real measure of optimism, the declining natural gas trend might be stopped and the production rate stabilized after 1980 as shown on Fig. 18. This estimate assumes that we will add 275 trillion cubic feet to reserves in the next 17 years. That is as much as we have added in the past 16 years. The dotted line shows the trend in production rate should no further reserves be added in the future.

Fig. 19 shows our estimate that coal production could conceivably increase to about 1.9 billion tons a year in 1990--over 500 million tons a year greater than our most probable projection, this would include accelerated coal production for coal gasification and liquefaction. It will require a tremendous effort--the growth must come largely from strip mining in the western states. The use of coal is also dependent upon stack gas scrubbing becoming commercially available at an early date to control air pollution.

As shown on Fig. 20, we increased our estimate of installed nuclear capacity to 500,000 megawatts in 1990 as compared to the 400,000 megawatts we think most likely. This is still 100,000 megawatts short of the Atomic Energy Commission "high" prediction.

One could talk at length about the manpower and material resource requirements, and the economic and social problems associated with such a nuclear energy program. They are tremendous. In any case, especially when

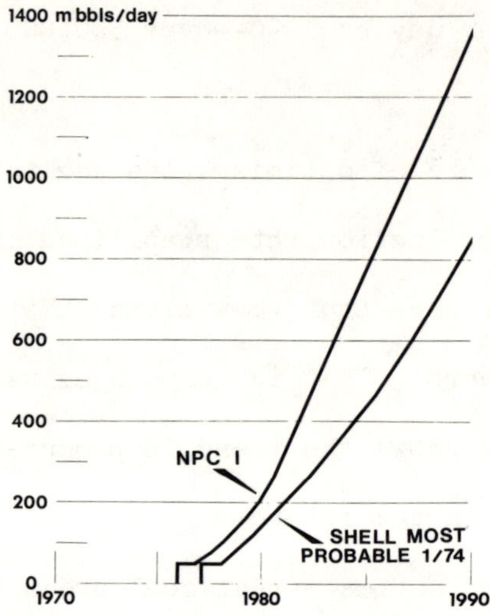

Figure 17

USA DOMESTIC CONVENTIONAL
NATURAL GAS PRODUCTION*

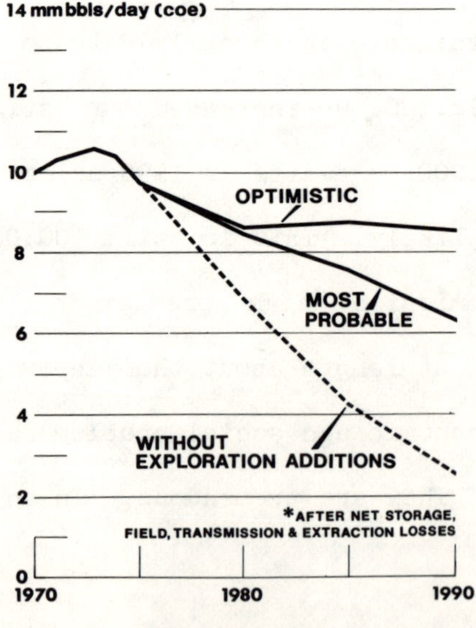

Figure 18

26

U.S. COAL SUPPLY

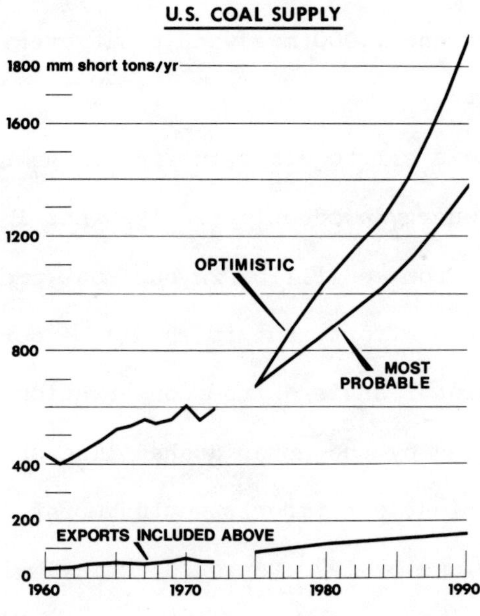

Figure 19

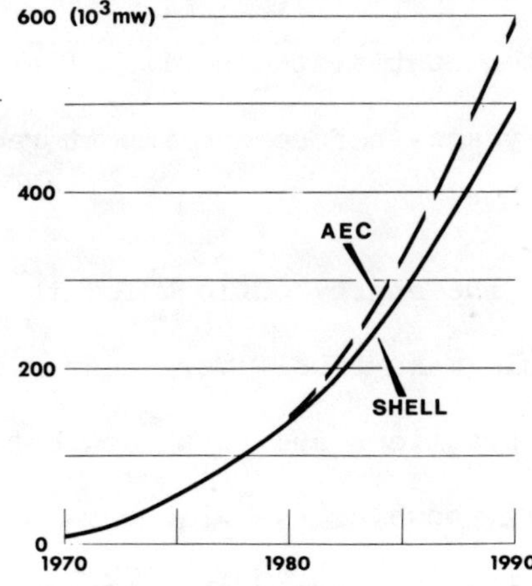

Figure 20

taking into account plant lead times now projected at 8 to 9 years--it will require an effort in scale that is difficult to comprehend. It is equivalent to completing one 1,000 megawatt plant every 10 days during each year of the 1980's.

When we add this all together in Fig. 21 we see the crude oil equivalent of domestic energy production in 1990 has increased to 57 million barrels per day from 46 million in our most probable forecast. This will take a fantastic effort, but it should be our national goal. The impact of the additional savings from conservation on this same supply picture is depicted by the upper dashed line shown on the chart.

With conservation, our imports would amount to only 7 million barrels per day by 1985 and by 1990 we would achieve virtual energy self-sufficiency. The cumulative volume of oil imported between now and 1990 will be reduced from 100 billion barrels to 45 billion barrels--a reduction of 55 percent. We think that the United States will be able to afford imports at this level--and, furthermore, that this volume is in a more reasonable relationship with the worldwide energy supply and demand situation. For the United States, the period of greatest stress will come in the next 5 to 8 years--before our import requirements begin to measurably taper off.

This, then, is the energy outlook for the United States as we see it. The country could make itself very nearly independent in energy with sound conservation practices and total dedication of human and material resources. More probably, however, we will remain dependent on imports to some significant extent for the foreseeable future.

USA
OPTIMISTIC
ENERGY SUPPLY
70 MMB/D COE
60
50
40
30
20
10
0
CONSERVATION
UNCONSTRAINED
DEMAND
NUCLEAR
IMPORTS
HYDRO &
GEOTHERMAL
COAL SYNCRUDE
COAL GAS
COAL
FOR DIRECT BURNING
SHALE OIL
NATURAL GAS
OIL & NGL
1965
1970
1975
1980
1985
1990

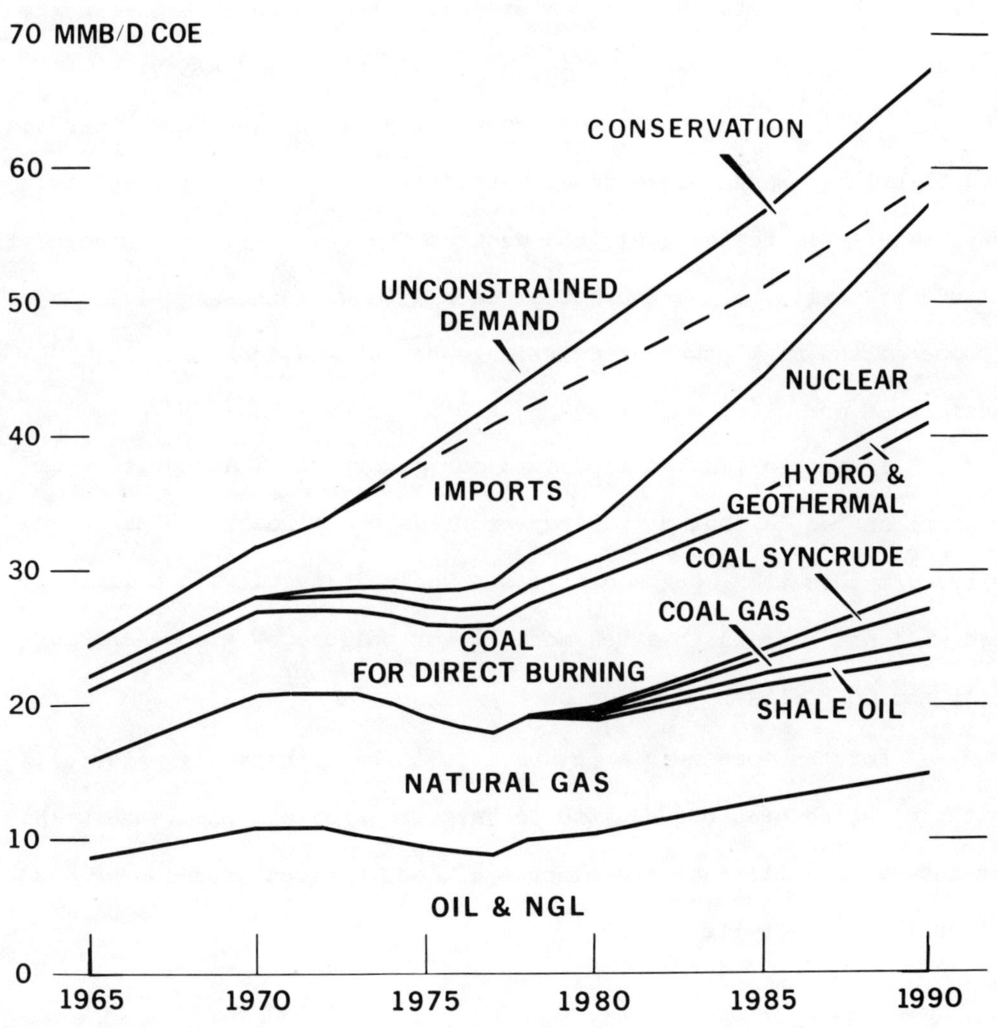

Figure 21

President Nixon has announced a goal of energy independence. We support such an objective. But we do not think it can be achieved by 1980. If it can be done at all, self-sufficiency will not be approached before 1990.

We should stop for a moment to realize what our country will have accomplished if we achieve the energy goals discussed today. It will mean that our nation has built a bridge to the future over a comparatively short period of time--time enough I hope to determine what our longer-range future will bring and how it will be dealt with.

Will it bring continued growth or does it mean some other state of equilibrium? We do not know--except that it will be different from today. What will be the goals and desires that evolve in the mind of the public? What will be the sources of energy to meet these goals? There are some of the most powerful political--social and technological questions of our time.

I want to make an appeal to our governmental authorities and the American people that our energy problems be approached with objectivity. At this time, our society as a whole is faced with a limitation which will not respond to a new morality or philosophy alone--and certainly not to partisan politics.

For the foreseeable future it will be a physical, material limitation which presents us with perhaps the greatest challenge we have ever known. It will take the efforts of each citizen--and courageous leadership--to meet it.

ENERGY RESEARCH AND DEVELOPMENT
ALTERNATIVES FOR FUTURE SUPPLY

Chauncey Starr[*]

Research and development in the energy industry is the key to meeting the continuously increasing demand under a continously decreasing availability of energy resources. Fuel resources are constrained by nature. They are non-uniform geographically, and are only partially accessible to man. They exist in two forms: the depletable fuels which represent nature's endowment to mankind, and the continuous or non-depletable resources originating in the Sun or the central heat of the Earth. It is our responsibility as a technological society to utilize these resources as effectively as possible, to preserve their value for future generations to the maximum possible degree, and to provide our energy services with the minimum of environmental impact to the biosphere.

Until recently, fossil fuels have been relatively abundant, easy to obtain from the confines of the Earth and have not presented problems of availability whenever needed. However, during the past few decades the exponential increase in demand has gradually revealed the future implications of a continuous drain upon the depletable resources. The economic pressure to utilize these natural fuels more efficiently has grown steadily, and this has been further compounded by a parallel growth in our awareness of the environmental impact of their use.

Thus, we now have accepted a spectrum of national objectives for technological development. These include the efficient utilization of depletable resources, increased emphasis on their replacement by continuous energy sources such as solar radiation wherever possible, the

---

[*] President, Electric Power Research Institute

use of these resources in a non-polluting fashion, and finally, improvement of the efficiency of the end-point use of various energy forms, and
in particular electricity, so that needed functions can be accomplished
with a minimum demand on the original sources.

It is now generally recognized that the mix of social amenities and
economic welfare of a large industrial society, which we commonly describe
as the quality of life, is heavily dependent on the continuous availability of energy. Human labor and energy machines are the basic ingredients for modifying nature's raw materials into those forms needed for
human existence and for the material needs of a social structure. In a
modern industrial society, energy is almost as essential as a food supply.
In fact, our food supply is itself dependent upon an intensive man-created
energy input. To provide an overview perspective of the role of energy
in our society, Fig. 1 lists the major uses of energy resources in the
U.S. Clearly, production of material goods and transportation represent
a combined use of roughly two-thirds of all our energy. It is in these
areas that the largest impact is felt either from energy constraints or
from technological improvements.

	PERCENTAGE OF ALL ENERGY USED IN U.S.
RUNNING INDUSTRY	37.3%
POWERING TRANSPORTATION	24.8%
HEATING HOMES AND OFFICES	17.9%
PROVIDING RAW MATERIALS FOR CHEMICALS, PLASTICS	5.5%
HEATING WATER FOR HOMES AND OFFICES	4.0%
AIR CONDITIONING HOMES AND OFFICES	2.5%
REFRIGERATING FOOD	2.2%
LIGHTING HOMES AND OFFICES	1.5%
COOKING FOOD	1.3%
OTHER USES	3.0%

Fig. 1 U. S. Energy - How It's Used

The total social cost of energy supply is a key component of the final
cost of all the end products resulting from this combination of machines
and labor. In a society in which all sectors have been encouraged to
expect steadily improving quality of life, the cost of energy may well
determine the social distribution of the amenities which our productive
systems produce. It is thus a function of research and development to
ensure, first, an ample availability of energy in forms most suitable for
energy machines, and, second, to make this energy available at a minimum social cost. We now recognize that the national accounting necessary
to determine these costs must include not only the direct investment of
capital goods and labor in the production of energy, but also the indirect cost of the environmental impacts on both the generation using
the energy and on future generations. These long-term burdens include
the future consequences of the depletion of limited natural resources,
irreversible or slowly reversible environmental changes, and the consequent adjustment in future life styles resulting from increased costs of
energy.

The combination of depletable and continuous resources can provide an
almost limitless energy supply, provided there are no constraints on
cost or time for development. Unfortunately, in the real world situation,
the issue of cost and availability are ever important because of their
constraining influence on the economic well-being of society, and the
dislocations they cause. The key issue is, then, how to allocate our
total national resources to provide technological options which may meet
our energy needs at an acceptable cost and in a timely fashion, so as to
remove such constraints on our society.

The energy issues are a part of the broader problem of meeting social
goals in a resource limited world, and thus the alternative approaches
to meeting energy needs often conflict for resources with other social
goals.

When comparing alternative technological systems, it may be helpful to
use a qualitative "figure of merit" such as the following:

$$\text{Technical Merit} = \frac{\text{End Performance}}{\text{Societal Cost}} \quad,$$

where the Performance is measured by a combination of availability and functional effectiveness; and where Societal Cost includes the resources used, both depletable and continuous (natural and man-made), and the cost of environmental and by-product effects.

Among the important man-made continuous resources are human labor and capital (i.e., the stored output of earlier labor - the accumulation of machines, supplies, operating systems, and similar inputs to a society's productive capabilities).

The characteristic impacts of resource constraint are different for depletable and continuous resources. Depletable resource use inherently creates an economic future burden that must be "present-valued" for planning purposes. This future burden results from the inevitable increasing unit cost as the resource is depleted, and the consequent impact on the future availability and utility of the technical system. The long-term effect of undesirable by-products must also be present-valued. Fossil fuel systems are an example of such issues.

In contrast, continuous sources do not necessarily create a future burden - except insofar as their use also creates long-lived undesirable by-products. Continuous resource availability is generally limited by the efficiency of the social system to produce the needed amount and by the limitations of technology to convert such resources into usable forms. The economic factors are primarily the "real time" costs, except for the long-term environmental impacts mentioned above, which need to be present-valued. Examples of such continuous sources are solar energy convertors, both physical and botanical and geothermal concepts. Such continuous sources have characteristic environmental effects involving land use, heat dissipation, and in the case of geothermal, chemical pollutants.

It is, of course, extremely difficult to quantify the true costs to society of any energy-generation system -- even those that have been in

operation for many decades. Only very long-time competitive operation
of alternative systems provides some insight into these costs. Yet,
for example, in mass transportation the comparative total costs of bus
and electric urban systems continues to be a subject of argument.
Nevertheless, it is instructive to attempt such an evaluation. National
planning for the very long term assumes a qualitative ordering of alternatives, even if this is done intuitively.

With above cautions, I have assembled in Fig. 2 a qualitative comparison
of the total costs of energy generation alternatives, both for the near
and long-term, with 1990 as the arbitrary break point. Near-term costs
are thus based on existent technology, and long-term on the assumed
success of R&D. Such estimates are heavily judgmental, even though they

	NEAR TERM (< 1990)				LONG TERM (1990 - 2010)			
FOSSIL FUEL	RESOURCE	CAPITAL	ENVIRON. & OTHER	WEIGHTED TOTAL	RESOURCES	CAPITAL	ENVIRON. & OTHER	WEIGHTED TOTAL
GAS	M	M	L	M	H	M	L	M
OIL	M	M	M	M	H	M	M	H
COAL	M	M	VH	H	H	H	M	H
NUCLEAR								
FISSION – LWR	L	M	L	L	L	M	L	M
FISSION – FBR	L	H	L	H	VL	M	L	L
FUSION (1973)	VL	(FEASIBILITY)	---	---	VL	VH	VL	H
SOLAR								
SPACE HEAT	VL	H	VL	M	VL	M	VL	L
ELECTRIC	VL	VH	L	VH	VL	VH	L	VH
WIND	VL	VH	L	VH	VL	VH	L	VH
GEOTHERMAL								
STEAM	VL	M	M	M	VL	M	M	M
WATER	M	H	M	H	L	H	M	H
ROCK	L	(FEASIBILITY)	---	---	L	H	M	H

QUALITATIVE SCALE: VERY LOW (VL), LOW (L), MEDIUM (M),
VERY HIGH (VH), HIGH (H)

Fig. 2 Comparative Costs of Alternative
Energy Generation Systems

are based on state-of-the-art assessments, and thus can only serve to
guide, rather than determine, policy and planning.

It is to be expected that the tabulation in Fig. 2 shows that the most
promising near-term technologies that might compete with fossil fuels
are nuclear (LWR) solar space heating, and geothermal steam. More

subject to argument are the long-term judgments. For example, assuming
both perform equally well, I believe the fast breeder will have a lower
total societal cost than fusion, primarily because of the assumed very
high unit capital cost for fusion. Recognizing the uncertainities of
such projections, they do stimulate attention to all the societal cost
factors that eventually must be considered.*

Because the technical, social, and economic frameworks in which future
decisions will be made are not clearly apparent to anyone today, the most
useful function for research and development is to provide a spectrum of
technical options for meeting the most probable future problems of energy
systems. From the experience of the past half century, it appears that
for the indefinite future the most desirable and versatile form of
energy for wide-spread public use is electricity. On the assumption that
the present growth in electricity use will, therefore, continue at the
expense of other energy forms, the research and development conducted in
the electrical utility industry may be a vital factor in determining the
availability of optional future energy systems. Fortunately, electricity
can be created by conversion of every known energy source, and thus the
growth of a national electricity distribution system does not inhibit
the development of alternative generation systems.

Most of us have been exposed to some of the stages through which a new
technological concept must pass before becoming commonplace in public
use. It is useful to enumerate these stages as shown in Fig. 3. The
first is that of scientific feasibility, involving both the analytic
development of the concept and its experimental verification. The second
is that of engineering feasibility, which moves from component development to subsystem development and to the total integrated performance
demonstration needed to establish the interactions of a complete system.
Commercial development then tests the concept against competitive alternatives. Finally, the technical system, although proven as a commercially

---

* A more detailed and independently developed table of this type was
  presented on pages 138-139 of the report "The Nations Energy Future",
  (WASH 1281), published by the U.S. AEC.

DEVELOPMENT PHASE	CONTROLLING PARAMETER

SCIENTIFIC FEASIBILITY
    ANALYTIC
    EXPERIMENTAL
	NATURAL LAWS

ENGINEERING FEASIBILITY
    SMALL-SCALE PILOT PLANT
    INTERMEDIATE DEMO
 	
    COMMERCIAL PLANT
	
	SUB-SYSTEM PERFORMANCE
	OPERATIONAL INTEGRATION OF
	COMMERCIAL SIZE COMPONENTS
	COST, PERFORMANCE, AND RELIABILITY

UTILITY INTEGRATION	COSTS OF PHASING IN MULTIPLE PLANTS, INCLUDING SUPPORTING SYSTEMS
SIGNIFICANT USE	COMPETITIVE ALTERNATIVES

Fig. 3 R&D Development Phase

desirable alternative, must now face the difficulties of integration
both physically and operationally into ongoing electric utility service
systems. If this hurdle is successfully achieved, we have reached the
final growth stage leading to substantial public use. The figure also
lists some of the criteria which have to be met at each stage. It is
certainly clear to all of us that a sequence of this nature is very
time-consuming. Occasionally, scientific discoveries, new technological
insights, and new economic conditions will speed up this process. Such
discontinuous events are certainly not predictable, even though history
has shown that they have a finite probability of occurring.

I have tried to illustrate in the next two figures the probable time
sequence for technical systems with which we have some familiarity.
Figure 4 shows the various stages for two of the hydrocarbon alternatives
now under commercial development. The first is the production of low
Btu gas from coal for direct use in utility boilers. The function of
coal gasification is to provide a process of removing coal impurities.
This system faces technological competition with improved combustion
methods in the broiler and with stack gas cleanup technologies. A second

37

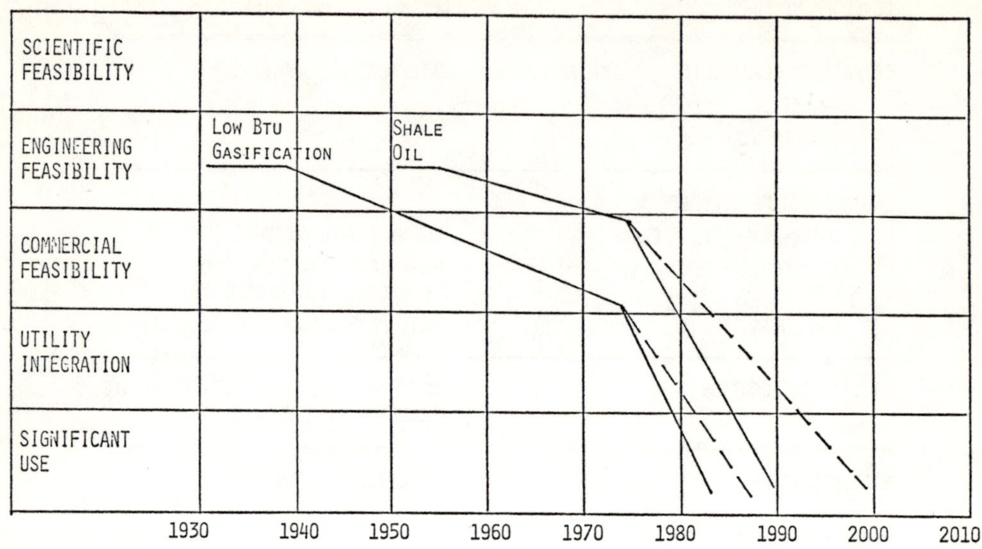

Fig. 4 Hardware Development Phases for
Hydrocarbon Alternatives

illustration is the development of shale oil as a fuel resource. Both
of these have been under study for several decades. The sharp acceleration in their development in the past year or two is, of course, a
direct result of the rapid increase in cost of the competitive natural
oil and gas which they are intended to replace. While such extrapolations into the future are certainly speculative, I believe that curves
illustrate the typical lead-time for such available technologies before
they can make substantial impact in public use.

Figure 5 illustrates in the same fashion the probable rate of development
of advanced generation alternatives, including conventional nuclear power,
the fast breeder, solar photovoltaic sources and fusion. I would certainly agree that all of these can have their rate of development accelerated by changes in national allocation of resources, by technological
break-throughs, and by changes of national policy. I doubt, however,
whether the rate at which these would become significant in end-use is
likely to be importantly altered by such changes. Money alone is not
sufficient. A proliferation of large-scale endeavors, pilot plants and

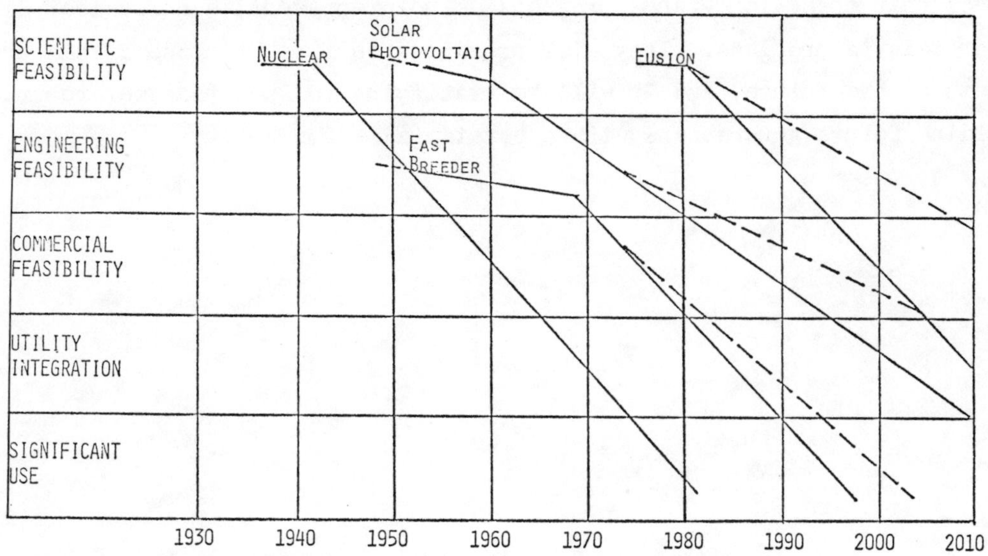

Fig. 5 Hardware Development Phases for
Advanced Generation Alternatives

the like, is no substitute for well-planned R&D spending directed at
advancing the state of the art and resolving key issues.

Of particular interest may be the estimated twenty-year differential
lead-time between the fast breeder and fusion. This has been a subject
of much public discussion, and certainly the projection presented here
represents a judgment which may not be shared by others. Nevertheless,
I believe the history of development of new technologies would support
this estimate.

Because of these very long lead-times required to bring new technologies
into use, if we expect to make a contribution to society's welfare some
decades hence, their development should be started now. It is obvious
that the more speculative developments will primarily benefit the next
or succeeding generations. If research is going to be called upon at
some future date to "pull a rabbit out of the hat" to meet some unforeseen
need, that rabbit will have to have been placed there by previous decades

of development. I do not believe we should be uncomfortable about the
fact that these long-range projects do not represent technological fixes
for today's problems. They will provide such fixes for the crisis problems of the future, and it will be gratifying to know that we are providing future generations with a beneficial endowment.

TIME FACTORS IN SLOWING DOWN THE RATE OF GROWTH OF DEMAND FOR PRIMARY ENERGY IN THE UNITED STATES *

Lester Lees , Mingin Philip Lo **

SUMMARY

The purpose of this report is to identify the time scales involved in slowing down the rate of growth of primary energy consumption in the U. S., as one component of an overall energy/environment strategy designed to limit the required volume of energy imports from overseas. Two important energy-consuming sectors of the economy are chosen as illustrative examples: (1) the "automobile" as a total system (25%); (2) space heating, air conditioning and water heating in the residential sector (22%). Efficient, light-weight vehicles are introduced into the automobile population by allocating an increasing percentage of new car production to such vehicles year by year until some fixed percentage is attained. Parametric calculations show that significant reductions in the annual rate of energy consumption by automobiles can be achieved if (a) the fuel consumption of efficient vehicles is 60% or less of "standard" vehicles; (b) the increment in percentage of new car production devoted to efficient vehicles is not less than 8% per year; (c) the efficient vehicles are "frozen" at not less than 80% or more of all new car production at the end of an eight to ten year period. In the residential sector the "turnover" rate is comparatively low, and the calculated reduction in annual energy growth rate produced by energy-conserving measures is modest, as expected, unless a "retrofit" rate of older living units of at least 2% per year can be attained.

These two components of an energy-conserving policy taken together would bring the growth rate in U. S. primary energy demand down from its present rate of 4.2% per year to about 2.8% per year by 1985. Reductions in the annual growth rate of the remaining 50% of U. S. primary energy consumption that seem quite feasible would bring the overall growth rate down to about 2.5% per year by 1985. If reductions in growth rate of this magnitude could in fact be achieved, energy imports would peak in the mid-1980s at a level no higher than about 60% above the present (1973) volume of imports. Incentives and disincentives designed to bring about this slowdown in the rate of U. S. energy consumption are discussed briefly.

* Supported in part by the National Science Foundation, Research Applied to National Needs (RANN), under Grant No. 29726, EQL Report No. 7
** Environmental Quality Laboratory, California Institute of Technology

1.   Introduction

Until quite recently it was generally assumed that the rate of supply of primary energy from domestic sources would always be adequate to meet the steadily rising rate of demand for energy in the U. S. The "energy crunch" of the 1970s shows that this assumption is no longer valid. During the decade from 1960 to 1970 total U. S. primary energy consumption grew at an annual rate of 4.3%,[1]* corresponding to a "doubling time" of about 16 years. In this same period domestic energy production grew at an annual rate of 2.6%, corresponding to a doubling time of about 27 years. The inevitable consequence of these mismatched time scales of domestic energy supply and demand was that the U. S. shifted its position from a net exporter to a net importer of energy in the late 1960s.

Any viable energy strategy for the next 25 years must contain a "mix" of the following sets of measures:

(a)   increasing the rate of imports of oil and liquefied natural gas (LNG);

(b)   increasing the rates of supply of primary energy from domestic sources, especially uranium and coal (including coal gasification and liquefaction), and, to a lesser extent, oil and natural gas;

(c)   developing new sources of energy, such as geothermal and solar energy (esp. solar/thermal conversion);

(d)   slowing down the rate of growth of energy consumption by improving efficiency, reducing "wasteful" practices, and shifting to less energy-intensive activities.

Each of these sets of measures has its own characteristic time scale, or doubling time, and these competing time scales will determine the nature of the "mix". These characteristic time scales are governed in turn by a complex set of logistical, technical, economic, environmental, land use, institutional and political-legal constraints. The authoritative study by the National Petroleum Council has already examined in detail the range of possibilities under option (b) above.[1,2] The purpose of the present report is to identify the time scales and magnitudes of option (d) — slowing down the rate of growth of energy consumption — by means of a few highly-simplified but typical illustrative examples. The results obtained are then examined for their impact on the required rate of energy imports and the required rate of domestic energy supply over the next 25 years.

----

* Superscripts denote references listed at the end of this report.

The question as to whether reductions of significant magnitude in the annual rate of growth in the demand for energy can actually be achieved over the next 10-15 years was addressed in the NPC demand projections[2] for 1985. In these projections the "low" estimate lies about 10% below the "intermediate" estimate. The recent Office of Emergency Preparedness report[3] on energy conservation projects a decrease of about 15% in energy demand by 1985 if all "mid-term" conservation measures are employed, and a decrease of about 22% if all "long-term" measures are utilized (Fig. IX-1, p. 59 of Reference 3). Beyond 1985 the various estimates of energy demand diverge somewhat, as expected, because of the difficulties involved in making such projections. For example, in 1995 the NPC "low" demand estimate is only 13% lower than the intermediate estimate, while the OEP estimated demand (extrapolated) is reduced by about 17% if "mid-term" conservation measures are employed, and by about 27% if all "long-term" measures are used.

In order to examine this question more carefully two important energy-consuming sectors of the economy are chosen in this report for purposes of illustration: (1) transportation (25%); (2) space heating, air conditioning and water heating in the residential sector (22%). The 25% of U. S. primary energy consumed in transportation is fuel alone, of which automobiles consume somewhat more than one-half, or about 14% of U. S. primary energy However the "automobile" considered as a total system all the way from raw materials through production, distribution, road-building, servicing and repair consumes 25% of all U. S. primary energy,[4] so the automobile is singled out for special attention in Section 2. "Efficient", light-weight vehicles are introduced into the automobile population by devoting an increasing percentage of new car production to such vehicles year by year until some fixed percentage is attained. The effect of this shift on the rate of growth of energy consumption is calculated over a 20-year period.

In Section 3 a similar study is carried out for the residential sector. Energy-conserving living units are introduced by increasing their percentage of new unit construciton year by year until some fixed percentage is attained.

These two sectors differ in one important respect: the annual "birth" and "death" rates of automobiles are relatively high (11% and 7% of the existing car population, resp.), so the time scale for making changes in the rate of growth of energy-consumption of automobiles is relatively short (of the order of 6-10 years), and the magnitude of the reductions in growth rate is significant. On the other hand the annual "birth" and "death" rates of residential units are relatively low (3% and 1% of the existing living units, resp.), so changes in the annual rate of growth in energy-consumption are correspondingly slower, and of smaller magnitude, unless a significant percentage of older structures can be "retrofitted" to conserve energy. These points are brought out by means of some simple illustrative examples in Section 3.

Finally, Section 4 examines the implications of the results obtained in Section 2 and 3 for energy imports, domestic energy supplies, and U. S. energy policy alternatives.

2.   Slowing the Rate of Growth of Energy Consumption by Automobiles

At any given time the automobile population is characterized by a wide "performance" spectrum in terms of miles per gallon, ranging at present from values as low as 7 miles per gallon to a high of around 28 miles per gallon. In order to bring out the importance of time scales in reducing the rate of energy consumption as simply as possible, we will work with a population consisting of only two classes of vehicles: (1) "standard" vehicles, all of which require $g_1(t)$ gallons per mile; (2) "efficient" vehicles, all using $g_2(t)$ gallons per mile, where $g_2 < g_1$. In this case the annual energy consumption rate, E, is given by the following expression:

$$\frac{E}{E_0} = \frac{m}{m_0} \frac{g_1}{(g_1)_0} \left[\frac{n_1}{n_0} + \frac{g_2}{g_1} \frac{n_2}{n_0} \right] \quad (1)$$

Here m is the average annual mileage per vehicle, n_1 and n_2 are the number of standard and efficient vehicles, resp., and $n = n_1 + n_2$ is the total number of automobiles at any given time. The subscript zero refers to the "base" year (say 1973) that is selected as the starting point for calculating the time history of the annual rate of energy consumption.

Suppose that all vehicles are standard vehicles at the beginning of the base year, i.e., $n_1 = n_0$ and $n_2 = 0$. In this simple example the process of introducing efficient vehicles into the automobile population is divided into two stages: In the first stage a steadily increasing percentage of new car production is assigned to such vehicles, e.g., 10% of all new cars during the first year, 20% the second year, etc., until the jth year ($j \leq 10$ in this example). After the jth year the percentage of new car production taken up by efficient vehicles is fixed at (10j)%. For simplicity the death rate of efficient vehicles is taken to be zero during the first j years, while the death rate of standard vehicles is some function d(t). Beyond the jth year the death rates of the two classes of vehicles are assumed to be the same.*

---

* This approach is easily generalized by allowing the percentage of new cars that are efficient vehicles to be equal to p the first year, 2 p the second year, etc., up to the jth year (jp $\leq$ 100), where 0 $\leq$ p $\leq$ 100. Beyond the jth year efficient vehicles are a fixed percentage jp of all new car production (Appendix).

At the present time the birth rate, b, and the death rate, d, of the total automobile population are 11% and 7% per year, respectively, so the total population is growing at an equivalent exponential rate of 4% per year. Taking these rates as fixed for the present, and considering the process as a continuous one, the differential equations describing the time history of the automobile population in this simple illustrative example can be written down and readily integrated to give the time histories of standard $n_1(t)$ and efficient vehicles $n_2(t)$. (See Appendix)

Actually the first and simplest case analyzed turned out to contain most of the "message". In this case the percentage of new car production allocated to efficient automobiles increases by 10% each year up to the tenth year, when these vehicles constitute 100% of new car production (Case 1, $j = 10$, $p = 10\%$). No new standard automobiles are produced thereafter. The automobile population history for Case 1 is shown in Figure 1. The population of standard vehicles peaks between the third and fourth years, when the birth rate of this class of vehicle is equal to its death rate. Beyond this point the population n_1 declines more and more rapidly, and, in fact, for $t \geqslant 10$, n_1 declines exponentially with time, because no new standard vehicles are being produced. The population of efficient vehicles builds up quite slowly at first (Figure 1), but increases rapidly beyond $t = 4$. By about the tenth year efficient vehicles account for one-half the total car population (Figure 1). If this process were to continue indefinitely with time, the build-up of efficient vehicles would approach the equivalent exponential rate of 4% per year.

At present the average mileage driven per automobile is increasing by 1% per year, i.e., $\frac{m}{m_0} = e^{0.01t}$. By adopting this rate in our calculations, and by utilizing the results for n_1 and n_2 in [Eq. (1)], the time history of the energy-consumption rate is calculated for various values of the fuel economy parameter $\frac{g_2}{g_1}$ between 0.50 and 1.0; the results are illustrated in Figure 2. For comparison the growth in $\frac{E}{E_0}$ is also shown for a car population consisting only of standard vehicles ($n_2 = 0$), i.e., $\frac{E}{E_0} = e^{0.05t}$.* At present the average gasoline consumption for U.S. automobiles is about 15 miles per gallon,[4] so values of g_1 of 0.750, 0.625 and 0.50 correspond to 20, 24 and 30 miles per gallon, respectively.

---

* According to Eq. (1) for various values of $\frac{g_2}{g_1}$ between 0.50 and 1.0 the value of $\frac{E}{E_0}$ at any given time can be found by means of a linear interpolation.

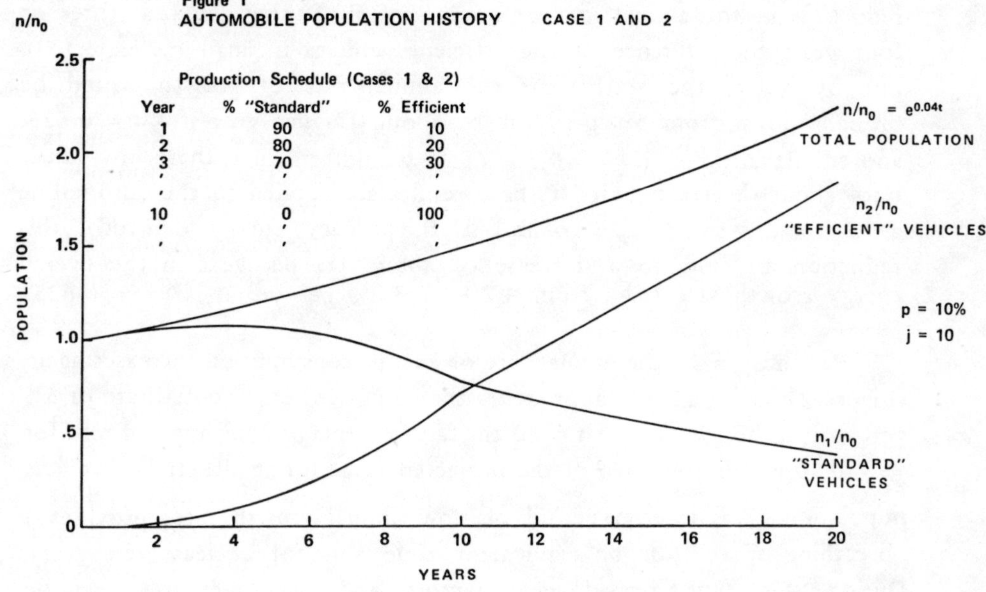
Figure 1
n/n₀
AUTOMOBILE POPULATION HISTORY   CASE 1 AND 2
2.5
Production Schedule (Cases 1 & 2)
Year   % "Standard"   % Efficient
1   90   10
2   80   20
3   70   30
10   0   100
2.0
n/n₀ = e^0.04t
TOTAL POPULATION
n₂/n₀
"EFFICIENT" VEHICLES
1.5
p = 10%
j = 10
1.0
POPULATION
.5
n₁/n₀
"STANDARD" VEHICLES
0
2   4   6   8   10   12   14   16   18   20
YEARS

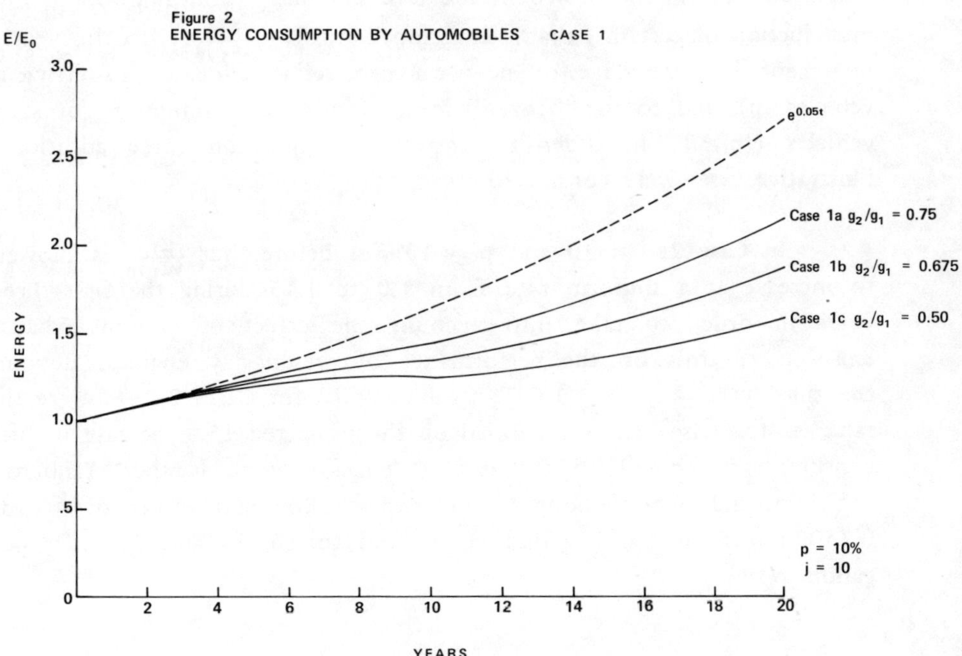
Figure 2
E/E₀
ENERGY CONSUMPTION BY AUTOMOBILES   CASE 1
3.0
e^0.05t
2.5
Case 1a g₂/g₁ = 0.75
2.0
Case 1b g₂/g₁ = 0.675
Case 1c g₂/g₁ = 0.50
1.5
ENERGY
1.0
.5
p = 10%
j = 10
0
2   4   6   8   10   12   14   16   18   20
YEARS

Values of $\frac{g_2}{g_1}$ in the neighborhood of 0.5 to 0.6 are required for this particular population time history in order to produce a substantial reduction in the annual *rate of growth* of E. During the first three or four years the influence of the efficient vehicles is small (as expected), but by about the sixth year the annual rate of growth would be brought down from 5% per year to about 1% per year if $\frac{g_2}{g_1}$ = 0.625, and to virtually zero if $\frac{g_2}{g_1}$ = 0.50, and it would remain at these low values until about the tenth year. If these results are applied to the automobile as a total system (25% of all U. S. primary energy demand), this reduction amounts to a decrease of about 1% per year in the overall energy growth rate (e.g., from 4.2% to 3.2% per year), if g_1 = 0.625.

Beyond t = 10 the annual rate of energy consumption increases again (Figure 2), and gradually approaches the equivalent exponential rate of 5% per year, although even at t = 20 the *level* of energy consumption rate for $\frac{g_2}{g_1}$ = 0.50 is still only 60% of the projected value for an all-standard vehicle population. This interim period of slow annual growth rates provides a "breathing space", or an equivalent time shift of at least ten years. During this time period new factors and new measures can be introduced to slow the overall rate of growth of the automobile population. (See Section 4)

The question naturally arises as to the sensitivity of the magnitude of the slowdown in the rate of energy—consumption to the introduction of certain exhaust emission controls $\left(\frac{g_1}{(g_1)_0}\right)$; to the yearly increment in percentage of new car production allocated to efficient vehicles (p), and to the "frozen" level of new car production for such vehicles [(pj)%]. In order to answer this question three additional illustrative cases were considered.

In Case 2, j = 10 and p = 10% as before, but $\frac{g_1}{(g_1)_0}$ is allowed to increase at a uniform rate from 1.0 to 1.15 during the first three years in order to take into account the effect of certain exhaust emission controls on the performance of present-day engines. Beyond the third year $\frac{g_1}{(g_1)_0}$ = 1.15. The values of $\frac{g_2}{g_1}$ for Cases 2a — 2c are the same as for Cases 1a — 1c. Based on the projected 15% increase in fuel consumption imposed by new exhaust emission controls, the "standard" vehicle would obtain about 12 miles per gallon and values of $\frac{g_2}{g_1}$ of 0.750, 0.625 and 0.50 would correspond to 16, 19 and 24 miles per gallon, resp.

Figure 3 shows the energy-consumption history for Case 2, and Figure 4 compares Cases 1c and 2c in order to illustrate the effects on the rate of energy-consumption of exhaust emission controls installed on both new and old vehicles. Clearly the situation is going to get worse before it gets better. Even with the introduction of the new, efficient vehicles the annual rate of energy consumption is higher up to the seventh year than the level reached by a 5% per year projected increase. However, the *rate of growth* of the annual energy-consumption rate is again reduced to about 1% per year between the sixth and tenth years if $\frac{g_2}{g_1}$ = 0.625.

Figures 5 and 6 show the effect of "freezing" the percentage of new car production devoted to efficient vehicles at 50% after the fifth year (j = 5), at 60% after the sixth year (j = 6), etc., for the particular case $\frac{g_2}{g_1}$ = 0.50, $\frac{g_1}{(g_1)_0}$ = 1.0, p = 10% (Case 3). These results show that significant reductions in the annual rate of growth of energy-consumption are achieved only if j $\geqslant$ 8, i.e., if the efficient vehicles are "frozen" at not less than 80% of all new car production after the eighth year.

Suppose that the process of introducing these efficient vehicles is slowed down so that some yearly incremental percentage of new car production less than 10% is employed (p $<$ 10%). The effects of this "stretchout" are shown in Figures 7 and 8 (Case 4, j = 10, $\frac{g_2}{g_1}$ = 0.50, $\frac{g_1}{(g_1)_0}$ = 1.0). Evidently significant reductions in the annual rate of growth of energy-consumption are attained only if the incremental percentage of new car production devoted to efficient vehicles is not less than about 8% per year over an eight to ten year period.

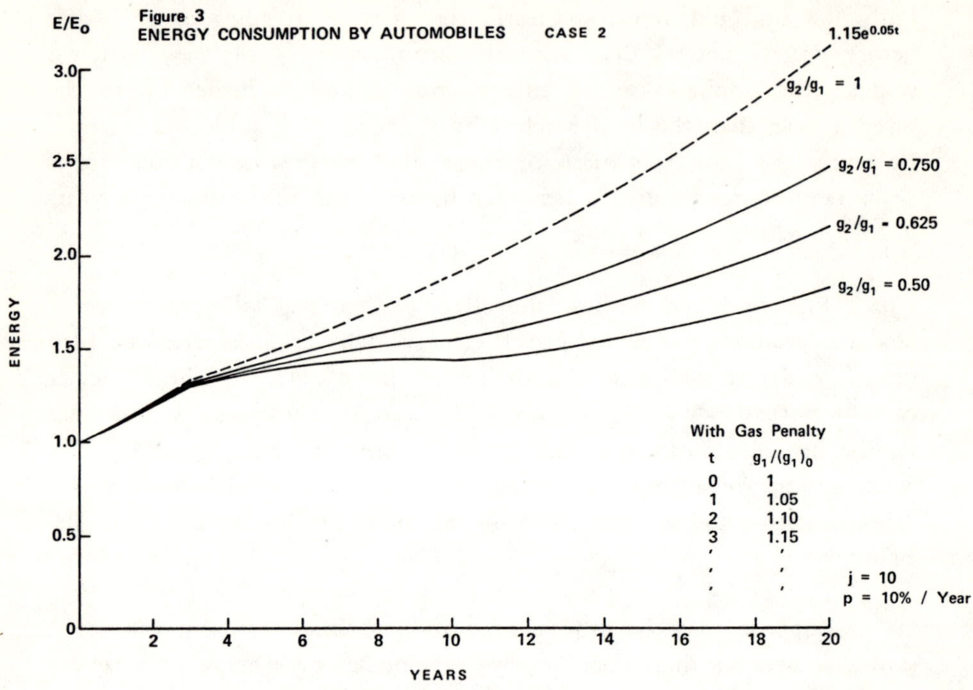

E/E_0
Figure 3
ENERGY CONSUMPTION BY AUTOMOBILES   CASE 2
1.15e^0.05t
g_2/g_1 = 1
g_2/g_1 = 0.750
g_2/g_1 = 0.625
g_2/g_1 = 0.50
3.0
2.5
2.0
1.5
1.0
0.5
0
ENERGY
With Gas Penalty
t   g_1/(g_1)_0
0   1
1   1.05
2   1.10
3   1.15
j = 10
p = 10% / Year
2   4   6   8   10   12   14   16   18   20
YEARS

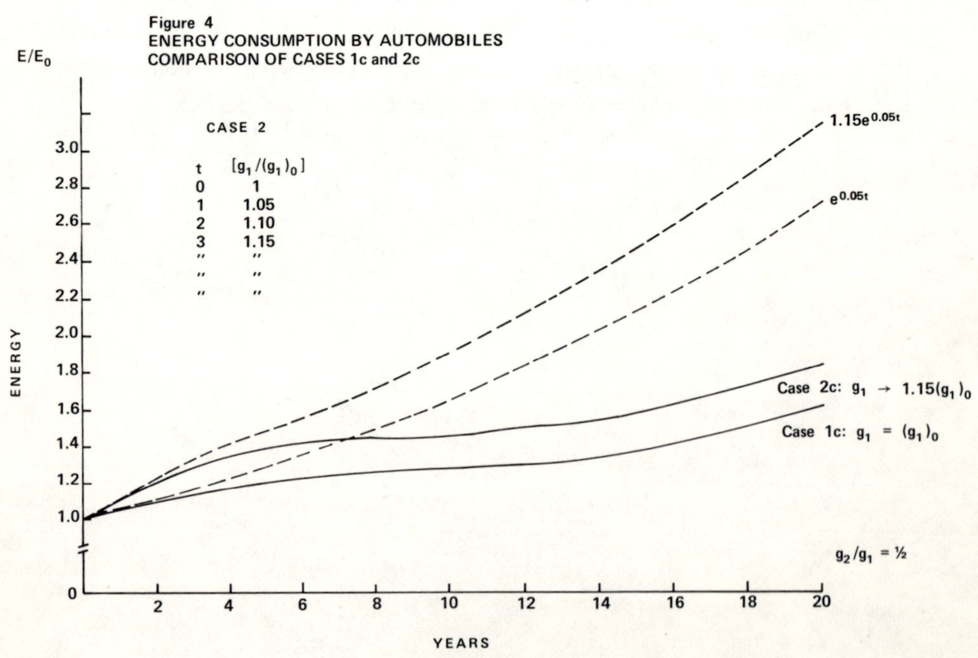

Figure 4
ENERGY CONSUMPTION BY AUTOMOBILES
COMPARISON OF CASES 1c and 2c
E/E_0
CASE 2
t   [g_1/(g_1)_0]
0   1
1   1.05
2   1.10
3   1.15
1.15e^0.05t
e^0.05t
3.0
2.8
2.6
2.4
2.2
2.0
1.8
1.6
1.4
1.2
1.0
0
ENERGY
Case 2c: g_1 → 1.15(g_1)_0
Case 1c: g_1 = (g_1)_0
g_2/g_1 = ½
2   4   6   8   10   12   14   16   18   20
YEARS

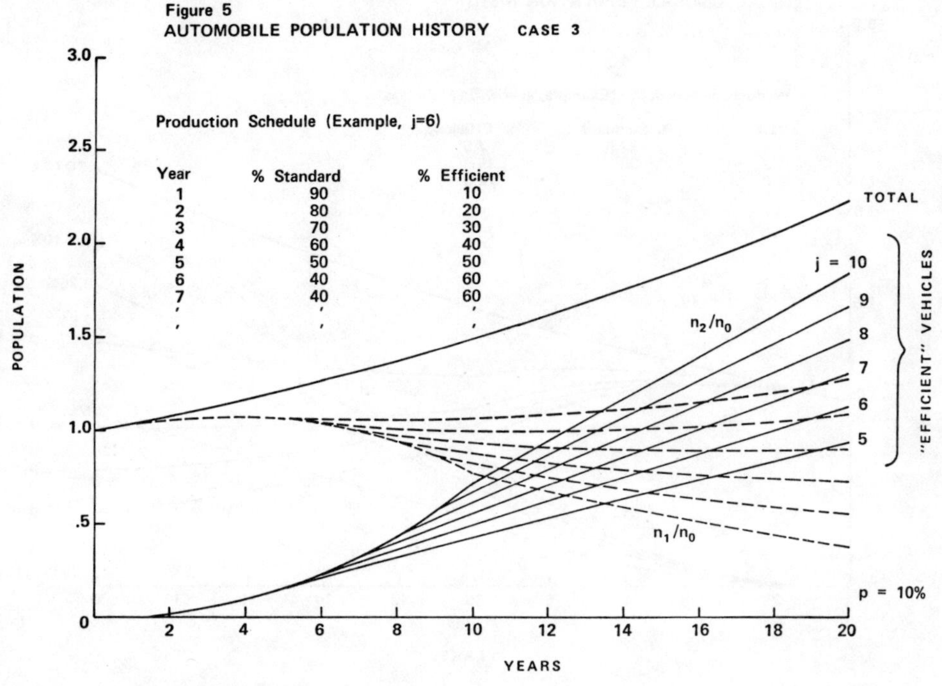

Figure 5
AUTOMOBILE POPULATION HISTORY   CASE 3
3.0
2.5
2.0
1.5
1.0
.5
0
POPULATION
Production Schedule (Example, j=6)
Year   % Standard   % Efficient
1   90   10
2   80   20
3   70   30
4   60   40
5   50   50
6   40   60
7   40   60
TOTAL
j = 10
9
8
7
6
5
"EFFICIENT" VEHICLES
n_2/n_0
n_1/n_0
p = 10%
2   4   6   8   10   12   14   16   18   20
YEARS

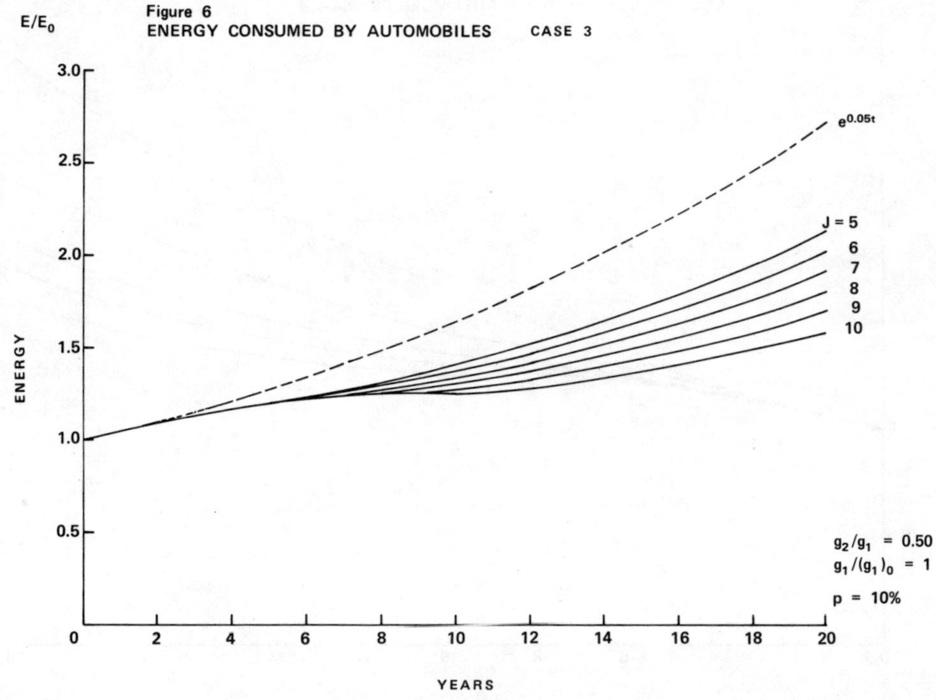

E/E_0
Figure 6
ENERGY CONSUMED BY AUTOMOBILES   CASE 3
3.0
2.5
2.0
1.5
1.0
0.5
ENERGY
e^{0.05t}
J = 5
6
7
8
9
10
g_2/g_1 = 0.50
g_1/(g_1)_0 = 1
p = 10%
0   2   4   6   8   10   12   14   16   18   20
YEARS

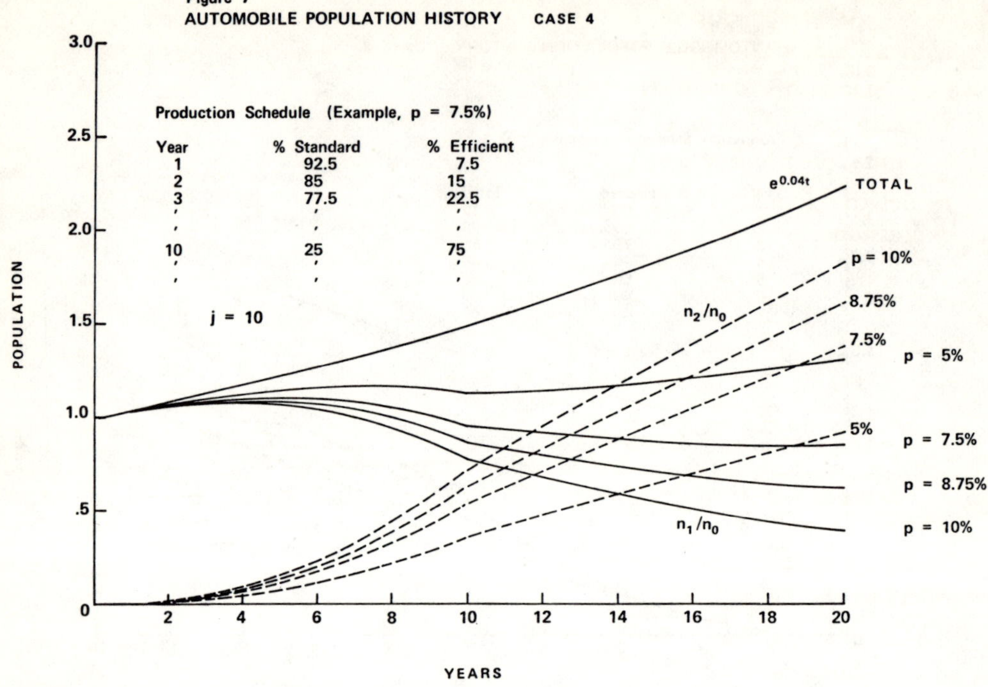

Figure 7
AUTOMOBILE POPULATION HISTORY   CASE 4
3.0
POPULATION
Production Schedule   (Example, p = 7.5%)
Year   % Standard   % Efficient
1   92.5   7.5
2   85   15
3   77.5   22.5
,   ,   ,
10   25   75
,   ,   ,
j = 10
2.5
2.0
1.5
1.0
.5
0
$e^{0.04t}$ TOTAL
p = 10%
8.75%
7.5%
p = 5%
5%   p = 7.5%
p = 8.75%
p = 10%
n_2/n_0
n_1/n_0
2   4   6   8   10   12   14   16   18   20
YEARS

E/E_0
3.0
Figure 8
ENERGY CONSUMED BY AUTOMOBILES   Case 4
ENERGY
2.5
2.0
1.5
1.0
0.5
$e^{0.05t}$
p = 5%
p = 7.5%
p = 8.75%
p = 10%
$g_2/g_1 = 0.5$
$g_1/(g_1)_0 = 1$
j = 10
0   2   4   6   8   10   12   14   16   18   20
POPULATION

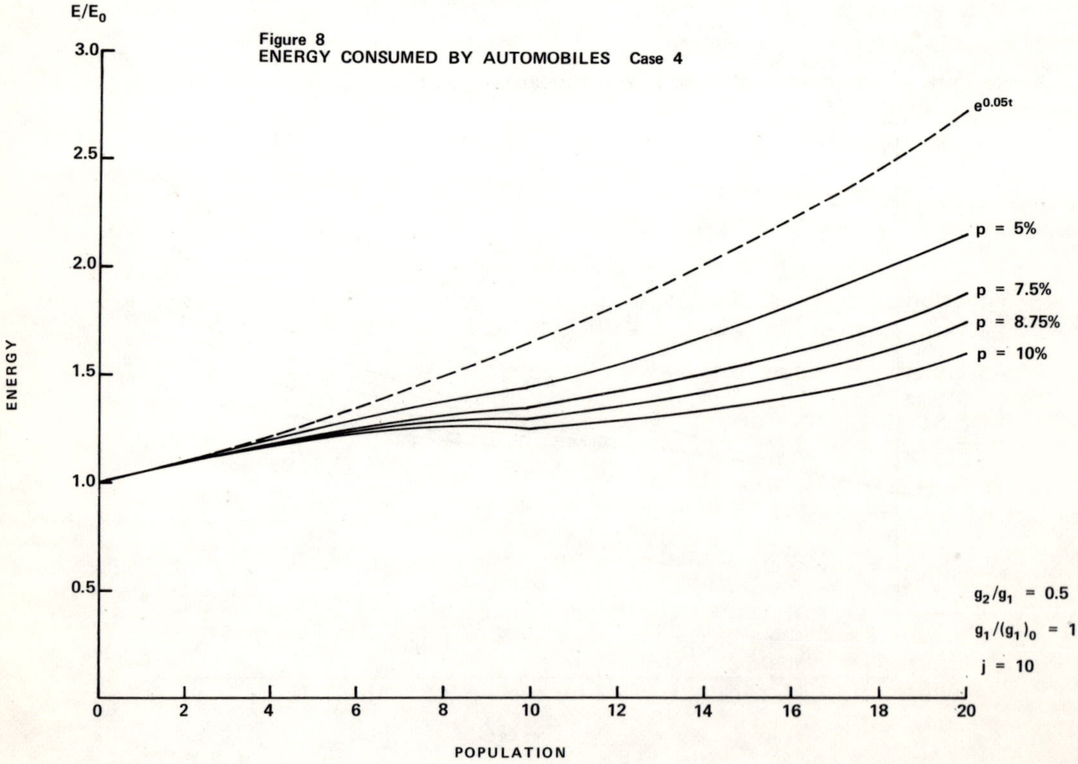

3.   Slowing the Rate of Growth of Energy Consumption in the Residential Sector

As in the case of automobiles, for simplicity we are going to work with a "bimodal" population of dwellings consisting of only two classes: (1) "standard" dwellings that consume e_1 units of primary energy annually on the average; (2) "energy-conserving" structures that consume e_2 units of primary energy annually on the average. With this simple model the annual rate of energy consumption E is given by the following expression:

$$\frac{E}{E_0} = \frac{e_1}{(e_1)_0}\left[\frac{n_1}{n_0} + \frac{e_2}{e_1}\frac{n_2}{n_0}\right] \quad (2a)$$

or,

$$\frac{E}{E_0} = \frac{e_1}{(e_1)_0}\frac{n}{n_0}\left[1 - \left(1 - \frac{e_2}{e_1}\right)\frac{n_2}{n}\right] \quad (2b)$$

where n_1 and n_2 are the number of standard and energy-conserving dwellings or living units, resp., and $n = n_1 + n_2$. The subscript zero again refers to the "base" year.

Suppose that all structures are standard in the base year, and that 10% of all new living units built in the first year are energy-conserving, 20% the second year, etc., up to the tenth year, after which time all new dwellings are energy-conserving. During the same period suppose that a certain percentage (100 R)% of standard living units are "retrofitted" annually so that they consume only e_2 units of primary energy per year.*

At present the annual rate of construction of new living units is about 3% of the existing population, and the demolition rate is about 1% per year. Considering all these processes as continuous the differential equations describing the time history of the population of living units can be formulated and integrated to obtain the time history of the living unit "population" (Appendix). In this simple example the assumption is made that no energy-conserving structures are demolished during the first ten years, but that after that time the demolition rate for these units is the same as for "standard" units.

---

* If more detail is warranted in the calculation, one could consider three classes of structures: (1) standard; (2) energy-conserving; (3) standard structures that are retrofitted.

In Figure 9 the population history of standard and energy-conserving living units is shown for retrofit rates of zero, 1%, 2%, and 3% per year, resp. (Also shown is the growth of the total population, $n = n_0 e^{0.02t}$). For $R = 0$ the population of standard dwelling places peaks between the sixth and seventh years, and then decays very slowly with time. The population of energy-conserving dwellings builds up very slowly, and even after 14 years, these units comprise only about 23% of the total population. On the other hand if the retrofit rate is 2% per year the build up of energy-conserving units is much faster, and these dwelling places amount to about 40% of the population after 14 years.

At present the annual consumption of primary energy per residential unit is growing at about 3% per year, or $\frac{e_1}{(e_1)_0} = e^{0.03t}$. Taking this growth rate as fixed for the present, the effects of introducing new energy-conserving structures and retrofitting old ones on total residential energy consumption rates can be calculated from Eq. (2a) or (2b); the results are shown in Figure 10 for $\frac{e_2}{e_1} = 0.50$.* As expected, with a zero retrofit rate the reduction in annual energy growth rate is modest, from 5% per year to about 4% per year for $t > 6$. Since the residential sector consumes about 22% of all primary energy, this reduction amounts to a reduction of about 0.22% per year in total primary energy growth rate.

However, if a retrofit rate of 2% per year could be achieved (for example), a reduction in growth rate in annual energy consumption in the residential sector from 5% per year to about 3% per year would occur, and this slowdown would account for a corresponding reduciton of about 0.4% per year in the growth rate for total primary energy. When combined with the introduction of efficient vehicles utilizing about 60% of the gasoline per mile that standard vehicles require (Section 2), the two components of an energy-conserving strategy taken together would bring the growth rate in the U.S. primary energy down from 4.2% per year to about 2.8% per year in the crucial time period $6 \leqslant t \leqslant 12$.

---

* According to Reference 3 values of $\frac{e_2}{e_1} = 0.50$ are achievable even with the application of present technology.

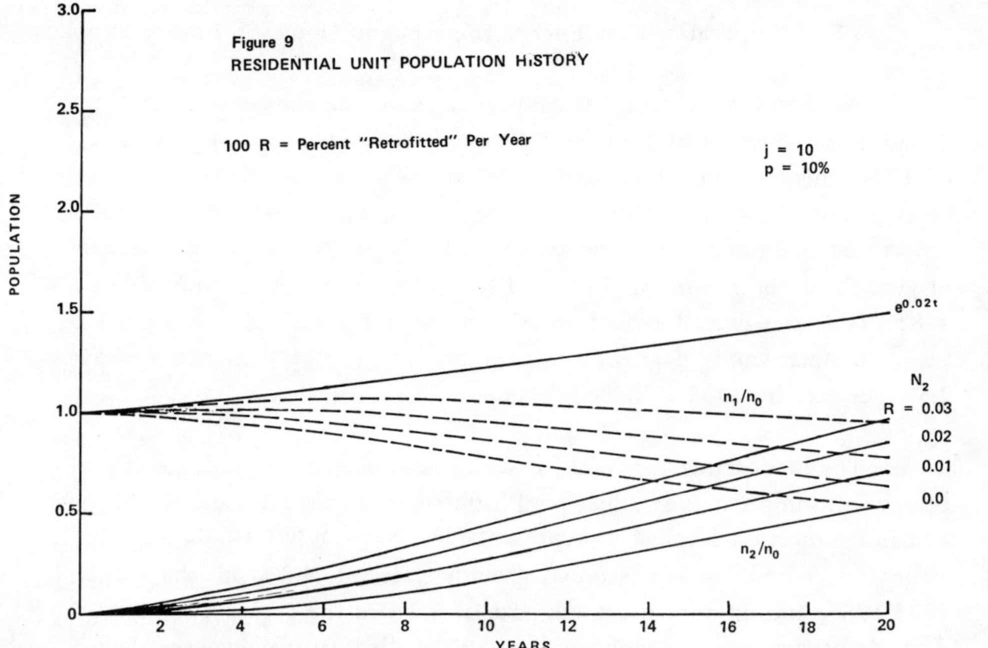

3.0
2.5
2.0
1.5
1.0
0.5
0
POPULATION
Figure 9
RESIDENTIAL UNIT POPULATION HISTORY
100 R = Percent "Retrofitted" Per Year
j = 10
p = 10%
$e^{0.02t}$
n_1/n_0
n_2/n_0
N_2
R = 0.03
0.02
0.01
0.0
ENERGY – CONSERVING UNITS
2 4 6 8 10 12 14 16 18 20
YEARS

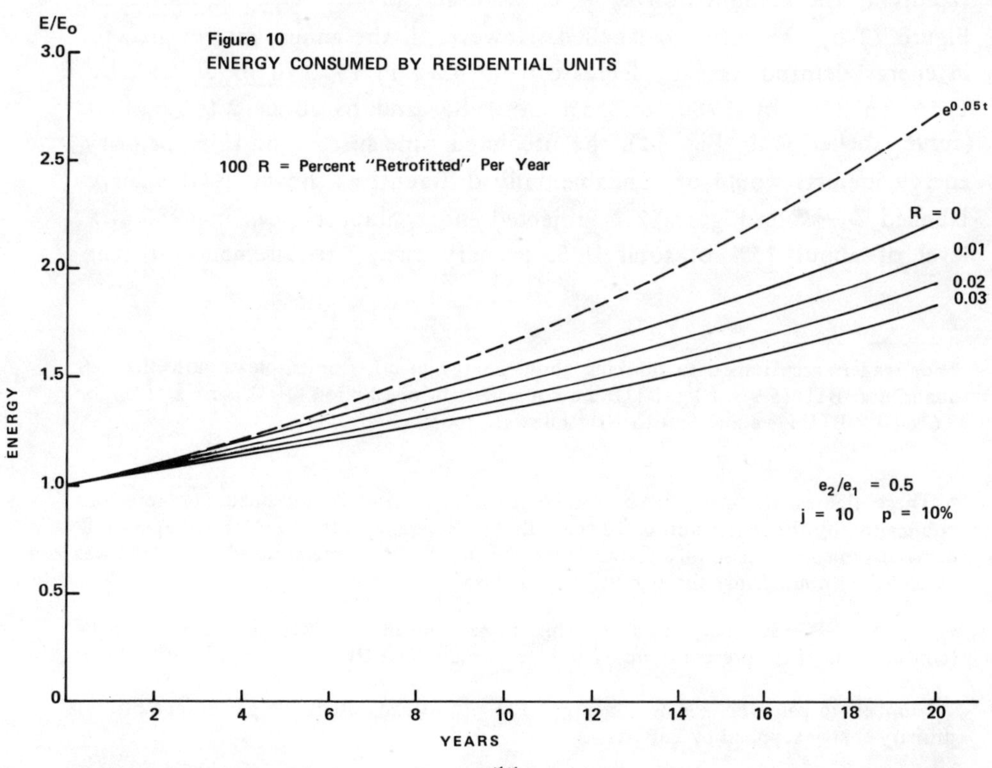

E/E_O
3.0
2.5
2.0
1.5
1.0
0.5
0
ENERGY
Figure 10
ENERGY CONSUMED BY RESIDENTIAL UNITS
100 R = Percent "Retrofitted" Per Year
$e^{0.05t}$
R = 0
0.01
0.02
0.03
$e_2/e_1 = 0.5$
j = 10   p = 10%
2 4 6 8 10 12 14 16 18 20
YEARS

4.   Implications of a Slowdown in the Rate of Growth of U. S. Energy Demand

4.1   Implications for Energy Imports and Domestic Energy Supplies

As shown by the highly simplified examples discussed in Sections 2 and 3, a reduction of 1.2% − 1.4% per year in the annual growth rate of U. S. energy demand is achievable by 1985, considering only space heating and the "automobile" as a total system. Judging by the OEP report[3] an additional reduction of about 0.4% per year in the annual rate of growth of the remaining 50% of U. S. primary energy is attainable by 1985. Thus the overall reduction lies in the range of 1.6% − 1.8% per year; in other words a growth rate in primary energy demand of about 2.5% per year by 1985 is indeed feasible.

The importance of such a slowdown in the rate of growth of energy consumption is graphically illustrated in Figures 11 and 12.* The demand projection labelled ① represents the NPC "intermediate demand" estimate,[2] based on an annual growth rate of 4.2% in the period 1971-1985, and an annual growth rate of 3.2% in the period 1985-2000.† The domestic energy supply projection labelled Ⓐ follows the NPC's "initial appraisal" of the annual rate of growth of about 2.6% through 1995.+ Obviously imports of primary energy would have to increase very rapidly if the combination ① − Ⓐ were actually to occur, as shown in Figure 12 by the curve so labelled. However, if the annual rate of growth in energy demand were to decrease from 4.2% in 1970 to about 3.5% by 1975, to 2.9% by 1980, to 2.5% by 1985, and to about 2% by 1990 (curve labelled ② in Fig. 11), the predicted time history of U. S. primary energy imports would be fundamentally different, as shown by the curve labelled ② − Ⓐ in Figure 12.‡ Projected energy imports peak in 1985 at a level of about 15% of total U. S. primary energy requirements at that

---

*For readers accustomed to thinking about energy in other units, please note that 5.9 quadrillion BTU (5.9 x 10^{15} BTU) are equivalent to one billion (10^9) barrels of oil; *or* 3.6 x 10^{15} BTU are equivalent to 10^{12} kilowatt hours.

† These figures are qualitatively similar to Figs. 1 and 2 contained in a previous publication by the senior author,[5] except that in Reference 5 the NPC "initial appraisal" demand projection based on an annual growth rate of 4.2% in the period 1971−1985 was extended without change through the period 1985−1995.

+ In the NPC Summary Report[2] this projection lies midway between Case IV (continuation of the present trends) and Intermediate Case III.

‡ Numbers in parenthesis shown in Fig. 12 (e.g., (36)) represent the percentage of total primary energy supplied by imports.

Figure 12
ANNUAL U.S. PRIMARY ENERGY IMPORTS AND TOTAL NUMBER
OF TANKERS REQUIRED (250,000 DWT) THREE PROJECTIONS
10^15 BTU
70
60
50
40
30
20
10
NUMBER OF TANKERS
OIL
LNG
630
105
540
90
450
75
360
60
270
45
180
30
90
15
(36)
(34)
(32)
(26.2)
(20.7)
(12.4)
1 A
2 A
2 B
(15)
(10)
1975
1980
1985
1990
1995
YEAR
Figure 11
ANNUAL U.S. PRIMARY ENERGY DEMAND AND
DOMESTIC SUPPLY — TWO PROJECTIONS
10^15 BTU
200
190
180
170
160
150
140
130
120
110
100
90
80
70
60
50
ENERGY
DEMAND
(NPC)
1
2
(2.02)
(2.22)
(2.5)
(2.89)
(3.44)
(B)
(A)
DOMESTIC SUPPLY
(INCLUDING ALASKA NORTH SLOPE)
(NPC)
1975
1980
1985
1990
1995
2000
YEAR

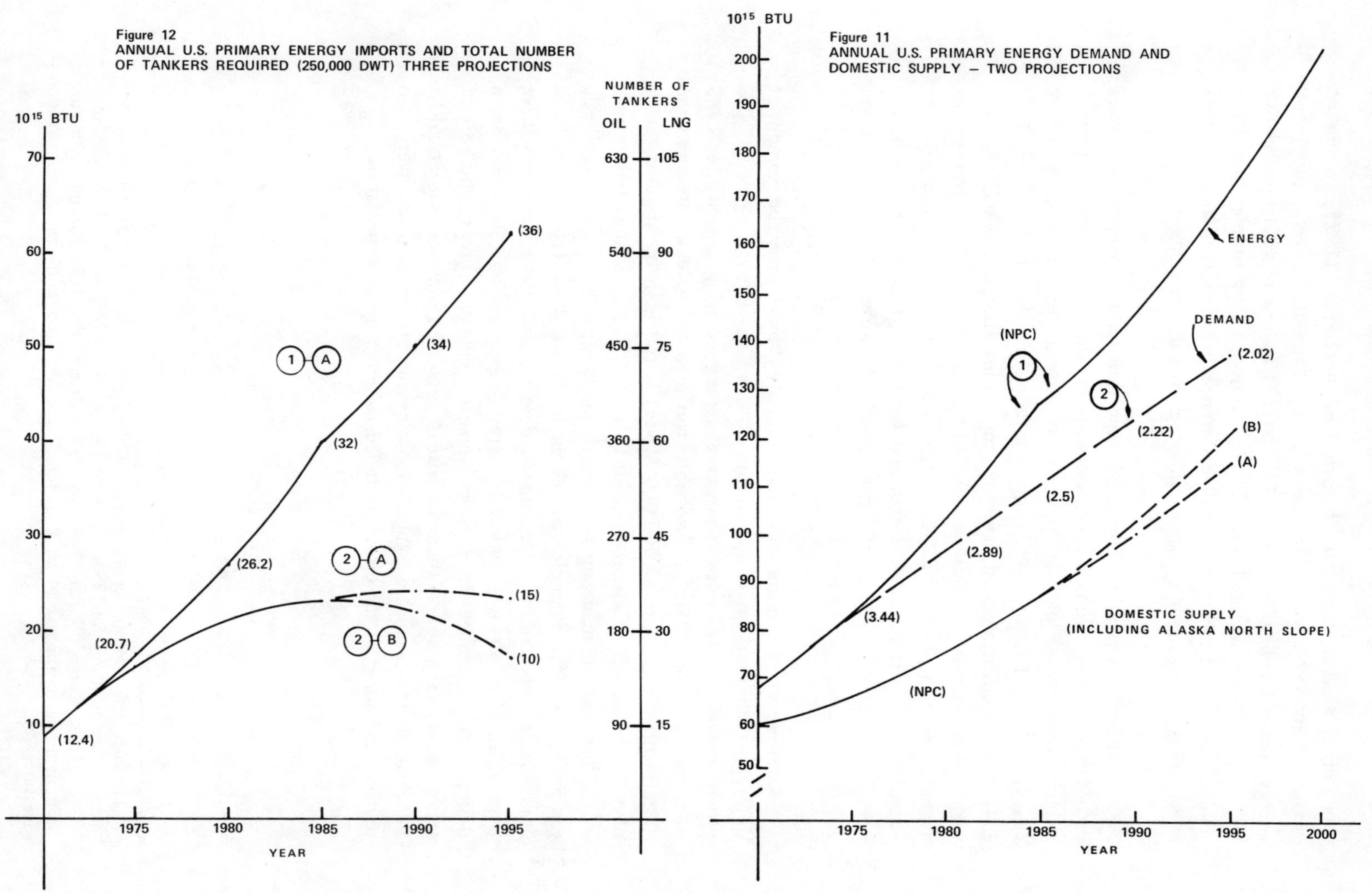

57

time, or at a level about 60% higher than the current (1973) volume of energy imports. Moreover, increases in the rate of domestic supply of energy (curve labelled Ⓑ in Figs. 11 and 12) have a much larger relative effect than if the demand curve ① is followed. This simple illustration reminds us once again that a small difference between two large numbers is very sensitive to modest changes in either of these two numbers.*

Although strenous efforts and important policy changes are needed in order to increase domestic energy supplies, the annual rate of increase required by curves Ⓐ or Ⓑ in Figure 11 is close to the intermediate supply Case III analyzed in the NPC study,[2] rather than the "high" supply Case I. In view of environmental, land use, capital investment and technical and logistical constraints, Case III is regarded by the authors of the NPC report as much more realistic than the "high" supply Case I. For example, the cumulative capital investment required by Case III is about 265 billion dollars (exclusive of electricity generation and transmission) in the period 1971-1985, as compared to 311 billion dollars required by Case I.

However the domestic energy supply "mix" may be somewhat different than that envisioned in the NPC study. Even the lowest NPC projection of installed nuclear power electric generating capacity for 1985 of about 240,000 MW(e) is probably too high by at least 20%, and the "intermediate" estimate of 300,000 MW(e) is too high by about a factor of 1.5. The deficit of about 4 quadrillion BTU's in annual energy supply would have to be made up by a much more rapid build-up of synthetic gas production (for example), as shown in Figure 3 of Reference 5. Such unavoidable uncertainties in estimated demand and domestic supply have large relative effects on required energy imports. Thus it would seem desirable to encourage diversity in domestic energy sources. The cost of this diversity may be less than the cost of increased imports that would be incurred if the available range of domestic energy sources were too narrow, and some of these sources failed to come up to expectation.

---

* The energy demand projection labelled ② in Fig. 11 lies about 12% below the NPC intermediate estimate[2] for 1985, and about 25% below the NPC intermediate estimate for 1995. Thus, curve ② lies about midway between the OEP demand estimates[3] corresponding to the use of all "mid-term" and all "long-term" energy conservation measures, resp.

4.2   Policy Implications

So much has been written recently about U. S. energy policy that only a few salient points are worth emphasizing here, pertaining especially to a slowdown in the rate of growth of energy demand. These points are concerned mainly with the complementary roles of pricing, taxation, incentives and regulations. These remarks are not meant to be difinitive, but are designed to stimulate discussion of policy alternatives.

Almost all recent energy studies agree that the unit price of energy in all forms is bound to increase substantially (in fixed 1973 dollars) over the next decade or two. But increases in unit prices may not be sufficient to reduce the rate of growth in energy demand in a timely fashion, because they affect operating costs much more strongly than "first costs", and because of well-known time lags in response to price changes. For example, the introduction of "efficient" automobiles at the desired rate (Section 2) might be encouraged not only by the predicted increases in the price of gasoline, but also by a "purchase tax" on passenger automobiles in proportion to their performance in gallons/mile,* and/or by specifications on fuel consumption in conjunciton with regulations on pollutant concentrations in automobile exhaust emissions. A "bonus" or "negative purchase tax" might be awarded to those makes of automobiles that do better than these standards.

Similarly, the desired rate of introduction of energy-conserving living units (Section 3) could be encouraged not only by the predicted increase in unit energy prices, but also by regulations adopted at local state and federal levels on the amounts of building insulation required. As our understanding of energy consumption levels of various building designs improves, these crude regulations might be replaced by regulations and incentives based on overall annual energy consumption in BTU per square foot, leaving the design details up to the ingenuity of architects and builders.

---

* **EPA** studies show that an automobile weighing 5000 pounds consumes about twice as much gasoline per mile on the average as an automobile weighing 2500 pounds.

In the next decade, however, even more difficult questions arise, as shown by Figure 2. If the growth rates of 4% per year in *total* automobile population and 1% per year in average mileage driven per vehicle persist, then the annual rate of energy consumption by automobiles begins to increase again after the tenth year, even if no new "standard" vehicles are produced after that year. The recent rapid growth in the number of "recreational vehicles" in the U. S. shows that a reliance on "saturation" to reduce these growth rates may prove to be illusory. In addition to pricing, taxation and regulation, it will probably be necessary to plan in the long run (t $\geq$ 10-12 years) for cities and towns in which energy-consumption is a primary consideration. Distances from home to work, shopping and recreation would be minimized, multi-passenger vehicles utilized wherever possible, and all structures would be designed to certain energy consumption specifications.

References

1. U. S. Energy Outlook: An Initial Appraisal, 1971-1985. Volume Two, Summaries of Task Group Reports, November 1971, an interim report of the National Petroleum Council.

2. U. S. Energy Outlook: A Summary Report of the National Petroleum Council, December, 1972.

3. The Potential for Energy Conservation: A Staff Study. Executive Office of the President, Office of Emergency Preparedness, October 1972 (Washington, D. C.).

4. Hirst, E.: Energy consumption in Transportation in the U.S., ORNL—NSF—EP—15, March 1972. (Work supported by NSF/RANN Program under NSF Interagency Agreement No. AAA—R—4—79.) See also Hirst, E. and Robert Herendeen, Total Energy Demand for Automobiles, SAE Paper #730065, presented at International Automobile Engineering Congress; Detroit, Mich., January 8—12, 1973.

5. Lees, L.: Why the Energy Crunch Came in the 70's. Cry California, Fall 1972, pp. 29-34 (published by California Tomorrow, San Francisco, Calif.).

APPENDIX

A.1   Automobile Population History

Phase 1: $0 \leqslant t \leqslant j$

During this phase the differential equations describing the time history of the automobile population are as follows:

$$\underset{\text{BIRTH RATE}}{\frac{dn_1}{dt}} = \quad b(1 - p't)n \quad - \quad \underset{\text{DEATH RATE}}{d \cdot n} \tag{A-1}$$

$$\frac{dn_2}{dt} = \quad (bp't)n - \quad 0$$

and

$$\frac{dn}{dt} = \quad (b - d)n \tag{A-2}$$

where $p' = 10^{-2} \; p$

According to the last of this set of equations, when b and d are taken as constants

$$n = n_0 e^{rt} = n_1 + n_2, \text{ where } r = b - d$$

By utilizing this expression for n in the differential equation for n_2, and integrating, one obtains

$$\frac{n_2(t)}{n_0} = \quad \frac{p'b}{r} \left[te^{rt} - \frac{(e^{rt} - 1)}{r} \right] \tag{A-3}$$

Also, $\quad \dfrac{n_1}{n_0} \quad = \quad e^{rt} \quad - \quad \dfrac{n_2}{n_0} \tag{A-4}$

Phase 2:

In Phase 2, $t \geqslant j$, and the birth and death rates of standard and efficient vehicles are "frozen" at their values for $t = j$.

The differential equations for this phase are as follows:

$$\frac{dn_1}{dt} = b(1 - p'j)n - d \cdot n_1 \tag{A-5}$$

$$\frac{dn_2}{dt} = (bp'j)n - d \cdot n_2$$

By integrating the differential equation for $n_2(t)$ one finds that for $t > j$:

$$\frac{n_2(t)}{n_0} = p'j \left[e^{rt} - e^{(bj-dt)} \right] + \frac{n_2(j)}{n_0} e^{-d(t-j)} \tag{A-6}$$

$$\text{Also, } \frac{n_1(t)}{n_0} = e^{rt} - \frac{n_2(t)}{n_0} \tag{A-7}$$

In all the calculations for automobiles made in this report $b = 0.11$ and $d = 0.07$.

When the birth and death rates are taken as constants, as in the present analysis, the governing parameters in Phase 1 ($0 \leqslant t \leqslant j$) are $\frac{pb}{b-d}$ (b–d) and j itself, while in Phase 2 the situation is more complicated and the parameters j, p, (b–d) and b, d all enter into the determination of the population history.

A.2 Population History of Residential Units

Phase 1: $0 \leqslant t \leqslant 10$

In this first period,

$$\frac{dn_1}{dt} = b(1 - p't)n - d \cdot n - Rn_1 \tag{A-7}$$

$$\frac{dn_2}{dt} = (bp't)n - 0 + Rn_1$$

$$\frac{dn}{dt} = (b - d)n \tag{A-8}$$

where $100R$ is the percentage of existing houses retrofitted each year. Again, $\frac{n}{n_0} = e^{rt}$, where $r = (b - d)$, provided that b and d are assumed to be constants.

By integrating the differential equation for $n_2(t)$ one gets

$$\frac{n_2(t)}{n_0} = \frac{p'b}{r+R} \left[te^{rt} - \frac{(e^{rt} - e^{-Rt})}{r + R} \right]$$

$$+ \left(\frac{R}{r+R} \right) \left[e^{rt} - e^{-Rt} \right] \tag{A-9}$$

and $\dfrac{n_1(t)}{n_0} = e^{rt} - \dfrac{n_2(t)}{n_0}$ $\tag{A-10}$

Phase 2: For $t < 10$, $(1 - 10p') = 0$ in this simple illustrative sample, so

$$\frac{dn_1}{dt} = -(d + R) n_1$$

and

$$\frac{n_1(t)}{n_0} = \frac{n_1(10)}{n_0} e^{-(d+R)(t-10)} \tag{A-11}$$

In this report $b = 0.03$ and $d = 0.01$.

SESSION II

REDUCING THE DEMAND: SHELTER AND
TRANSPORTATION

Chairman: Roger W. Sampson

Sr. Staff Eng., Management Plans &
Proposal Operations, Advanced Systems &
Technology, McDonnell Douglas Astro-
nautics Co., Huntington Beach, Ca.

ENERGY USAGE IN THE HOME –
CONSUMPTION AND CONSERVATION

R. B. Rosenberg*

Energy usage in the home accounts for about 20%
of all energy consumed in the United States each
year. Moreover, the residential sector is the
second-fastest growing energy market. An effective
program of residential energy conservation therefore could have a significant impact on our future
energy needs. One use area dominates residential
energy consumption: space conditioning (heating
and cooling), which used 61.2% of the energy consumed in 1968. This area then offers the greatest
potential for making a major impact on our national
energy demand. This paper describes some of the
more promising methods for realizing significant
energy savings and the problems associated with
their application.

INTRODUCTION

From the consumer's point of view, at least, there can be no argument that

the current energy shortage is real. Gasoline shortages are widespread;

bulk propane is unavailable, even to priority customers in some areas; and

many workers are being laid off because of a lack of fuel. Considerable

disagreement exists, however, on whether the current shortage will persist

for a significant period. Against this backdrop of uncertainty, the U.S.

Government, industry, and private citizens must make some major energy

policy decisions that will have far-reaching (and possibly very expensive)

consequences.[†]

---

*Vice President, Engineering Research, at the
Institute of Gas Technology, IIT Center, Chicago
60616, (312) 225-9600.

[†] For detailed information on energy supply, the reader is referred to other
IGT publications.[1,2]

One energy policy question concerns the importance of energy conservation.
Fortunately, energy conservation makes technical and economic sense,
regardless of whether the energy shortage persists. If the shortage persists,
conservation will be an absolute necessity in order to spread our limited
resources over all those utilization areas we deem vital to our way of life.
If, on the other hand, the shortage lessens, conservation will decrease costs
and preserve a reasonable portion of our limited resources for the next
generation. This latter rationale for conservation may sound like an appeal
to "motherhood and the flag," but it is much more than that. Preservation of
some of our resources actually means maintenance of a sufficient inventory
to justify continued production, thereby avoiding another crisis.

ENERGY CONSERVATION

Let us now consider where and how much energy we can conserve. Based
on studies at the Institute of Gas Technology (IGT), it is technically feasible
and economically viable to conserve significant amounts of energy in every
major consuming sector. For example, in the industrial sector (which
consumes 41.2% of all U.S. energy[3]), under the sponsorship of the American
Gas Association IGT is preparing a series of seminars on rapid-payout
techniques that can be applied immediately to conserve energy. For two
major utilities, we are developing new, energy-conserving industrial process
equipment, as well as in-plant process evaluations. In the transportation
sector (which accounts for 25.2% of the total U.S. energy consumption[3]), we
are evaluating alternative engine-fuel combinations for maximizing economy
and fuel availability, as well as the use of hydrogen as a more efficient aircraft
fuel. In the commercial sector (which consumes 14.4% of all U.S. energy[3]),
we are developing new heating/cooling equipment and a variety of other
techniques. The remaining 19.2% is consumed by the residential sector,
and it is energy conservation in the home that is the subject of this paper.

RESIDENTIAL ENERGY CONSUMPTION

Before considering how energy conservation can be applied to residences,
let us examine how energy is used. As shown in Table 1, the largest single
application by far is space heating, which accounts for 57.5% of all energy
consumed in the residential sector. Space heating and air conditioning
combined consume more than 60% of the residential total. As a percentage
of total energy used in the U.S., residential space heating accounts for 11.5%.

Table 1

RESIDENTIAL ENERGY CONSUMPTION BY END USE[3]

End Use	1968 Consumption (10^{12} Btu)	% of Residential Total	% of National Total
Space Heating	6,675	57.5	11.0
Water Heating	1,736	14.9	2.9
Refrigeration	692	6.0	1.1
Cooking	637	5.5	1.1
Air Conditioning	427	3.7	0.7
Clothes Drying	208	1.8	0.3
Other	1,241	10.6	2.1
Total	11,616	100.0	19.2

The second-largest category of residential energy consumption is water heating (14.9% of the residential total), followed by refrigeration (6.0%) and cooking (5.5%). All other uses account for the remaining 12.4%. This latter category comprises at least 27 different uses, as shown in Table 2. The largest of these (lighting) accounts for less than 1% of the total national energy.

On the basis of these data, apparently the area with the greatest potential for savings is space conditioning (heating/cooling). Of course, the sum of many small savings can be significant. Nevertheless, our greatest emphasis should be in this area because it is here that we can reap the greatest rewards.

THE COMPLEXITIES OF ENERGY CONSERVATION

A note of caution must be interjected at this point. Every method that is suggested to conserve energy may not actually do so. Moreover, predictions of the amount of energy that can be conserved by various techniques frequently are grossly overestimated because proponents of specific concepts often adapt too simplistic a viewpoint and fail to recognize the "systems" nature of residential energy consumption. An excellent example is the case of the pilot light.

Many persons have estimated the number of pilot lights in operation and, after multiplying this figure by an average energy input per pilot light, have predicted major energy savings if pilot lights are eliminated. This position is absolutely wrong because it is far too simplistic. Consider the pilot light on the water heater. The energy input to this pilot light is less than or equal to the heat loss from the storage tank. Thus, if the pilot light were eliminated, the main burner would have to operate more frequently to maintain the desired

Table 2
"OTHER" USES OF ENERGY[3] *

End Use	Annual Consumption (10^{12} Btu)	% of Residential Total
Lighting	412	3.5
Television Sets	352	3.0
Food Freezers	220	1.9
Clothes Driers	208	1.8
Irons	88	0.8
Frying Pans	66	0.6
Automatic Coffee Makers	57	0.5
Radios	52	0.5
Bed Coverings	43	0.4
Washing Machines	41	0.4
Dishwashers	36	0.3
Fans (Circulating)	34	0.3
Heaters (Portable)	32	0.3
Vacuum Cleaners	26	0.2
Toaster	23	0.2
Dehumidifiers	15	0.1
Broilers	15	0.1
Hot Plates	14	0.1
Clocks	10	0.1
Humidifiers	7	0.1
Mixers	7	0.1
Shavers	5	< 0.1
Food Disposals	4	< 0.1
Hair Driers	3	< 0.1
Blenders	3	< 0.1
Knives (Carving)	1	< 0.1
Toothbrushes (Automatic)	1	< 0.1

---

*The total "other" consumption disagrees somewhat with that in Table 1 because of different primary information sources.

water temperature. This situation would essentially offset any saving realized by eliminating this pilot light (allowing for differences in the heat-transfer effectiveness between the main burner and the pilot light).

The pilot light on the range is a more complex situation. For a humidified home during the heating season, the pilot light represents a 100% efficient space heater, compared with the conventional furnace **with** a rated efficiency of 75-80% and an actual operating efficiency (sometimes called seasonal efficiency, as discussed later under efficiency definitions) of about 65%. For a nonhumidified home during the heating season, the effective efficiency of the pilot light is decreased to about 90% because credit cannot be taken for the heat added as latent heat in the water from combustion. In either case, eliminating the pilot light <u>increases</u> the energy required for heating. On the other hand, the input energy of the pilot light is wasted during the nonheating months. And, if the house is air-conditioned, this waste is compounded by the increased energy required to cool the home.

Therefore, determining the amount of energy that can be conserved is a complicated analysis that must consider — among other things — relative heating efficiencies, the number of air-conditioned homes, and the ratio of heating to nonheating hours per year.

Marketing is another factor that further complicates evaluation of the pilot light on the range. The elimination of the pilot light and its replacement by an electric igniter will significantly increase the cost of today's range (probably by about $20-$30 for the oven and top burners). This increase could upset the competitive balance between gas-fueled and electric ranges. Because independent studies[3,4] have shown that electric ranges use more primary (source) energy than gas ranges, by a factor of about 1.6, an increase in the proportion of electric range sales will increase U.S. energy consumption.

This discussion should not be construed as an argument in favor of pilot lights. Reasons for advocating their elimination do exist; however, conservation may not be one of them. In fact, when all factors are considered, eliminating pilot lights may well increase total energy consumption. Therefore, great caution should be exercised in evaluating the potential "savings" to be realized by any energy conservation concept.

A POSITIVE APPROACH TO ENERGY CONSERVATION

As stated earlier, energy conservation in space conditioning (heating/cooling) represents the greatest potential. We discuss two basic approaches here: decreasing heat losses from buildings and improving space-conditioning equipment.

Many persons have suggested various means for decreasing heat losses, such as changing roof (and house) colors, redesigning windows, appropriately locating the house to minimize wind effects on infiltration, increasing humidity, and so on.[5] Of these various techniques, the one with the greatest effect is increasing the amount of insulation. Already, we are witnessing efforts in this direction, for example, the changes in FHA standards for wall and ceiling insulation. The problem here is that these standards, as well as many other proposed conservation techniques, are applicable only to new construction. They do nothing to decrease the energy now being consumed in the almost 69 million existing residences (in 1971).

The flywheel effect of these existing units is enormous. For example, if energy consumption in all houses built in 1971 (a record year) were cut by 25%, total consumption in the residential sector would be decreased by less than 1%. On the other hand, if this energy conservation were applied to as few as 3.5% of the _existing_ residences, we would achieve more than from conservation in _all_ new homes for that year. Of course, we should not ignore energy conservation in new construction because it will continue to pay off year after year as more homes are built. Nevertheless, we cannot afford to ignore energy conservation in existing structures; a great potential for savings exists here.

One study[6] has estimated that 40-45 million of the 65 million homes in the U.S. were built prior to World War II. As shown in Table 3, building standards prior to 1940 included little or no insulation. Even in the 1950's, insulation standards were minimal. Assuming, therefore, that 50% of the existing homes have little or no insulation, we estimate that 2.4 trillion CF*

---

*Savings per year arbitrarily assumed to be 74,000 CF to account for
 warmer climates.

of natural gas could be saved annually if all these homes were retrofit for full insulation. This represents a potential saving of more than 10% of <u>all</u> the natural gas used for all purposes in the U.S.

Table 3

EFFECT OF INSULATION ON HEATING REQUIREMENTS*

Insulation	Annual Heat Required (1000 CF) of natural gas	Annual Savings (% of maximum input)
None (Typical Pre-1940 Construction)	270	--
3 inches in Attic Plus Storm Windows and Doors (Typical 1950's)	180	33
Full (6 inches in Attic and ~ 3 inches in Walls Plus Storm Windows and Doors)	122	55
Annual Savings (Maximum)	148	55

*Calculated for a typical 1400-square-foot ranch-style house in a cold climate for a heating season.

Table 4 compares the cost-effectiveness of the various energy conservation techniques. Installing 3 inches of ceiling insulation is the <u>most</u> cost-effective and provides the <u>largest</u> energy saving. Adding another 3 inches in the ceiling is next most cost-effective, but provides the smallest incremental energy saving.

Table 4

COST-EFFECTIVENESS OF ENERGY CONSERVATION TECHNIQUES

	% of Total Energy Savings	Retrofitting Insulation Cost*	% of Total Energy Saving/ $ of Cost
No Insulation	0		
3 inches in Attic	47	$105[†]	0.45
6 inches in Attic (Addition of 3 inches)	7	$ 70[†]	0.10
Storm Windows and Doors	13-26	$250-300	0.05-0.09
3 inches in Walls	33-20	$720	0.05-0.03

*Installed by consumer on a do-it-yourself basis.

[†] Based on average data.

A positive program for insulating existing homes has been undertaken in the
State of Michigan. In this program, two Michigan utilities are actively
promoting insulation sales through advertising and mail campaigns. In
addition, they are offering to arrange for installation and to finance it
through the gas bill. Because payout periods for the homeowner can be
very short (2.15 years if he installs the insulation himself), the program is
proving to be attractive.

According to preliminary reports, in the service territory of the Consumers
Power Co., insulation sufficient to provide attic insulation for 7065 homes
was sold by all outlets in December and January. In the service territory
of the Michigan Consolidated Gas Co., insulation sufficient for 14,600 homes
was sold. The annual potential for energy saving from these homes could
be as high as 1.7 billion CF of natural gas (1.7 trillion Btu).* This is an
indication of what can be achieved in a positive program.

Table 4 also illustrates that retrofitting <u>wall</u> insulation can contribute a
significant energy saving, but is not cost-effective. Here, then, is a
technical challenge: to develop new materials or new installation techniques
so that this major conservation potential can be economically justified.

IMPROVED HEATING/COOLING EQUIPMENT

As mentioned earlier, another way to achieve a major decrease in the energy
used for space conditioning is to improve the heating/cooling equipment.
Because state-of-the-art equipment is already good, the development of a
major improvement will not be easy.

Many sins are committed in the name of efficiency when different heating/
cooling systems are compared. To avoid this, we will adopt the following
terminology.

- <u>Design or Rated Efficiency</u>. This refers to an efficiency measured under
  specified conditions for rating purposes. In general, these conditions
  correspond to steady-state operation at design temperatures. On this
  basis, electric-resistance-heating systems are 95-100% efficient; gas-
  fueled equipment must be at least 75% efficient for approval; and oil-fired
  equipment will be about 75% efficient.

---

*Neglecting the energy cost to make, market, or install the insulation.

- <u>Seasonal or Utilization Efficiency.</u> In actual practice, equipment must operate in a cyclic mode at off-design conditions. The seasonal efficiency is usually based on measurements in actual homes and includes inefficiencies associated with case losses and air flow through the draft diverter. In addition, the condition (maintenance) of the system and the accuracy of its sizing relative to the house heat load will affect seasonal efficiency. On this basis, most gas-fueled equipment will be found to be 60-65% efficient. Oil-fueled furnaces usually are lower, unless they are well maintained, because oil has a greater tendency to soot.

- <u>Overall Efficiency.</u> This efficiency term accounts for differences between primary and secondary energy sources. For example, natural gas and oil are primary energy sources, whereas electricity is a secondary source (i.e., it must be generated from a primary source). The national average efficiency of electric power generation (including distribution and transmission losses) is about 30%. Thus, the overall efficiency of an electric-resistance space heater is less than 30%.

- <u>Resource Utilization Efficiency.</u> This efficiency measures the actual amount of natural resources that must be used to provide 1 unit of useful energy. It includes the inefficiencies in producing, transmitting, and distributing our primary energy sources. Based on data compiled by the Council for Environmental Quality, a gas furnace has a resource utilization efficiency of about 40%, an oil furnace will be 20% or less, and an electric-resistance space heater will be between 10% and 20%, depending upon the primary energy source to the power plant (oil, coal, nuclear, or natural gas).

The <u>design</u> efficiencies for electric-, gas-, and oil-heating systems afford only small room for improvement. For example, if the rated or steady-state efficiency of the gas furnace (normally 80% but at least 75%) approaches 90%, the water of combustion condenses, and this corrosive water greatly decreases equipment life. Thus, the potential for improvement in the type of equipment is limited. Similarly, electric-resistance equipment is highly efficient. Improvements can be made in the efficiency of electricity production or in the applications of gas and oil furnaces, but again, the expected energy saving is small.

The improvement that offers the greatest potential for energy conservation is a practical, cost-effective heat pump. These units can be either electrically driven or heat-actuated (gas- or oil-fueled). The reason for current interest in the heat pump is its very high seasonal efficiency.

Heat-pump performance is measured by a COP, or coefficient of performance, which is defined as the heat transferred by the heat pump divided by the energy input to drive the system. This figure can exceed 1.0 without violating any laws of thermodynamics because the energy input term actually is pump work, whereas the heat transferred and supplied to the home comes from the ambient air. Considered on this basis, the gas furnace has an operating COP of 0.6-0.65, and the electric-resistance heater has an overall COP of 0.3. Heat pumps have demonstrated COP's of more than 2.0 for heating and up to 1.2 for cooling. Thus, the energy requirements for heating have the potential to be cut by a factor of 2 or 3.

HEAT-PUMP SYSTEMS

A heat pump is a combination of a prime mover and a refrigeration cycle. Thus, many different hardware combinations and fundamentally different thermodynamic cycles are involved. Any one or more of several developments could provide the improved heating/cooling equipment.

Currently, the only heat pump available for residential applications operates with an electric-motor-driven Rankine refrigeration cycle for heating and cooling. This system has two fundamental problems. First, the capacity of the machine decreases as the outside temperature decreases. Thus, this heat pump has its lowest output when heat is needed most. To overcome this, electric-resistance heaters with an <u>overall</u> COP of 0.3 are used; this largely offsets the heat pump's energy-conserving potential. Second, frost builds up on the outside coils and the electric energy used for defrosting is wasted. One study[7] estimates an <u>overall</u> COP for heating of 0.75 for these electric units. The estimated cooling COP is 0.6.

Heat-actuated heat pumps can overcome these fundamental problems in several ways, for example, by using a different cycle. The following systems of heat-actuated heat pumps are being or have already been tested:

- Stirling-engine-driven Rankine cycle
- Rankine-engine-driven Rankine cycle
- Otto-engine-driven Rankine cycle
- Brayton-engine-driven Brayton air cycle
- Brayton-engine-driven Rankine cycle.

The performance (COP) of heat pumps will vary because of cycle or equipment differences or differences in ambient temperature, installation, or maintenance. For the above systems, COP's of 1.4 to 2.0 have been reported for heating and COP's of 0.7 to 1.2 for cooling.[7] Most of these were measured under laboratory conditions. However, in one notable study, a seasonal COP of 2.2 was demonstrated for three heat-actuated systems operating in homes. One of these was modified to include an advanced control system, and a seasonal COP of 4.3 was achieved.

A second means of overcoming the problems with electric systems is to use the waste heat from the prime mover in a heat-actuated system to provide supplemental heat, when necessary, and to accomplish defrosting. Recent work at IGT has indicated a major potential in saving energy for this approach.

Heat-actuated heat pumps can be retrofit into existing structures to provide both heating and cooling. They also can be used in new construction. Thus, they offer a major potential for energy conservation. Most of the current development is being sponsored by the American Gas Association. However, a much larger effort is justified and should be undertaken.[7]

CONCLUSIONS

Energy conservation is important, and many cost-effective methods are available by which it can be accomplished. Some of these methods save only small amounts of energy, but taken collectively, they can be significant.

This paper has highlighted two methods for conserving energy in residential applications: increased insulation and heat pumps. These methods offer the potential for major energy savings, and they can be retrofit to existing homes so that energy conservation can be accomplished rapidly. Some equipment is available, so we can start today. However, research can greatly expand the potential of these energy conservation methods.

REFERENCES CITED

1. H. R. Linden, "Review of World Energy Supplies," SNG Symposium I. Papers on Substitute Natural Gas From Hydrocarbon Liquids. Institute of Gas Technology, Chicago, 1973, pp. 15-53.

2. H. R. Linden, "SNG in the U.S. Energy Balance," Part 1, Gas, vol. 49, July 1973, pp. 29-33; Part 2, ibid., August 1973, pp. 65-66.

3. Stanford Research Institute, Menlo Park, Calif., Patterns of Energy Consumption in the United States, for the Office of Science and Technology, Executive Office of the President. U.S. Government Printing Office, Washington, D.C., January 1972. Stock No. 4106-0034.

4. E. Hirst and J. C. Moyers, "Efficiency of Energy Use in the United States," Science, vol. 179, March 30, 1973, pp. 1299-1304.

5. Building Environment Division, Center for Building Technology, Institute for Applied Technology, National Bureau of Standards, Technical Options for Energy Conservation in Buildings. Technical Note 789. U.S. Government Printing Office, Washington, D.C., July 1973.

6. R. J. Littin, "Impact of Insulation in Conservation," Paper presented at Gas Conservation Seminar, O'Hare International Tower, Chicago, October 2, 1973.

7. D. Kennedy, "Energy Conservation...Some Wheat, Some Chaff," A.G.A. Monthly, vol. 55, February 1973, pp. 4-25.

8. R. A. Conden, Consumers Power Company, Jackson, Mich., private communication, Jan. 29, 1974.

EVOLUTION AND EFFECTS OF ENERGY CONSERVATION IN THE TRANSPORTATION SECTOR

Richard L. Strombotne[*]
Bunli Yang[+]

The transportation sector of the U.S. accounts for 25% of the energy consumed by the U.S., and it accounts for more than half of the U.S. annual petroleum usage. With the present and growing problem of energy (especially petroleum) availability, one of the important ways to bring supply and demand into balance is through energy conservation. This paper discusses a variety of possible ways to reduce the relative energy consumption of the transportation sector, with emphasis on highway vehicles, provides estimates of energy savings and points out the important factors that must be evaluated along with potential savings of energy. Changes in the design of automobiles to obtain greater fuel economy appear to be a promising approach.

INTRODUCTION

Since the start of the Arab oil embargo in October 1973, the United States has been going through withdrawal pains. Frustrated motorists

---

[*]Assistant for Physical Sciences, Office of the Secretary
  U.S. Department of Transportation

[+]Transportation Analyst
  U.S. Department of Transportation

waiting in long lines for gasoline, truckers finding high prices and
partial fill-ups at truck stops, bus and train riders in cars now
suddenly filled to capacity, and owners of large cars who sell at a
loss to buy smaller cars are among those feeling the pains of about
a 16% shortage in oil supplies. The providers and users of transportation services are quick to feel the effects of reduced oil supplies
because the Nation's transportation vehicles and systems rely almost
completely on the high energy density, easily distributed products of
petroleum. As a result, the transportation sector is conserving
energy by necessity.

This paper discusses briefly the energy use of the transportation
sector and the basic methods for energy conservation. It gives some
estimates of energy savings due to several specific energy conservation actions and the time phasing of conservation actions and effects.
The strategies to induce energy conservation are discussed. Next it
provides estimates of energy savings over time due to changes in
highway vehicles and the highway vehicle fleet. It describes the
improvements in automobile fuel economy that appear to be possible
through the application of relatively mature technology. Finally, it
notes the short-term effects of energy conservation upon consumers
and industry and suggests some possible long-term effects.

TRANSPORTATION ENERGY USE

Until recently the transportation sector consumed directly about 25%
of the total U. S. energy budget and had been projected to continue to
consume about that same share for the next several decades. [1,2]
It uses more than half of the petroleum consumed by the United States.
Highway vehicles account for about 75% of the direct energy consumption by the transportation sector and aviation accounts for about 10%.
Rail, pipeline, waterborne traffic, and miscellaneous users split the
remainder. [1]

In 1972, more than 100 million passenger vehicles and about 21 million trucks traveled almost 1,300 billion miles and consumed over 105 million gallons of gasoline and diesel fuel at an average overall fuel economy of just over 12 miles per gallon (mpg). The passenger car fleet averaged about 13. 5 mpg. This information from the Federal Highway Administration illustrates the large numbers associated with any rigorous discussion of either the transportation sector or national energy demand. The total volume and modal distribution of transportation services are closely related to the demographic and economic characteristics of the country. The dominant pattern of life in the U. S. with residential areas away from central business districts and dispersion of family members has evolved with a level of personal mobility not achieved elsewhere. The data of the 1972 National Transportation Report account for (both historically and in projection) the influence on transportation demand of such factors as: population, income, population distribution patterns, quality of transportation, and the cost of transportation relative to other costs.[4] In brief, it projected passenger travel expenditures to increase at a rate of 4. 3% annually for the two decades between 1970 and 1990. There is also much information available about the use of energy by the various modes of transportation, including their total fuel consumption, ton-miles or passenger-miles delivered, and relative energy efficiency.[1,5] It is sufficient to point out here that aviation generally requires more fuel per unit of service than the surface transport modes and that the automobile and truck normally use more energy per unit of service than public transport modes as for example, bus and train. In some specific situations, however, these relations are not true.

CONSERVATION METHODS

There are only a few methods, well known, to conserve energy in the transportation sector.[1,5,7] The following equation provides a focus for discussion.

$$E_p = \sum_i \frac{(PM)_i}{u_i\, e_i},$$

where E_p is the total energy consumed (in gallons) by the passenger transportation modes i, $(PM)_i$ is the number of passenger miles, u_i is the average vehicle occupancy in passengers per vehicle, and e_i is the average vehicle efficiency in vehicle miles per gallon of the mode i. The first method is to reduce the total demand for transportation services. Substitution of telecommunications for transportation, land use policies to permit walking instead of driving, better planning of trips to make one trip do the work of several are often cited as specific ways to reduce demand. Events of recent months have shown that simple lack of fuel or rationing will also effectively reduce total travel. Indeed, there was 13.7% less automobile travel on the first two Sundays in December when gas station operators voluntarily closed on Sundays and people took into account the sudden unavailability of gasoline. The second method is to shift demand from less efficient to more efficient modes. Shifts from air to any form of surface transportation, shifts from auto to intercity bus or train, to public transportation, or to bicycles are examples in this case. The third method of energy conservation is to increase the average vehicle occupancy factor u_i, the average passengers per mile, in any mode of transportation. The best example of this method is car pooling. As of last December, more than 125 organizations were using the Federal Highway Administration's car pool matching computer program. The fourth method is to improve the average vehicle efficiency or fuel economy to obtain more miles per gallon. Better maintenance and operations practices will improve the efficiency of existing vehicle fleets, whether they are aircraft, ships, trains, buses, trucks, or cars. For highway vehicles, operational changes include reduced speed limits, improved traffic flows, and smooth driving without jack rabbit starts, while keeping engines in tune and tires

correctly inflated are obvious examples of good maintenance
practice. It has been estimated that average fuel consumption of
cars could be reduced by $1\frac{1}{2}\%$ by increasing the regular engine
maintenance frequency to every 6, 000 miles from every 12, 000 miles.
New additions to the vehicle fleet may intrinsically be more efficient
than the average vehicle in the fleet and thereby increase the overall
efficiency. Besides the shift to small cars that is now occurring in
the automotive industry, cars of the future may have better fuel
economy due to improvements or modifications to the engine,
modifications to the rest of the car, or the use of a new more efficient
engine.

The use of fuels derived from nonpetroleum based energy resources
conserves petroleum, if not energy. In the future, gasoline or
distillate from coal or oil shale is likely to find increasing use.
Methanol, although in short supply as a fuel for transportation now,
may supplement the liquid hydrocarbons. Hydrogen, which is a
chemical way of storing energy derived from nuclear power or
other energy resources, may turn out to be advantageous as a fuel,
but has severe cost and storage problems as a fuel for transportation
and may find limited application. The various alternative fuels for
transportation have received much attention and will not be discussed
further here. [1,7]

CONSERVATION STRATEGIES
Consider next the strategies that are available to public policy
makers in order to achieve the energy savings of conservation
measures. [8,9] The conservation strategies run from minimal
involvement of the government to a high level of involvement. The
wait-and-see approach lets the existing forces and marketplace
pressures sort out the relative priorities for fuel use for both
supply and demand. In a highly dynamic complex situation this
approach has the advantage that matters are not made worse by

taking the wrong action. It is not likely to prevent gross inequities for some groups arising from the rapid change in relative economics and availability of fuels. The recent truckers' strike provides an example of the dangers of letting matters drift.

A second strategy is to inform the public about the nature of the problem, of the available alternative courses of action, and the effects of those alternatives so that they can make informed choices. The next stage is the exhortation of the public to do the right thing. This technique may be quite effective when a relatively small number of people or organizations are involved. Last fall, the airlines voluntarily cut back their cruising speeds after being asked to do so. When the large and diversified population of users and operators of cars and trucks are the ones who must be persuaded to do the right thing (assuming someone knows what that is), then one can reasonably expect only a limited and delayed response. Yet last December, the Gallup Poll reported that three-quarters of households had turned down their thermostats and more than 60 percent of those interviewed said they were driving more slowly.   Continued voluntary adherence to lower levels of comfort and to reduced speed limits is problematical.

The government can also intervene in the marketplace by taxing or providing subsidies. Gasoline is now taxed both by state and Federal Governments, not to control its use, but to raise revenue. The price elasticity of demand for gasoline is relatively low, at least for small price changes. Since the most effective way to obtain a desired response is to reward it whenever it occurs, people who buy or use cars with good fuel economy should be rewarded, and conversely they should receive some kind of penalty when they buy or use cars with poor fuel economy. The main issue is whether the greater annual fuel costs for a gas guzzler than for a less thirsty car are sufficient to reduce the fuel consumption of cars to meet the demands of public policy or whether greater incentives are needed.

These incentives could take the form of a fuel consumption penalty
tax levied on cars at the time of their first sale or an annual tax. A
tax on some attribute of a car which correlates with fuel economy,
such as weight or horsepower or engine displacement, will affect those
attributes, but does not necessarily affect fuel economy. Correlation
is not causation and a tax on engine displacement may reduce the size
of engines used, but still permits the use of inefficient engines. There
does not appear to be any ground swell for taxation as a way to conserve
energy in the transportation sector, perhaps because of concern about
the regressive nature of some of the proposed taxes.

Direct regulation of the activities of business or the general public,
puts the government in most intimate involvement with the affairs of
citizens. Still there are degrees of regulation and infinite variations
upon regulations. Several variations on the theme of fuel economy
regulation follow. A bill to require labeling of automobiles as to
their fuel economy has been proposed as a substitute for the present
voluntary program. Mandatory labeling presumably would give the
public the kind of information it needs to be able to make better
choices among similar models with different fuel economy
characteristics and also would give manufacturers incentive to
improve the fuel economy of their products. A section of the Energy
Emergency legislation (S. 2589) passed by both the Senate and the
House of Representatives (in somewhat different forms), calls for
the Department of Transportation and the Environmental Protection
Agency to jointly study and report within 120 days the practicality of
fuel economy standards that would require a 20% improvement in the
fuel economy of the entire annual production of each automobile
manufacturer starting with the 1980 model year. Another bill,
S. 2176, passed by the Senate and pending before the House, calls
for the Secretary of Transportation to establish a minimum fuel

economy standard within 18 months that would apply to 1978 model year cars. These different approaches to accomplish the same purpose have quite different ramifications.

Rationing of gasoline is an issue worth much discussion all by itself. The long lines of cars at gasoline stations in the eastern cities where one or two hour waits are common to get $3 worth of gasoline (at the time of this writing) impose a rationing system where the criteria are stamina, patience, luck, and favorable working hours. Several states and the District of Columbia now have an even-odd rationing system keyed to the parity of license plates and the calendar in an attempt to impose a kind of order on the scramble for gasoline. By comparison with this situation, there is much to be said for the rationing contingency plan proposed by the Federal Energy Office which would give every licensed driver the chance to buy a certain amount of gasoline each month and to either obtain money for his excess ration coupons or obtain excess coupons from others, that is, if he is willing to pay the going rate.[12]

In general, the following points can be made about the conservation strategies and their variants. The main criteria for judging the conservation strategies are or should be: administrative feasibility, equity, costs and benefits, time to implement, enforcement procedures, adaptability, relations to other national policies, and effectiveness. Some strategies may be particularly effective in dealing with temporary situations or transition periods. Rationing belongs in this category. On the other hand, some strategies, such as the marketplace strategy, are likely to be more effective over the long haul.

ENERGY SAVINGS

Whatever strategies are actually adopted, it is useful to examine the potential savings that would result from specific actions taken by governments, industry, or consumers. Clearly some particular energy conservation actions offer much more opportunity to save energy than others, some can be put into effect more rapidly, some will have greater public acceptance, and some will be more cost effective. For rational discussions of public policy the results and effects of transportation energy conservation actions must be considered so that the relative advantages and disadvantages of each may be compared and the most suitable combinations of actions may be supported.

Table 1 lists ten specific transportation energy conservation actions and estimates of the resulting energy savings as a percentage of the total transportation energy used in 1970.[7] Reference 7 describes the actions in greater detail and gives the basis for the estimates. Of the ten listed, the first two have the largest percentages of energy savings. The first assumes that the composition of the automobile fleet changes to a 50/50 mixture of cars with 13.1 and 22 mpg from the composition 90/10 which prevailed in 1970. The second action listed assumes a 30% reduction in the fuel consumption of half of all highway vehicles, not just cars, as the result of producing more efficient highway vehicles. It leads to a savings of 11.5% of the total transportation energy. The remaining actions on the list, which represent some of the more commonly discussed ways to conserve energy in the transportation sector, and involve modal shifts, car pooling, reduced speed limits, and reduced congestion, generally have energy savings of about 3% or less. Savings of a percent or so can be important in a time of crisis and each of these actions can contribute its share of energy savings. Note that some of the actions involve the same group of people, namely, urban commuters, and to some extent are mutually exclusive. A more recent report has refined and confirmed these estimates of energy savings and has evaluted the listed actions with respect to other important considerations, including cost differentials, time to implement, travel time, environmental quality, safety, and implementation strategies.[11]

TABLE 1

SELECTED TRANSPORTATION ACTIONS
AND ESTIMATED SAVINGS AS PERCENTAGE
OF TOTAL TRANSPORTATION ENERGY

Number	Action	Total Transportation Energy Savings (%)
1	Convert 50% of Passenger Car Population to Small Cars (22 mpg)	9.0%
2	Introduce in 50% of Highway Vehicles a 30% Reduction of Fuel Consumption	11.5%
3	Eliminate 50% of Urban Congestion	1.1%
4	Achieve 50% Success in Limiting Highway Speed to 50 mpg	2.9%
5	Persuade 50% of Urban Commuters to Car Pool	3.1%
6	Shift 50% of Commuters (to & from City Centers) to Dedicated Bus Service	1.9%
7	Shift 50% of Intercity Auto Passengers to Intercity Bus & Rail, Evenly	3.0%
8	Sift 50% of Intercity Trucking to Rail Freight	3.4%
9	Shift 50% of Short Haul Air Passengers to Intercity Bus	0.29%
10	Persuade 50% of the People to Walk or Bike up to 5 Miles, Instead of Driving	1.6%

TIME FACTORS

The general public, industry, and the National economy are reacting
to the shock of reduced petroleum imports. The shock has not
affected each household or each business to the same extent nor at
the same rate. Each of the affected groups will react at its own
pace. The independent truckers were among the first to be affected
and to take action, appropriate or not, to get some kind of relief. At
this writing, motorists in some states have to wait hours for gasoline.
A year from now and five years from now the effects of this energy
crisis may well be continuing for some groups. An understanding
of the time scales by which different groups can respond or in which
different actions can take place is an essential ingredient in making
judgments about desirable policies to pursue. For example, the
available information about car use says that while the average age
of the car on the road is about five years, the life of the average car
is approximately ten years, and that a typical car accumulates almost
half of its lifetime mileage in about four years.[12] This means that
the effect of an innovation that appears in all cars in the same model
year would be distributed over half of the total annual vehicle miles
traveled within four model years. A truck or bus may remain in
service for 15 years, however, so innovations in trucks will not
propagate so rapidly as innovations in cars. For aircraft, the commercial planes of 1985 are on the drawing boards today and are relying
on today's proven technology. Policy formulation, as well as research
and development, for the different transport modes should proceed in
full recognition of these different time scales and their relations
to time scale for change in the energy and fuels supply industry,
the supplier industries, and other important elements of the U.S.
and world economics.

HIGHWAY VEHICLE EFFICIENCY

Improvements in the efficiency of highway vehicles offer one of the
most hopeful techniques for conserving energy in the transportation
sector, as Table 1 shows. The technical staff at Transportation
Systems Center in Cambridge, Massachusetts, as part of the

Department of Transportation's Automotive Energy Efficiency
Program, has calculated the gasoline savings that would result
from improving automobile fleet efficiency in several scenarios.[10]
These scenarios assume a 25% reduction in vehicle miles traveled
(VMT) in 1974 as compared with 1973 and a steady 4% annual
increase in VMT thereafter.

The baseline scenario assumes that the average fuel economy of the
car fleet remains at 13.5 mpg. A second scenario assumes an
accelerated fleet change to small automobiles having 20 mpg fuel
economy and that the conversion is complete by model year 1977. In
this case, fuel savings amount to 18 billion gallons per year by 1980.
The third scenario assumes the introduction of fuel conserving
vehicle changes over the next few years by such techniques as shift
to radial tires, changed axle ratio, and improved transmission,
reducing aerodynamic drag, and weight reduction to increase overall
fuel economy from 13.5 to 16.5 in the 1978 models. The fuel savings
would amount to 9 billion gallons by 1980. The fourth scenario
assumes that modifications are made to car engines over the next
few years to increase the fuel economy of 1978 models to 16 mpg.
By 1980 the annual fuel savings amounts to 7.5 billion gallons. In
the fifth scenario, a new engine, probably a stratified charge or
diesel type, that delivers fuel economy of 18 mpg was assumed to
be phased into production in all models over the five years starting
in 1979. Although fuel savings would only be about 1 billion gallons
in 1980, they would increase to 12.5 billion gallons by 1985 in this
scenario. The final scenario assumed a large increase in vehicle
utilization, with average occupancy of cars going from 1.6 persons
to 2.5 persons for urban commuters, modal shifts, some trips
foregone, etc., and estimated a fuel savings of 7 billion gallons in
1976 and 16 billion gallons in 1980. This scenario is the only one in
which VMT did not follow the baseline assumptions.

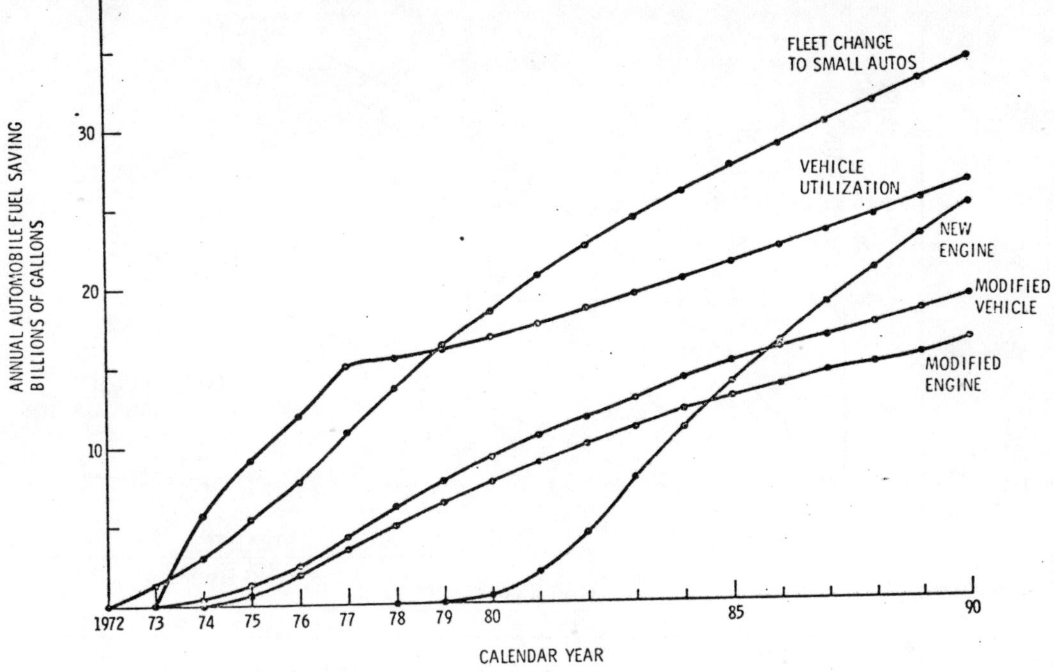

Fig. 1   National Automobile Fuel Savings for Various Conservation Options. See text for discussion.

Figure 1 shows the time development of the fuel savings for each scenario.   They show that between now and 1990 approximately three years consumption of gasoline by cars could be saved by (1) shift to small cars, (2) increased vehicle utilization, or (3) modified vehicle and modified engine, if the assumed changes are implemented. These scenarios are not predictions of what will happen, of course, only of what might be the fuel savings under the various assumptions. Nevertheless, they indicate the different time scales for response of the existing fleet, of the evolving fleet, and of a fleet with a new engine.

FUEL ECONOMY TECHNOLOGY

The fuel savings calculations show that technological changes to improve the fuel economy of production cars in the next few years,

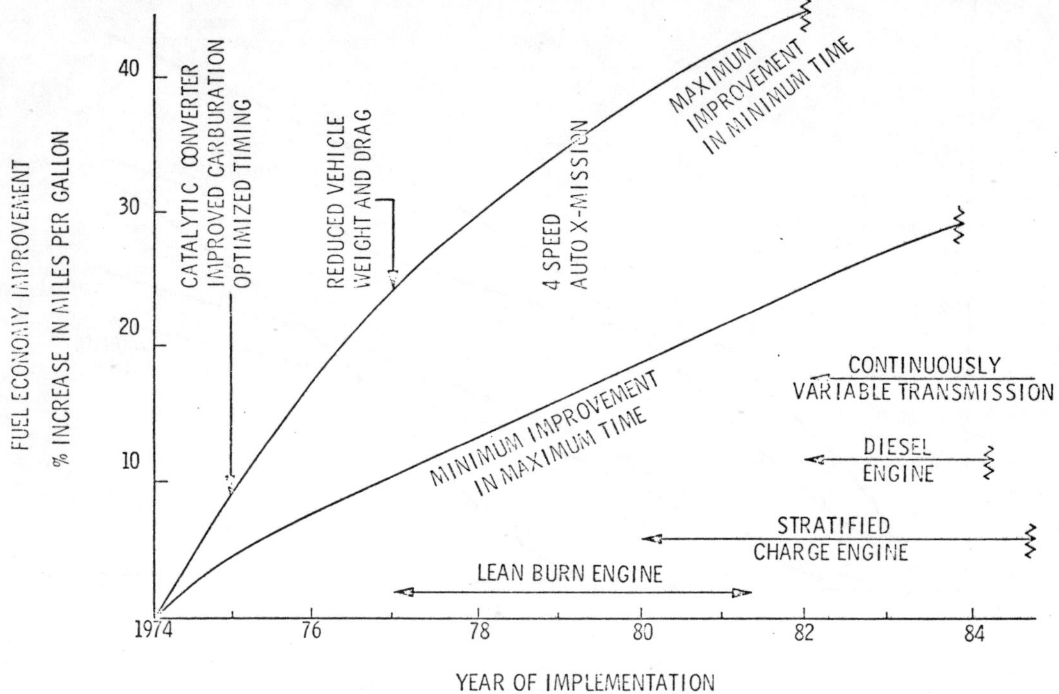

Fig. 2   Potential Percent Improvement in Fuel Economy
vs. Time of Implementation. See text for discussion.

if they are feasible changes, could lead to significant fuel savings.
The three main areas of fuel economy technology are: reduction of
engine fuel consumption, improved matching and coupling of the
engine and the road load through the drive train, and changes to
vehicle structure to reduce weight and aerodynamic drag. Potential
improvements in fuel economy technology must be judged by their
related effects on emissions control, first cost and life cycle cost,
performance requirements, customer acceptance, lead times and
development status. Staff at the Transportation Systems Center
have conducted a preliminary assessment of the potential fuel
economy technology for single car models over the next few years.
This assessment is still highly qualitative but represents the types
of devices, fuel economy improvements and the timing of first
production that are being widely discussed within the automotive
industry. Figure 2 displays the range of percentage improvement
in fuel economy that may be feasible from 1974 to 1984 and

identifies specific technologies and the time of their first use.
These improvements apply to vehicles that are introduced in the
year shown. By 1984, the potentially feasible (but not fully
evaluated) improvements could increase fuel economy to 30 to
45 percent. By 1980, fuel economy might improve by 18 to 36
percent with the same caveats. These results are generally
consistent with the assumptions of the two scenarios about modified
vehicles and modified engines. In making this assessment,
emission standards were considered to be controlled at the levels
of the 1976 Federal interim standards (0. 41 g/mi HC, 3. 4 g/mi CO,
and 2. 0 g/mi NO_x), acceleration capability was maintained
approximately equivalent to the corresponding 1973/74 vehicles,
no deviation from present safety standards was permitted, changes
from present life cycle cost were minimized as much as possible,
and there were no significant departures from present vehicle
designs.

The 1975 model cars with catalytic convertor, improved carburetion,
and optimized ignition timing show a fuel economy improvement of
about 5 - 10 percent. The 1977 models, with reduced wieght and
drag and a lean burn engine show a potential fuel economy improvement of 10 to 22 percent in this analysis. The lean burn engine
operates with a relatively high air/fuel ratio, particularly at part
load, to obtain its good fuel economy. By 1979, improved transmissions, such as a four speed automatic or a three speed with
overdrive, may lead to a further increase in fuel economy of about
10 percent. For these transmission types, the issues relate to
customer acceptance of changed engine operating regimes and shift
characteristics rather than technical or commercial feasibility.
Stratified charge or light-weight diesel engines or continuously
variable transmissions may appear early in 1980's to give a

further improvement in fuel economy but must have more
development before they are ready for production.

EFFECTS OF CONSERVATION

The effects upon consumers and industry of energy conservation
measures and of the strategies employed to conserve energy within
the transportation sector are as diverse and complex as the users
of energy.   The effects are occurring now, will persist for many
years, and undoubtedly will have lasting consequences for U. S.
society.

Short-Term Effects

Many of the short-term effects are visible now.   Some have been
discussed in previous sections.   The newspapers tell the stories
daily.   They tell of one or two hundred thousand autoworkers laid
off as a result of the most dramatic turn around in the attitudes of
the car buying public in several decades.   They tell about the
occasional fistfights and shooting that have happened at gasoline
stations.   They tell of the plight of the owner of the Poodle Parlor on
Wheels who misses appointments to trim the toenails of poodles
because he spends so much time getting gas or because the customers
are out getting gas themselves.   They tell of the people who have
waited two hours for their turn to get gasoline only to be turned
away because the station is closing early or is out of gasoline.   The
good news is that reduced travel has meant fewer accidents and
fatalities.   Moreover, in those states where the speed limit has
already been lowered to 55 mph there has been a further significant
reduction in accidents and fatalities.   In all, traffic accidents killed
about 1,000 fewer people in 1973 than in 1972, and most of the
reduction was in December of 1973.   Businesses that rely on the
convenience and mobility of the ubiquitous automobile, such as
drive-in movies, drive-in restaurants, businesses along highways,

the tourist industries are not seeing their usual numbers of
customers and some will go into bankruptcy. Crimes associated
with the energy crisis also make the news: the man critically ill
because he swallowed some of the gasoline he was siphoning, the
fuel trucks hijacked, the con man who received three dollars from
each of the drivers waiting in line for gasoline by saying that he was
doing it to speed up the line at the pumps, the gasoline dealers who
put an extra tax on the gasoline they sell, etc. The public blames the
oil industry for the energy crisis and as a result, there have been
and will be investigations by Congress of the activities, policies,
economics, and practices of the oil industry on a wide scale. The
short-term effects are of all kinds; economic, social, technological,
and political. They touch all of the modes of transportation
although the emphasis here has been on the highway modes.

Long-Term Effects

A look to the future is necessarily speculative. Nevertheless, some
suggestions and comments about long-term effects of transportation
energy conservation are offered here. With a long-term change in
the availability and price of fuel for cars, trucks, planes and trains,
the public, in response, is likely to make long-term changes of its
own. The shift to small cars or to cars with good fuel economy will
probably continue, but not at its present intensity. Since the U. S.
is a mobile Nation and one fifth of the households move each year,
even though the stock of housing changes on a time scale of decades,
within a relatively short time may people can relocate to be closer
either to their work or to their shopping areas or both. This
relocation will reduce the need to travel. There will be more
demand for public transportation. It is already in evidence. The
number of cars per family may drop permanently. De Toqueville's
description of U. S. society one and a half centuries ago, when

compared with U. S. society, shows an underlying stability of institutions and manners that can be expected to continue into the long-term.   The long-term social effects will be of a more superficial nature but may well lead to dramatic changes in courting practices, living habits, and in what people do for recreation, for example.   The proportion of disposable income that the U. S. spends on transportation, which has been 14-16 percent for several decades, will probably not go far out of that range, but the allocation of that disposable income between maintenance of the national automobile fleet, purchase of new additions to that fleet, and operation of both will probably change.   Clearly, in a world where fuel is expensive, there is greater value to a car with good fuel economy than has been true in the past.   The prices of such cars and the choices people make will reflect that difference in value.

There are some important interactions among fuel use, emissions standards, safety features, economics, and mobility as they relate to the automobile that must be recognized in considerations of long-term effects.   The level of emission control that Congress establishes for oxides of nitrogen has direct implications for fuel economy and indirect implications for the range of engine technologies that may be economical in the future.   The reason is simple but has complex qualifiers which will not be discussed here.   The high temperatures that improve engine efficiency also result in emissions of oxides of nitrogen (NO_x) and the production of NO_x increases more rapidly with temperature than does engine efficiency.   A standard for NO_x that is too high may impair the health of many, while a standard that is too low will lead to reduced fuel economy.

Automobile safety features of recent years have saved lives, but they have also added weight to cars and thereby have increased fuel consumption.   Small cars are generally less safe than big cars involved in the same types of accidents, but they have better fuel economy than

large cars, and in the future their safety characteristics can be
significantly improved by comparison with the average small
car on the road today. The growing share of small cars in the
automobile fleet may have substantial safety implications. These
complex interactions need to be sorted out and examined in a
consistent way so that an informed public can make choices among
policies affecting highway vehicles with greater understanding for
the consequences.

SUMMARY

The transportation sector uses about one quarter of the Nation's
energy for a great diversity of essential and nonessential purposes.
There are good opportunities for conservation of significant amounts
of energy both in the short-term and long-term; with many of those
opportunities focused on highway vehicles. Public policy makers
have a wide range of strategies to consider and use in promoting
energy conservation in the transportation sector which range from
informing the public of the issues to regulating the choices of daily
living. Institutional measures which force either more effective use
or less use of existing transportation facilities or both have to carry
the burden of the short-term energy shortages, but many of these
measures are at high social cost. Technological changes to
vehicles offer the prospect of conserving energy within a few years
and for the longer term. The evolution of energy conservation
measures to cope first with the immediate crisis and then to the
more stable long-run situation leads to a wide variety of
economic, social, technological and political effects both in short-
term and long-term. It is necessary to keep opening up new
energy conservation options through research and development.
And we scientists and engineers who work in this field need to have
a realistic appreciation of the requirements for successful
applications in the large consumer market of transportation.

<u>ACKNOWLEDGEMENT</u>

The authors gratefully acknowledge the many valuable discussions with colleagues within and outside of the Department of Transportation. Particular appreciation must go to Messrs. R. Husted and M. Miller in the Office of the Secretary, A. French of the Federal Highway Administration, and A. C. Malliaris of the Transportation Systems Center. The views expressed in this paper are those of the authors and do not necessarily represent official policies or positions of the U. S. Department of Transportation.

REFERENCES

1. "Research and Development Opportunities for Improved Transporation Energy Usage," Summary Technical Report of the Transportation Energy R&D Goals Panel, Department of Transportation, September 1972 (Available from the National Technical Information Service, PB 220612)

2. Walter G. Dupree, Jr. , and James A. West, "United States Energy through the Year 2000," U. S. Department of the Interior, Washington, D. C. , December 1972

3. "Estimated Motor Vehicle Travel in the United States and Related Data--1972," Table VM-1, Preliminary, November 1973, Federal Highway Administration

4. "1972 National Transportation Report--Present Status-- Future Alternatives," U. S. Department of Transportation, Office of the Secretary, Washington, D. C. , July 1972

5. Eric Hirst, "Energy Intensiveness of Passenger and Freight Transport Modes, 1950-1970," ORNL-NSF-EP-44, Oak Ridge National Laboratory, Oak Ridge, Tennessee, April 1973

6. Richard A. Rice, "Energy Efficiencies of the Transport Systems," SAE Paper 730066

7. A. C. Malliaris and R. L. Strombotne, "Demand for Energy by the Transportation Sector and Opportunities for Energy Conservation," Paper at Intersociety Conference on Transportation, September 1973, ASME--73-ICT-87

8. Robert K. Whitford and Frank L. Hassler, "Some Transportation Energy Options and Trade-Offs: A Federal View," Transportation Systems Center, U. S. Department of Transportation, Cambridge, Massachusetts, 8 January 1974

9. Joel Darmstadter, "Energy Options: Limiting Demands," Outlook for Energy Conference, December 1972, Sponsored by the Upper Midwest Council, Minneapolis, Minnesota

10. Federal Register, Vol. 39, No. 11, pp. 2066-2068

11. "Transportation Energy Conservation Options," Report No. DP-SP-11, Transportation Systems Center, U. S. Department of Transportation, Cambridge, Massachusetts, October 1973

12. "Nationwide Personal Transportation Study," A series of reports--Federal Highway Administration, U. S. Department of Transportation, Washington, D. C.

HUD TAKES TOTAL ENERGY CONCEPT
ONE STEP BEYOND*

Gerald S. Leighton[+]

This paper presents a total energy package concept which supplies the normal range of utilities, such as electricity, air-conditioning, and heating, but also provides for the treatment of water, the processing of solid and liquid wastes. In such an integrated system more than half the normally wasted energy can be recovered or diverted for heating, air-conditioning, domestic hot water, and for water and liquid waste treatment. At the same time fuel requirements are reduced.

A pilot combined-package utility plant, called MIUS (Modular Integrated Utility System), has been developed.

---

* This paper appeared in full in the January 1973 issue of Air-conditioning & Refrigeration Business", Industrial Publishing Co., Division of Pittway Corp.

+ Utility Systems Program Manager, Office of Research and Technology, Housing and Urban Development

SESSION III

REDUCING THE DEMAND: INDUSTRIAL ENERGY
CONSERVATION

Chairman: Martin Meyerson
            Director, Corporation R&D Planning,
            Baltimore, Maryland

INDUSTRIAL ENERGY CONSERVATION OVERVIEW

Martin Meyerson[*]

Industry consumes approximately 43% of the energy used in the United States, excluding raw materials used (e.g. petrochemical feedstocks), which amount represented about 15 million barrels of oil equivalent per day in 1972.

Regardless of the improvements made in energy supplies, industry can and must reduce its energy consumption, without reducing its production, as one of the most vital elements in meeting America's energy challenges.

This overview of industrial energy consumption briefly surveys the energy intensive industries of cement production, steel and refractories production, and aluminum production as an introduction to the subsequent more detailed papers on those industries. It suggests that savings on the order of 10 to 20 percent in energy consumption are possible through the application of existing economically justifiable techniques. Additional savings are possible with some production process changes.

After the excellent presentations this morning on national energy issues, my task this afternoon is to try to put a few of our industrial energy issues into perspective, before the more detailed industrial energy conservation papers are presented in my session. To do this I would like to portray those issues in the oversimplified manner depicted in Figure 1. Here we see the three major topics of energy supply, energy demand, and energy opportunities which challenge a typical corporation. These topics raise the following major issues:

---

* Director, Corporate Research and Development Planning, Martin Marietta Corporation, Baltimore, Maryland

1.   To continue profitable production, can we obtain adequate energy supplies?

2.   To improve profitability and aid energy availability, can we reduce our energy demand?

3.   To improve growth and diversification, can we participate in the expanding energy-oriented market?

Each of these energy-oriented issues can be answered in the affirmative or negative considering the constraints of time, cost, and our national and industrial policies, plans and operations.   In the short period that I will spend with you this afternoon, I'll touch on a few of the near term impacts, and then conjecture somewhat on some possible longer term impacts of some of the issues associated with energy availability and conservation.   Following me in this session will be experts on energy conservation in the metal processing, chemicals, utilities and cement industries.

The suggestion that savings of 10 to 20 percent in industrial energy consumption are possible, comes from the recent experience of many companies.   A graphic depiction of this fact is noted in Figure 2, which represents the energy conservation experiences of several different types of industrial operations in 1973.   In these particular cases energy cost savings varied from about 11% to about 44% per year[1], or an average of about 22% annually.

Having thus depicted an overview of general energy conservation possibilities, I will spend the remainder of my time discussing Cement, Steel and Refractories, and Aluminum energy issues.

CEMENT

As I noted earlier, industrial energy consumption amounts to approximately 43% of total national energy consumption, and the stone, clay, glass, and concrete products sector consumes almost 10% of industrial energy.   Of this amount, about 43% is in cement production, hence cement production accounts for nearly 2% of total national energy consumption[2], and about 1.5 billion dollar sales split among about 51 companies.

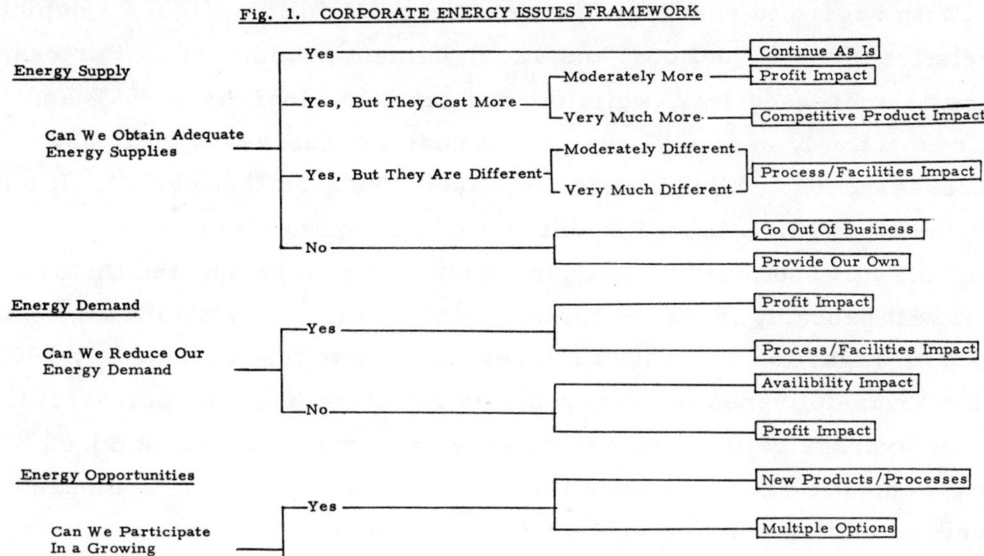

Figure 2. TYPICAL INDUSTRIAL ENERGY CONSERVATION COST SAVINGS

Industry Type	Total Annual Energy Cost - $M	Energy Conservation Annual Cost Savings $M	%
Basic Chemicals	5.5	2.4	44
Textiles	0.9	0.3	33
Agricultural Chemicals	1.7	0.3	18
Oil Refining	10.3	1.1	11
Intermediate Chemicals	13.2	1.9	14
Food Processing	1.1	0.3	27
Pulp & Paper Mill	5.3	1.7	32
Rubber & Tires	2.9	0.5	17
Average	5.1	1.1	22

Source: "Energy Conservation in the Processing Industries,"
a presentation at the McGraw Hill conference on
"How To Conserve and Manage Energy in the Process
Industries," in Washington, D. C., by Mr.
Alfred E. Waterland, E. I. DuPont de Nemours & Co.,
29 January 1974.

With regard to energy utilization and availability, Figure 3 depicts the relative trend of fuel cost and use in cement production[3]. For example coal costs increased 16%, while oil and gas costs increased 156% and 143% respectively in two decades. But coal use decreased 12%, and oil use decreased 8%, while gas use increased 20% over that period. It should be obvious that gas will be difficult and more expensive to obtain in the future, oil will continue to increase significantly in price, and the use of coal will probably increase again if adequate transportation is available.

With regard to costs it is interesting to note that in the early 1900's crude oil was delivered to cement plants for about 85 cents per barrel, when the average selling price per barrel of cement was about $3.00. Today, domestic crude oil sells for about $5.00 per barrel, compared to an average selling price of cement of just a bit less than that per barrel. Hence the cement industry had to increase its productivity to account for the 600% increase in the cost of one of its basic fuels, with less than a 50% increase in the price of a barrel of cement.

Figure 4 depicts some of this productivity improvement in terms of the thermal energy input to produce a barrel of cement, which has improved 24% over about 25 years, using the same basic production process[3]. You will see later that aluminum and steel have also improved productivity and conserved energy over the past decade or two - aluminum, by using the same basic process, and steel by changing its process.

Modern cement production facilities operate at between 950,000 to 1,100,000 BTU/bbl for long wet kilns, and between 750,000 to 850,000 BTU/bbl for long dry kilns. But European technology is using kilns with traveling grate preheaters, known as Lepol kilns, which use between 500,000 to 580,000 BTU/bbl, a significant reduction below current U.S. technology[4]. However, since this is a semi-dry process, the raw material has to be nodulized before being fed onto the traveling grate, and the gases leave the system at approximately 250-300°F and cannot be used for further drying. Further, the current European product will not meet existing U.S. standards for strength and alkalinity.

Figure 3. Relative Trend From 1945 To 1967
Of The Average Costs And Use Of The
Fuels Used In Producing Cement

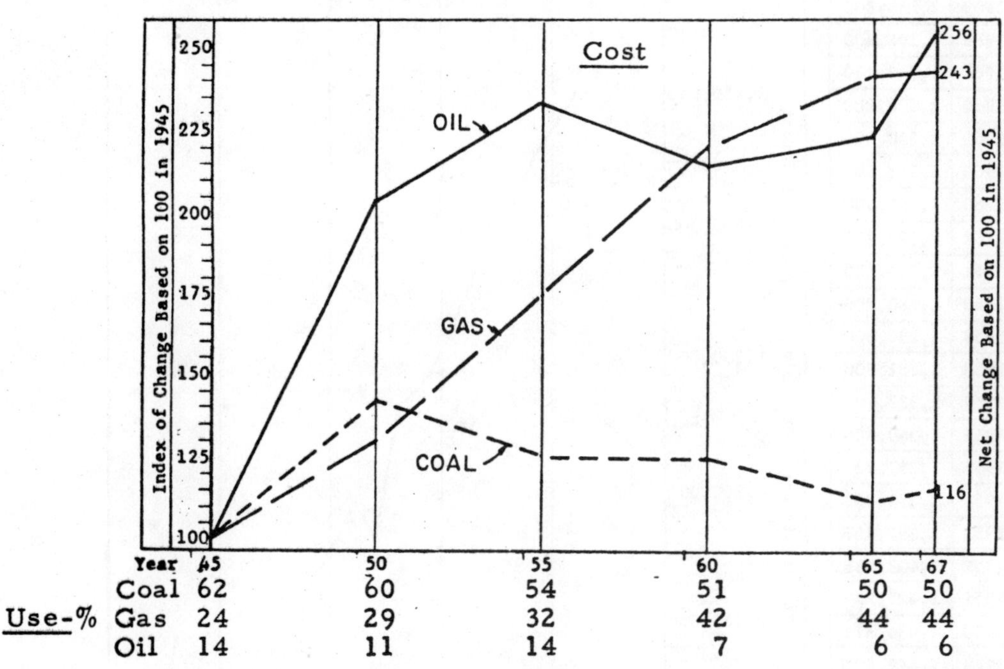

Source: "Burning and Fuels," F. S. Tripp, PCA, Sept. '71

Figure 4. Average Heat Required for Producing One Barrel of Cement

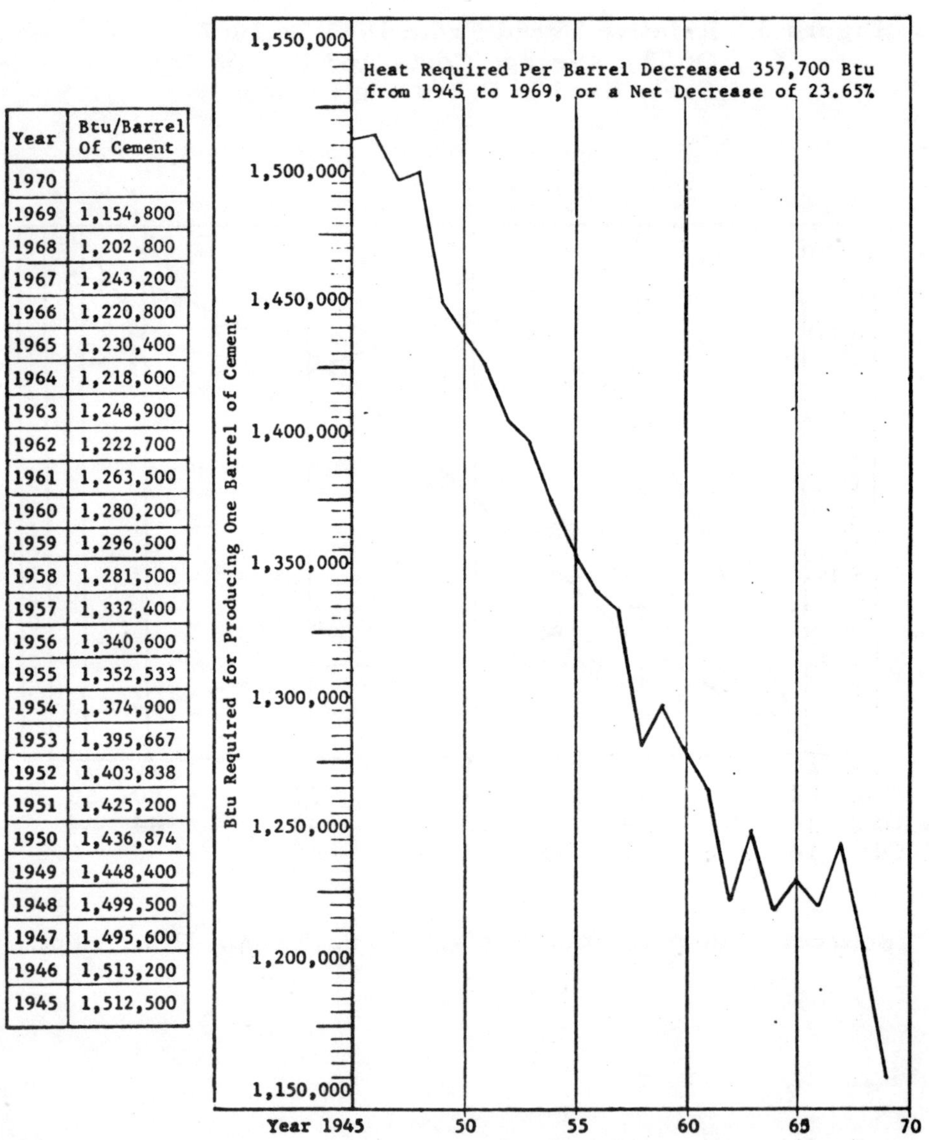

Year	Btu/Barrel Of Cement
1970	
1969	1,154,800
1968	1,202,800
1967	1,243,200
1966	1,220,800
1965	1,230,400
1964	1,218,600
1963	1,248,900
1962	1,222,700
1961	1,263,500
1960	1,280,200
1959	1,296,500
1958	1,281,500
1957	1,332,400
1956	1,340,600
1955	1,352,533
1954	1,374,900
1953	1,395,667
1952	1,403,838
1951	1,425,200
1950	1,436,874
1949	1,448,400
1948	1,499,500
1947	1,495,600
1946	1,513,200
1945	1,512,500

Source: "Burning and Fuels," F. S. Tripp, PCA, Sept. '71.

110

The difference between a heat consumption of 550, 000 BTU/bbl
for an efficient preheater kiln and 1, 100, 000 BTU/bbl for a typical wet
process kiln could mean a production cost difference of 17 cents/bbl
of clinker produced[5], when fuel costs less than 50 cents per million
BTU's. With fuel costs increasing to $1. 00 per million BTU's these
days coupled with shortages, the trade-offs between energy conservation
and its impact on profitability, on the one hand, and product performance
and requirements on the other hand need to be constantly reevaluated.

You will hear more detailed information on cement production and
energy conservation later in this session, hence I will not cover them
further here.

Another important consideration of energy cost increases for cement,
and possible concomitant price increases per barrel is to look at its final
applications, and potentially competitive products for those applications.
Figure 5 shows the additional cost for each construction application for
each one percent increase in the current average price of cement[6]. Here we
see that the cost of a typical $30, 000 private home would rise only $3 with
each percentage increase in the cement price. Only $79 would be added
to the cost of a $500, 000 commercial building; $3 to the cost of a $25, 000
condominium unit; $280 to the $400, 000 cost of building one mile of
highway; and finally, $60 to the cost of building a $360, 000 school. Thus,
even if the cost of fuel doubles, as it has in some cases, it would have a
minor impact on the application cost of cement, but it could affect the
profitability of individual cement producers.

The reason that cement prices have such little effect on total construction costs is that in nearly all cases the material is a minor item
of cost, compared to labor and other costs. Yet without this material,
Portland cement, construction of many types would come to a virtual
standstill. Hence, competitive construction products would have a hard
time displacing cement and concrete.

In summary, it would appear that the following energy issues are
important to the cement industry:

Figure 5. <u>IMPACT OF EACH 1% INCREASE IN CEMENT PRICES</u>

<u>Would Increase By</u>

$30,000 single-family home $3

$500,000 commercial building $79

$25,000 apartment or condominium unit $3

$400,000 one-mile highway $280

$360,000 school $60

Source: "Testimony Presented at the Price Commission
Public Hearings on the Cement Industry,"
6 October 1972, R. D. MacLean, President,
Portland Cement Association.

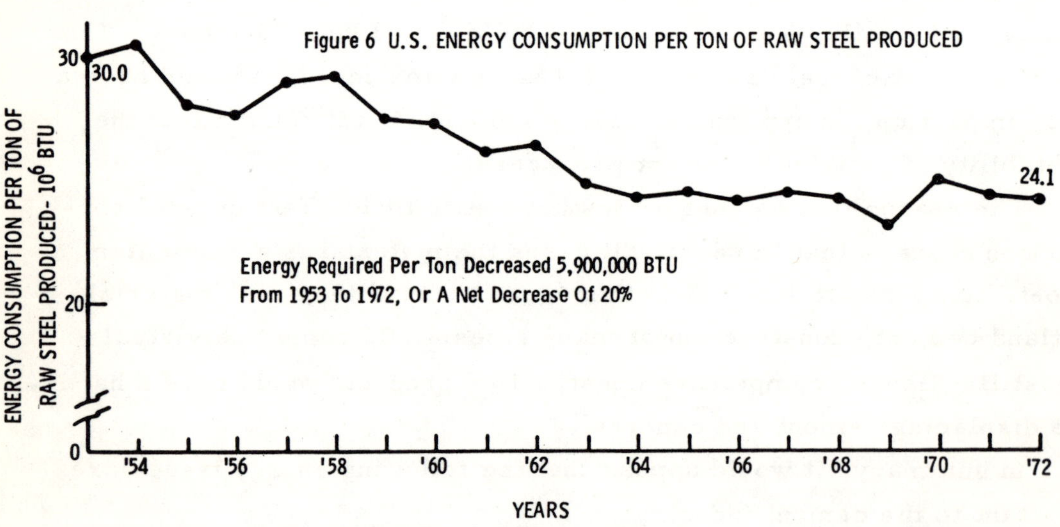

SOURCE: American Iron & Steel Institute raw data.

1. While many cement companies have provided modern production facilities that could inherently operate reasonably efficiently on a variety of fuels, national fuel allocation could adversely affect capital equipment availability and cost, and hence cement production. The switch from oil to coal is expensive.

Improvements in energy conservation on the order of 10% have already been accomplished in many plants, and more is possible with more emphasis on efficiency of heat exchangers and grinding, and improved equipment maintenance, etc. Significant improvements beyond that amount are possible only by using new process techniques and equipment (e.g. suspension preheater kiln, high alkali cement, etc.) which could affect product performance. Trade-offs between product performance requirements and energy conservation need to be constantly reevaluated.

2. Energy costs per barrel of cement produced are typically 14% of the average selling price, and are rising.

While these costs are small when considered in the perspective of the use of cement in construction applications, and would not connote more use of competitive construction materials, it could affect profitability among cement producers. Hence energy conservation may again be the only viable option in this area.

STEEL and REFRACTORIES

About 6% of total U.S. energy is used by the steel industry which has shown marked improvement in energy efficiency as the use of the Basic Oxygen steel making furnace is expanded. Figure 6 depicts an energy efficiency improvement of about 20% to produce a ton of raw steel from 1953 to 1972, where the Basic Oxygen Process is currently used for about 50% of total primary raw steel production, as noted in Figure 7. Widespread adoption of the Basic Oxygen Process could ultimately cut steel's energy consumption by nearly 50% or 3% of total U.S. energy - an amount saved which is half-again larger than the cement industry's total energy consumption.

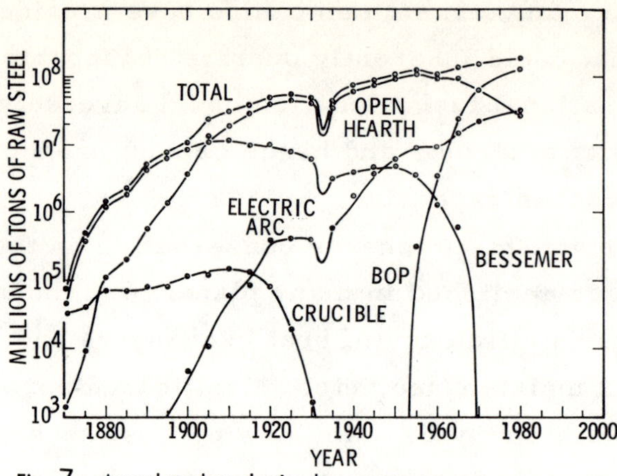

Fig. 7. Annual steel production by process.

Source: "Steelplant Refractories in the Seventies,"
K. Kappmeyer and D. H. Hubble,
U. S. Steel Corp., Ceramic Bulletin
Vol. 51, No. 7, 1972.

Fig. 8. Impact of Basic Oxygen Furnaces on Energy Consumption in
the Lackawanna Plant of Bethlehem Steel Corporation

	April, 1963		May, 1967	
Steel Production	Ingot Tons		Ingot Tons	
Open Hearth	495,193		95,260	
Basic Oxygen	0		412,155	
	495,193		507,415	
Energy Consumption	10^6 BTU	10^6 BTU/Ton	10^6 BTU	10^6 BTU/Ton
Heavy Fuel Oil	412,011	0.84	0	0.00
Petroleum Tar	311,017	0.63	73,888	0.15
Coal Tar Pitch	363,389	0.73	152,444	0.30
Coke Oven Gas	511,990	1.03	143,237	0.28
Mixed Gas	57,564	0.12	27,225	0.05
Sub Total	1,661,981	3.35	396,794	0.78
Steam	203,039	0.41	98,746	0.19
Electric Power	22,618	0.05	40,089	0.08
Gross Power	1,887,638	3.81	535,629	1.05
Waste Heat Used to Generate Steam	(377,341)	(0.76)	(88,083)	(0.17)
Net Total	1,510,296	3.05	447,546	0.88

Source: Rand Corporation, Interim Report: The Growing Demand for
Energy, April 1971.

Figure 8 graphically depicts the change in energy consumption in steel making in one plant, as a result of the shift from open hearth to basic oxygen production. Here it can be seen that total energy consumption decreased dramatically from about 3 million BTU/ingot ton for the open hearth to less than 1 million BTU/ingot ton for the basic oxygen process[7].

The increasing demand for steel, the shift in the basic steel making process, and the pressure of increased foreign competition have combined to cause the U.S. steel industry to operate its processes at somewhat elevated temperatures, and to attempt to obtain more and more heats of steel from its furnaces before major overhaul. This, in turn, has required better performance from refractory products, where 50 to 60% of total production is used in making steel. Figure 9 notes the increase in steel making furnace lining life, as a result of better refractories. These better refractories resulted from a shift from products based on natural raw materials (e.g. silica) to ones made from synthetic materials (e.g. magnesia, high alumina, zirconia), and in the shift to higher purity and higher bulk density refractory products.[8]

Producing these refractory products at somewhat higher temperatures in the same process has required somewhat more energy input per ton of product produced.

For example, calcining in a rotary kiln the magnesium oxide resulting from the reaction of brine and dolime, to produce 88% magnesite now takes about 16 million BTU/ton compared to about 14 million BTU/ton ten years ago. And the fuel costs per ton of magnesite produced have increased from about $5/ton in 1963 to almost $9/ton at this time.

To produce the basic periclase refractory grain used to make brick to line steel making furnaces requires a two step process - the first to calcine magnesium hydroxide blended with the proper calcium hydroxide and silica materials, and the second to hard burn the briquettes resulting from the first step in the final grain. One part of the refractories industry uses a dual rotary kiln process which requires about 12 million BTU/ton for the first process step, and about 9 million BTU/ton for the second process step for a total of 21 million BTU/ton.

Another part of the refractories industry uses a Herreschoff furnace, requiring about 10 million BTU/ton for the first process step, and a shaft kiln, which requires about 2 million BTU/ton for the second process step, for a total of 12 million BTU/ton, thus saving about 9 million BTU/ton over the dual rotary kiln approach. But even this more efficient approach has seen fuel costs increase 50% from about $6/ton in 1970 to almost $9/ton at this time.

Like cement, refractories represent a minor cost in the production of steel or in the final applications of steel products. Open hearth processes use about $2 worth of refractories per ton of raw steel produced - basic oxygen furnaces use less than $1[9]. Thus increases in energy costs could impact profitability more than potential competitive refractory products.

Major process changes (e.g. nuclear reactors, plasma technology) in the refractories industry are unlikely because the market is stabilized and capital investments would be prohibitive. But energy conservation techniques on furnaces and kilns are being implemented, and will continue to be expanded.

As in the case with the cement industry, fuel availability is an important issue, and fuel shortages or allocation could cause potential loss of production. Coal is predominately used to fire the rotary kilns to produce magnesite, and, on balance should be readily available. However, natural gas is the primary fuel in the production of periclase, and its curtailment has already caused a shift to #2 fuel oil. Shortages or allocation of this fuel oil could cause potential production losses, unless new supplies are provided.

In summary, it would appear that the following energy issues are important to the refractories industry:

1. While some refractories companies have provided modern production facilities that would operate efficiently on a variety of fuels, national fuel allocation could potentially affect production. Again the switch from gas or oil to coal is expensive.

2. The use of the Herreschoff furnace in combination with a shaft kiln in lieu of dual rotary kilns for the production of periclase refractories is a major energy conservation approach.

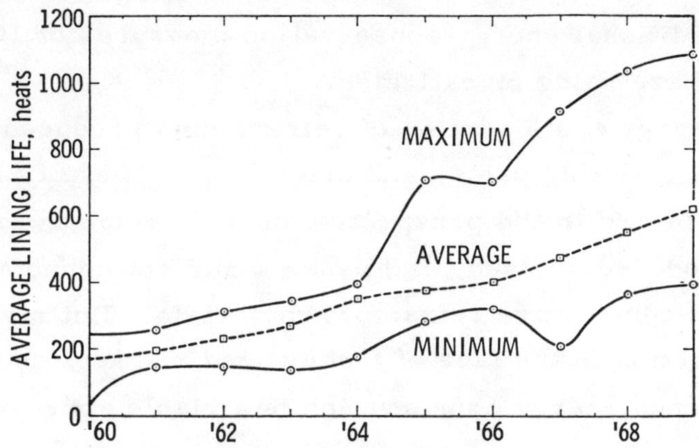

Fig. 9. BOP lining life for indicated years.

Source: "Steelplant Refractories in the Seventies,"
K. Kappmeyer and D. Hubble, U.S. Steel
Corp., Ceramic Bulletin Vol. 51, No. 7,
1972.

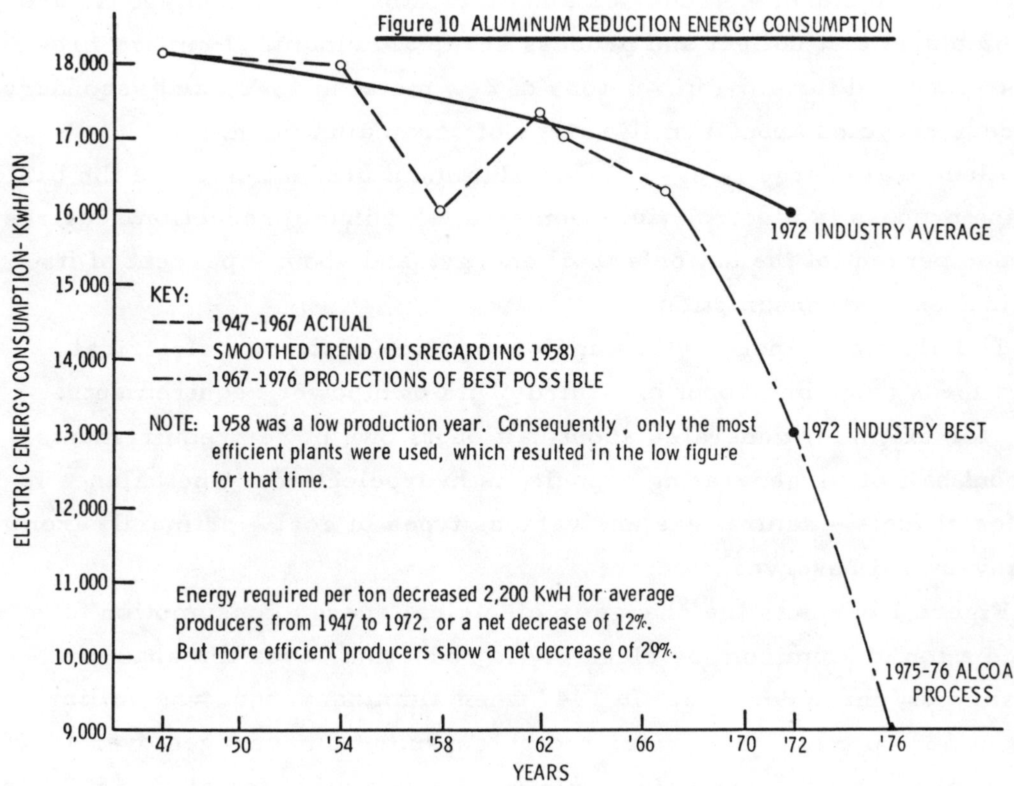

SOURCE: 1. Aggregate U.S. data, Census of Manufacturers.
2. "The Growing Demand for Energy," Rand Corp. April 1971.
3. Statement by Dr. E. Walker, Alcoa before Joint Hearings
on Conservation and Efficient Use of Energy,
12 July 1973.

Additional energy conservation measures of 10% or more
are possible and are being investigated.

3. Energy costs per ton of refractories produced are typically
16% of the average selling price, and are rising. These costs are very
small when considered in the perspective of their relation to the cost of
a ton of steel produced or used, and hence would not connote more use
of quite different competitive refractory materials. But rising costs for
refractory production in the face of a stabilized market, could affect
profitability. Hence energy conservation is a viable approach in this area.

ALUMINUM

The aluminum industry in the U. S. encompasses 12 integrated producers of primary metal, hundreds of sheet, extrusion and other mill
products manufacturers, foundries and fabricators of finished goods, and
about 85 plants that collect and process scrap aluminum. Primary producers accounted for 4. 1 million tons of new metal in 1972, and secondary
producers recycled about 1 million tons of scrap aluminum.

Electrical energy is essential to aluminum production since the basic
smelting process is electrolytic. Domestic aluminum production requires
about one percent of the nation's total energy, and about 4 percent of its
electrical energy consumption.

The aluminum industry is a major supplier of its own electrical
energy needs providing about one-third of its own power requirements.
Alcoa, for example, generates about half of its own power requirements,
and about half of its generating capacity is hydroelectric. The balance is
from fossil fuels - natural gas and various types of coal - primarily from
company-owned reserves.

Figure 10 depicts the change in electrical energy consumption to
produce a ton of aluminum by electrolytic reduction, over the span of
approximately three decades. In 1947 when aluminum ingot was selling
for about $300 per ton, electrical energy consumption averaged over 18,000
KWH per ton. In 1972 when aluminum ingot was selling for about $500 per ton,
electrical energy consumption averaged from 16,000 KWH per ton, a net

decrease of about 12%. But the most modern smelters require only 13,000 KWH per ton, hence the net decrease is about 29%. Alcoa's recently announced chloride process[10] promises to reduce electrical energy consumption to about 9,000 KWH per ton by 1975, a reduction of 30% below the most modern smelters of today. You will hear more about this later.

It might be of interest at this point to briefly compare the energy requirements for the production of several primary metals, and especially their recycle, to consider aluminum, and its energy requirements, in a larger perspective. Figure 11 summarizes the energy requirements for the production and recycle of magnesium, aluminum, iron, copper and titanium. It is seen that for production, in diminishing order of energy intensiveness, titanium requires about 126,000 KWH/ton on an equivalent coal energy basis; magnesium, about 91,000 KWH/ton; aluminum, about 51,000 KWH/ton; copper about 14,000 KWH/ton; and iron, about 4,000 KWH/ton from present sources[11].

It is also seen that for recycle, or reprocessing magnesium, aluminum, iron, and copper all require substantially less energy (about 1500 KWH/ton) than that for production, while titanium requires moderately less (about 39,000 KWH/ton). With about 20% of total aluminum output recycled in 1972, at energy consumption requirements 95% less than that for new metal production, it is seen that this results in substantial energy conservation.

For particular applications, such as steel and aluminum cans the ratio of input energy requirements changes dramatically when recycle is considered. For example, for a single 12 ounce beverage can, steel cans with 30% recycle, require 5400 BTU/can from mining, through transport, to manufacture. Aluminum cans, with about 88% recycle require only 3600 BTU/can for the same operations[12].

To complete this part of the comparative analysis, it is noted from Figure 11 that future energy requirements for titanium will increase about 20% because ilmenite will be increasingly substituted for scarce high-grade rutile ores. Future energy requirements for magnesium will remain essentially constant because of its infinite resource in seawater. Use of

Figure 11. Summary of the Energy Requirements for the Production and Recycle of Metals

Metal	Present Source	Equivalent Coal Energy in KWH/ton of metal	Future Sources	Reprocessing
Magnesium	Sea water	90,821		1395
Aluminum	50% bauxite	51,379	30% bauxite 59,615 Clays 65,972 Anorthosite 72,356	Al scrap 1300-2000
Iron	High grade hematite Magnetic taconite	4,270 4,656	Specular hematite 5,135 Non-magnetic taconites 5,273 Iron Laterites 6,268	Iron & steel scrap 1,240
Copper	1% sulfide ore	13,532	0.3% sulfide ore 24,759	98% Cu scrap 635 Impure Cu scrap 1,555
Titanium	High grade rutile ore Ilmenite rocks Ilmentite beach sands Ferruginous rocks	126,115 149,440 153,055 152,813	High alumina clays 156,400 High Ti soils 206,075	Ti scrap 39,000

Source: "Energy Expenditures Associated With The Production and Recycle of Metals,"
J. Bravard, H. Flora, C. Portal, Oak Ridge National Laboratory, ORNL-NSF-EP-24
November 1972.

lower grade bauxite ores in the future will increase aluminum energy requirements by only about 16%, but use of high aluminum clays and anorthosite will result in 30 to 40% higher energy requirements. For copper production the quality of mined ore is falling and the use of lower grade ores could raise energy requirements as much as 83%. High grade hematite ores for pig iron production are virtually exhausted, but magnetic taconites increase energy requirements by only about 9%. Thus iron and steel will remain the lowest users of energy per ton of raw metal product in the future, but recycle will change that balance somewhat.

Modest process improvements in aluminum reduction and fabrication have been the main reason for the reduction of energy requirements in the past. Increased recycling of aluminum, use of waste gas heat, and improved capital equipment and maintenance programs are continuing to be expanded and will be discussed in more detail in a later paper.

In summary, it would appear that the following energy issues are important to the aluminum industry:

1. Electrical power shortages in the Pacific Northwest was a most critical energy issue in 1973, and was only resolved by additional rainfall early in 1974.

Should hydropower shortages again occur, and supplemental thermal power be purchased, if it is available, it would add several cents per pound to the cost of aluminum ingot.

2. Reduction energy costs per ton of aluminum ingot produced are typically 11% of the average selling price, and are rising. These costs are not large when considered in the perspective of most aluminum applications, with the possible exception of containers. Here recycle is important. Expanded recycle of aluminum will further reduce the impact of its rising energy demands and costs.

3. Energy conservation techniques are being utilized to reduce energy requirements by 10% or more.

As Alcoa's chloride process matures, it might be considered for wide application if its electrical energy conservation promises of 30% below today's most modern smelters are found to be possible in full scale operations.

CONCLUSIONS

1. Cement, steel and refractories, and aluminum production represent energy intensive operations in industry incurring a cost of between 10 and 15% of the unit average selling price of their respective products.

The cost of energy is increasing as a fraction of operations cost and, while competitive products do not appear to be a major threat, profit impact is a continuing important consideration.

Hence expanded energy conservation measures in the near term, and potential new processes in cement and aluminum production in the longer term could be important to ameliorate profit impact.

2. National fuel allocation could affect production of cement, steel and refractories, and aluminum depending upon the scope and type of allocation. Shifting from one primary fuel to another is costly.

REFERENCES

1. "Energy Conservation in the Processing Industries," a presentation
    at the McGraw Hill conference on "How To Conserve and
    Manage Energy in the Process Industries" in Washington, D. C.
    by Mr. Alfred E. Waterland, E. I. DuPont de Nemours & Co.,
    29 January 1974.

2. "The Potential For Energy Conservation - A Staff Study,"
    October 1972, Executive Office of the President.

3. "Burning and Fuels," September 1971, Mr. Fred B. Tripp,
    Portland Cement Association.

4. "Technical & Economic Aspects of Long Dry Cement Kilns vs
    Suspension Preheater Kilns," Dr. Gird A. Schroth, Fuller
    Company, 6-8 December 1971, Rock Products Seminar.

5. "Cement Plant Design Economics," Mr. Joseph P. Wyner, Bendy
    Engineering Company, 6-8 Dec. 1971, Rock Products Seminar.

6. "Testimony Presented at the Price Commission Public Hearing on
    The Cement Industry," 6 October 1972, Mr. Robert D. MacLean,
    President, Portland Cement Association.

7. "The Growing Demand for Energy," Rand Corporation, April 1971.

8. "Steelplant Refractories in the Seventies," Messrs. K. Kappmeyer
    and D. Hubble, U.S. Steel Corp., Ceramic Bulletin, Vol. 51,
    No. 7, 1972.

9. "Impact of Changing Technology on Refractories Consumption,"
    Bureau of Mines Information Circular IC8494, 1970.

10. "Statement by Dr. Eric A. Walker, Alcoa, before the Joint Congressional
    Hearings on Conservation and Efficient Use of Energy, 12 July 1973.

11. "Energy Expenditures Associated With The Production and Recycle of
    Metals," J. C. Bravard, H. B. Flora II, C. Portal, Oak Ridge
    National Laboratory, ONRL-NSF-EP-24, Nov. 1972.

12. "Recycling Can Cut Energy Demand Dramatically," P. R. Atkins,
    Alcoa, Engineering and Mining Journal, 1973.

ENERGY CONSERVATION IN PRIMARY METALS PROCESSING

Allen S. Russell[*]

Current technology for converting ores to useful metals was developed largely when energy was plentiful. Energy was an important, but not overriding, consideration. Metallurgical plants generally were built where power could be readily provided. Industry and energy development grew simultaneously. Coal mining and coking were an integral part of the steel industry; power dams were built by the early aluminum industry and later the great hydropower developments sought out this steady load. Metallurgical processes, designed to minimize cost, did not necessarily try to minimize energy consumption, because power was relatively inexpensive and abundant.

Nevertheless, years before the use of energy became such a dominant national concern, the metals industry was working on processes to improve energy utilization. The current shortage has emphasized the need for such energy reduction programs. This paper summarizes some industrial accomplishments and current challenges in producing primary metals with less energy.

ENERGY REQUIREMENTS FOR THE PRIMARY METALS INDUSTRIES

To identify energy usage and put economizing practices in perspective, the procedures for energy calculation will be illustrated by values for aluminum.

Total energy to produce aluminum and convert it to usable forms necessarily includes energy for mining, refining, smelting, melting, ingot casting, mill fabrication, transportation in-house and outside the plants, space heating in the plants, air conditioning, lighting, and the collection, treatment and disposition of waste. Also, it includes the energy content of all supporting materials such as caustic used in refining, and pitch and coke employed in smelting. The values in this paper do not include the energy content of capital equipment: buildings, process vessels, furnaces or rolling mills. Nor do they include the requirements for general administration, sales, research and development, or other staff

* Associate Director for Production R&D, Alcoa Laboratories, Center, Pa.

functions. Human energy is omitted by convention, whatever its magnitude.

All energies are expressed in terms of BTUs used for the operation or
inherent as fuel in the product consumed. Hydroelectric power is converted
to BTUs by multiplying kilowatt hours by the theoretical factor 3413 BTU/
KWH. Electrical power generated from fossil fuel stations (which operate
at about 33% efficiency) is converted to BTUs by multiplying by 10,342
BTU/KWH. The sum of the BTU and converted KWH requirements is the total
energy used.

Although this paper focuses attention on the energy balance in the U.S.
aluminum industry, the industry is multinational. For the U.S. aluminum
industry nearly all of the mining and one-third of the refining of bauxite
is done outside the country. The listed values are the total energy requirements for aluminum smelted in the U.S. regardless of the location of the
mining and refining operations. They do not include the energy for imported
aluminum, 800,000 tons in 1972.

The total energy to produce one ton of aluminum equivalent at major steps
in its production is listed in Table 1.

Table 1

ENERGY PER TON OF ALUMINUM

| | THERMAL | + | ELECTRIC | = | TOTAL |
| | | | Equiv. | | |
	Millions BTU	KWH	Millions BTU		Millions BTU
Mining, Ore Purification (a)	26	440	4*		30
Metal Production (a)	24	16,500	133**		157
Mill Processing (b)	26	1,660	15*		41

   *1/6 of this energy comes from hydropower, 1 KWH = 9,187 BTU
 **1/3 of this energy comes from hydropower, 1 KWH = 8,033 BTU

(a) per ton primary metal
(b) per ton metal shipped

The thermal energy to dig, dry and transport the 4 to 5 tons of bauxite
treated for ultimate production of one ton of aluminum, and to refine the
bauxite to purified aluminum oxide, is 26 million BTU. The electric
energy component adds 4 million equivalent BTU, principally from the energy
equivalent of the caustic employed in ore purification, for a total of 30

million BTU per ton of equivalent primary aluminum.

Metal production, the electrolytic reduction of aluminum oxide to aluminum, has the largest energy demand. For the industry, one-third of the energy for the electrolytic process itself comes from hydropower. The thermal component is largely the energy equivalent of the carbon anodes consumed during the electrolytic reduction of aluminum oxide. To this is added the fuel required to bake this carbon.

The sum of the energies for mining, ore purification and reduction of the ore to primary metal is 187 million BTU/ton of primary aluminum.

The estimated material balance for the composite primary and secondary U.S. aluminum industry is shown in Fig. 1 as a flow diagram.

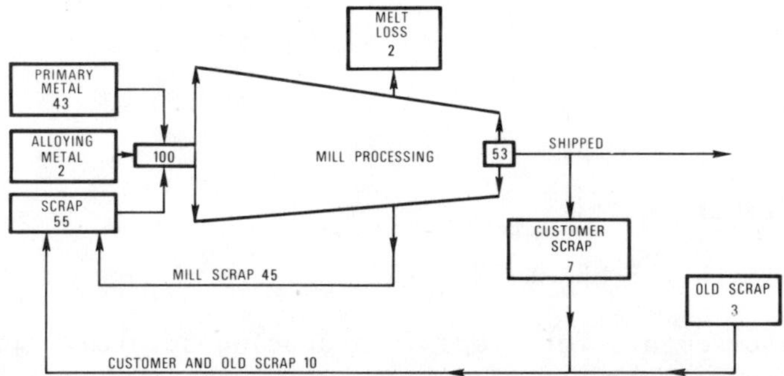

FIGURE 1 — U.S. ALUMINUM INDUSTRY FLOW DIAGRAM

Scrap makes up 55% of the aluminum charged to a melting furnace in this composite picture. Forty-five percent of the charge is mill scrap generated in-plant from ingot rejections, ingot scalpings, trimmings, and the like. Scrap returned by customers and "old scrap" such as beverage cans, auto parts and used appliances exceeds 10% of the charge.

There are losses at every step that nibble away the amount of product. The increased unit energy because of these losses has already been in-corporated in the values for mining, refining and smelting. But once the primary metal is combined with recirculated scrap, melt losses and mill scrap must be considered separately.

When aluminum melts it oxidizes and forms a surface skim of finely divided oxide and metal. This is periodically removed and treated to recover metal, but much of the skim burns. The resulting loss of aluminum is equivalent to about 2% of the metal melted and represents a loss of about 5% of the

primary metal and alloying elements. The net result of melt loss and
scrap recycle is 53 tons shipped for 100 tons of melt containing 45 tons
of primary and alloying metal.

The energy requirement per ton of aluminum shipped for major steps of the
process is listed in Table 2. Two-thirds of the energy is required in
producing the metal from its oxide. The total energy requirement per ton
of aluminum shipped is 200 million BTU. About 190 million BTU of this
energy is domestic.

Table 2

ENERGY FOR PROCESSING ALUMINUM

	Million BTU / Ton Treated	Tons Treated / Tons Shipped	Million BTU / Ton Shipped
Mining, Ore Purification	30 [a]	45 / 53	26
Metal Production	157 [a]	45 / 53	133
Mill Processing	41	1	41
(a) per ton primary metal			200

ENERGY REQUIREMENTS IN THE METAL INDUSTRY

Overall energy requirements for steps in producing several major metals are
found in Table 3. The data for steel are 1972 values from Bouman[4] Those
for copper are from Kellogg[7] The value for zinc was taken arbitrarily between the extremes given by Morgan[10] 31 million BTU/ton for the blast furnace
and 56 million BTU/ton for the horizontal retort. Those for magnesium are
from Bravard[5] adjusted for chlorine credits, those for titanium are from Bravard[5]

Table 3

ENERGY FOR METAL PROCESSING

	Million BTU per Ton of Metal					
	Steel[4]	Aluminum	Copper[7]	Zinc[10]	Magnesium[5]	Titanium[5]
Mining, Ore Purification [a]	1 [c]	30	42 [b]	--	171*	2
Metal Production [a]	15 [c]	157	34 [b]	40	118**	490
Mill Processing [b]	17	41	26	--	--	--

*$MgCl_2$ from seawater
**Adjusted for chlorine credits

(a) per ton primary metal, (b) per ton metal shipped, (c) per ton raw steel

For steel, "metal production" is used to include cokemaking, ironmaking and steelmaking operations (including scrap recycle); in other words all the energy from processed ore to raw steel. Energies in mill processing steel are about the same as for its production. As was true for aluminum, these energies cannot simply be added because of the large influence of losses in processing and of scrap recycle which constitutes about 45% of the metallic charge for steelmaking.

Copper mining requires more energy than steel because the ore is lean. A 0.8% copper content was the basis for the calculations. With magnesium, the "mining and refining" category is applied to the extraction of puri-fied magnesium chloride from seawater.

Table 4 lists the total energy requirements for these metals, calculated as the product of 1972 shipments times energy per ton shipped.

Table 4

ENERGY THROUGH MILL PROCESSING

	Primary Metal Smelted in United States					
	Steel	Aluminum	Copper	Zinc	Magnesium	Titanium
1972 Shipments[1,13] (Million Tons)	92	5.1	2.0	0.8	0.1	0.01
Million BTU Per Ton Shipped	40[4]	200	102	40*	289*	492*
Total Energy Quadrillion (10^{15}) BTU	3.7	1.0	0.2	0.03	0.03	0.01

*Mill processing not included

Bouman lists the total energy used by the steel industry in 1972 as 3,647 trillion BTU. For 92 million tons of shipments this is 40 million BTU per ton. Luerssen's 1973 value[8] agrees approximately with this figure when the electrical energy is converted to BTUs with 33% efficiency. For direct comparison with the value for aluminum the steel data would have to be increased to include transportation, refractories and the energy for processing ores, which American companies obtain from sources outside their United States facilities. It is interesting that the 5.2×10^{10} KWH electrical energy used by the steel industry in 1972[6] is quite comparable to that used by the aluminum industry.

127

Because of its large tonnage, steel is the largest overall
energy consumer of the metals. The aluminum industry requires
about one-fourth the energy used by the steel industry. The other metals
play decidedly less significant roles in the total energy picture. Hence
for this purpose it is not too important that the listed energies do not
include fabrication for zinc, magnesium, and titanium.

ENERGY SAVINGS BY SUBSTITUTING ONE METAL FOR ANOTHER

For the most part, metals have similar energy requirements on an equal
volume basis, an approximate but still inadequate overall comparison when
one metal is substituted for another (Table 5). The energy for mill
processing is not included for titanium, magnesium, and zinc so that the
listed values for these metals are low.

Table 5

UNIT VOLUME ENERGY FOR PRIMARY METALS
Thousand BTU/Cubic Inch

Titanium	Copper	Aluminum	Magnesium	Steel	Zinc
40+	16	9.8	9.1+	5.7	5.2+

+Mill processing not included

The energy required to make a cubic inch of aluminum mill product exceeds
that for mill steel by decidedly less than two-fold. Clearly the traditional
factors of cost, strength, modulus, corrosion resistance, strength to weight
ratio, etc., will dictate the proper choice of a metal because the energy
requirements for the fabricated item do not differ strongly.

Metal substitution does not appear to promise significant energy savings
in metal production energies per se. But proper metal utilization does offer
real promise in decreasing the nation's total energy needs.

For example, in electrical transmission, there is opportunity for a great
overall energy saving by using thicker conductors which decrease electrical
resistance. As another example, the use of lightweight aluminum in transportation can return more than tenfold in gasoline savings the initial
extra energy expenditure to produce the light metal.

In metal fabricating plants a number of procedures can help reduce energy consumption. Savings of 5 to 20% are being made by turning down standby furnaces; careful adjustment of the fuel-air ratio in furnaces; automatic combustion controls; frequent cleaning of burners, furnace coils and heat exchangers; improvements in burner and damper design; prompt replacement of faulty valves; avoiding blow-off of excess steam; and operating "as well as we know how." One steel fabricating plant decreased fuel consumption 20% after the manager requested a daily report of meter readings, and employees knew their use of energy was being monitored. Inland Steel finds a computerized plant energy model to be useful.

Most large users of energy have set up plant and company-wide teams to conserve energy. On a national scale these efforts are coordinated through the National Industry Energy Conservation Council of the Department of Commerce.

It is often feasible to use shorter heat treating cycles. This is an important endeavor in all plants particularly in conjunction with computer control. More accurate temperature sensing devices improve control of sintering and melting operations. Since electrical power shortages may be simpler to cope with than those involving oil or gas, heat treating will turn to induction and other electrical-based energy sources.

As with all other industry, appreciable savings result from reduced lighting levels, less air heating in winter, less air cooling in summer, and more use of car pools and public transportation.

An important means of reducing energy per unit of production is to increase the output of units by making sure more of the product meets specifications.

Implementing these immediate steps gives promise of meeting the President's goal of 5% reduction in energy consumption this year. Additional action is required for further savings.

SCRAP RECYCLING

Table 6 shows that important energy savings for aluminum, copper and steel result from recycling. In the aluminum industry it takes only 5% as much energy to recycle scrap to

molten metal as to make the metal from its ore. Of course, the energy
consuming steps involved in producing finished goods from molten metal
are the same for melted scrap (of the right composition) as for virgin
metal. The overall degree of energy saving thus depends on the product.
However, the energy content of aluminum can sheet, assuming normal
recycling of mill and customer scrap, is about cut in half if 60% of the
cans are recycled.

Table 6

ENERGY TO PROCESS AND MELT SCRAP[5]
(% of Energy Required for Primary Metal)

Aluminum	Copper	Steel
5	11 -15	25-50

As this paper has shown, 55% of the average melt for the composite aluminum
industry is scrap. (The use of averages should not disguise the remelt of
a great deal of this scrap by the secondary producers.) In addition to
mill scrap, 875,000 tons of customer scrap, usually clean, segregated by
alloy and easily collected, was handled in 1972. The outstanding problems,
belatedly receiving national attention, are with "old" scrap - some 250,000
tons of aluminum compared to an estimated 1.8 million tons available[3] in
1972. Of course, aluminum in many of its applications--architectural,
electric transmission, and airplanes, for example--has a long life
so that relatively little scrap is available from these sources
today.

The best known source of "old" aluminum scrap is the aluminum can. In 1973,
2 billion of these were returned to collection centers for recycling.
Reynolds Metals Company estimates it recycles 25% of the cans it makes.
The overall industry rate is one aluminum can recycled for every six used.
The latest directory of reclamation centers for aluminum cans lists over
1,200 centers in 43 states, up from 285 centers in 20 states two years ago,
and more collection centers are being established.

There are many opportunities to improve the recycle system--collection,
sorting, shredding, shipping, cleaning, and remelting. Can scrap forms
more skim than thick scrap or ingot when it is melted. Certain pigments

and lacquers on the cans accentuate this oxidation so that up to 1/5 of
the metal might be lost as skim by conventional melting methods. Alcoa
recently announced a multi-million dollar can reclamation facility with
special furnaces to minimize the melt loss problem.

Table 7 shows shows the use of customer scrap and "old scrap" as a percent
of shipments for steel, copper, and aluminum.

Table 7

USE OF CUSTOMER AND OLD SCRAP, 1967 [11]
(% of Shipments)

Steel	Aluminum	Copper
31	18	50

But successful as the can recycle program is, 1.5 million tons
of aluminum are still lost each year in garbage and refuse. At $200 per
ton, aluminum is by far the most valuable material found in municipal
waste. Studies suggest that reclaiming this aluminum can generate sufficient revenue to make total waste recycling economical.

The aluminum industry is involved in a variety of development programs to
establish systems for reclaiming aluminum from waste in concert with other
industries and government agencies and organizations like the National
Center for Resource Recovery.

The aluminum recovered from garbage and refuse, even if the separation
process frees it of other materials, is usually a mixture of alloys.
Alloy 5182 used for the aluminum can end contains about 4.5% magnesium,
0.3% manganese; alloy 3004 used for the can body has 1.0% magnesium and
1.3% manganese. In reasonable quantities, mixed scrap can be fed into
the main aluminum stream. The secondary industry employs mixed scrap in
some casting alloys with wide composition limits. But the large scale
utilization of mixed scrap requires economic procedures to remove
selectively the alloying elements. These procedures are being
developed. Recently Alcoa offered for sale a new economical
process for removing magnesium from aluminum alloys without

generating toxic or corrosive fumes. The process also recovers a market-
able grade of magnesium chloride.

Incidentally, if garbage is incinerated prior to separation, the resulting
alloy contains so many impurities that no economic procedure has been
developed to recover the aluminum in the purity required for most of its
applications.

Recycling of old scrap deserves more attention on a national scale. Table
8 shows total energy savings equalling nearly one percent of the nation's
total energy could result from recycling steel and aluminum. In this
calculation Miller's[9] value of energy savings of 13 million BTU per ton of
steel scrap recycled is adopted. The technological problems in using this
scrap are capable of solution. The remaining problems are social, political
and economic and may be solved in part by such methods as tax credits, lower
rates for transporting scrap or other economic incentives.

Table 8

ANNUAL ENERGY SAVINGS FROM RECYCLING "OLD" SCRAP

	Steel	Aluminum
Million Tons Available	27[11]	1.5[3]
Quadrillion (10^{15}) BTU Saved	.35[9]	.27

WASTE HEAT RECOVERY AND USE

In a recent Steel Industry Report to the National Petroleum Council, H.
E. Miller[9] said, "Major emphasis in the integrated [steel] plant should be
placed on the collection and more efficient use of the by-product fuel
generated by coke ovens and blast furnaces..." The wisdom of this statement
is evident from the fact that the energy in the auxiliary fuels arising
from the production and use of coke totaled .58 quadrillion BTU in 1972[6], nearly one percent of total U.S. energy use. The level of bleeding is said
to be zero or quite low in the large integrated plants but this gas could
be wasted in small integrated steel plants. The problem of this use is the
very low energy content (approximately 95 BTU/cu ft) of this fuel.

Miller[9] has further pointed out, "Flue gases exiting from a high temperature
steel heating furnace at 2400°F carry with them 55-60% of the available
heat from the combustion of fuel. Two methods of heat recovery are
predominant: recuperators to preheat combustion air and waste heat
boilers to generate steam. The latter has a potential efficiency of 80%
compared to 60% for recuperators."

Experiments are underway in Alcoa to utilize exhaust stack heat from
aluminum melting furnaces to distill pure water from finishing or
lubricant emulsion streams while concentrating the chemical residue for
reuse or disposal.

TECHNOLOGICAL DEVELOPMENTS TO DECREASE ENERGY FOR STEEL

Tenenbaum and Luerssen[12] have said that over one-third of the energy is
lost because of the sequential nature of steel manufacturing technology.
Continuous casting of slabs, billets, or blooms results in increased
yield and elimination of reheating prior to primary rolling, thereby
consuming about one-half the energy required for conventional processing.
Continuous casting is being adopted by the industry as rapidly as the
development of equipment, operating practices and capital will permit.
Twenty-three million tons of continuous casting is estimated in the
United States in 1974[12].

Basic oxygen steelmaking was developed by the steel industry over the
last decade for its cost savings. It is important to our present
thinking that it uses much less energy than the open hearth, even though
because of its lower scrap tolerance the energy per ton of steel actually
appears to favor the open hearth. With basic oxygen steelmaking the
excess scrap is handled by electric arc furnaces. The values in order
of magnitude are basic oxygen 0.7 million BTU/ton steel; open hearth 3.4
million; and electric furnace 6.6 million[6].

In 1973 the basic oxygen furnace turned out 55% of the 150 million tons
of steel made in this country. Open hearth production declined to 27%
and electric furnace production rose to 18%[2]. In recent years scrap
preheating techniques have increased the scrap consuming capability
of the oxygen furnace. Gas collection hood systems have been developed
to recover the off-gas for its energy potential. Future steelmaking
expansion is predicted to be with a combination of oxygen furnaces and
electric furnace capacity[12].

PROCESSES TO DECREASE ENERGY FOR OTHER METALS

The most promising new processes,   employing oxygen enrichment, save up
to half the energy now consumed in smelting copper.[7]

A new electrolytic process is said to be under development that markedly
reduces energy requirements for magnesium.

The zinc industry is tending toward electrolytic zinc production. This
has an energy efficiency of about 25% in contrast to the disappearing
horizontal retort process whose energy efficiency is only about 5% and
the vertical retort at 10% energy efficiency. Also in the picture is
electrothermic smelting at 25-30% energy efficiency and the zinc blast
furnace, although the latter is not in use in the United States.[10]

REFINING ALUMINA

From an energy standpoint, the Bayer Process for alumina refining is a
series of heat exchangers operating on a mammoth scale. Bauxite is treated
with hot caustic solution under pressure to dissolve the alumina values.
This solution is filtered and cooled in stages, allowing steam to evaporate
in heat exchangers of successively lower pressure and temperature. When
the solution has cooled enough, aluminum hydroxide is precipitated. This
product is filtered from the cooled caustic solution which then recirculates back through the heat exchangers until it is once again hot enough
to extract alumina from a new batch of bauxite.

More efficient heat exchangers, sophisticated computer analyses and improved control have decreased the heat requirements dramatically.

Once the aluminum hydroxide is precipitated, it is washed, filtered and
calcined to anhydrous aluminum oxide. This calcination has customarily
been in a rotary kiln. Culminating several decades of development, fluid
flash calcination is a proven energy saver. The Alcoa developed flash
calciner system utilizes fluidized bed and dispersed phase technology to
improve heat exchanger and reduce radiant losses. Full-scale systems now
in operation calcine metal-grade aluminum oxide with a net heat requirement of 1400-1500 BTU/lb. The extensively utilized rotary kiln calcining
system uses upwards of 2000 BTU/lb of aluminum oxide. This advance in
the area of calcining can reduce the energy requirement for primary
aluminum production by nearly 1%.

SMELTING ALUMINUM

In 1973, Alcoa announced a significant accomplishment by disclosing that
after 15 years of effort and expenditures exceeding $25 million, it had
developed a revolutionary new way of producing primary aluminum. The
implications of the new Alcoa Smelting Process are enormous and the
coincidence of timing most fortuitous in this era of cutbacks in energy
usage.

We have seen already that smelting requires 2/3 of the total energy for
production of aluminum mill fabricated products. Three decades ago,
smelting required 12 KWH/lb in addition to the energy content of the carbon
anodes. Now the average electrical consumption is just over 8 KWH/lb and
modern cells can operate at 6.5 KWH/lb. As new capacity is installed,
there will be a dramatic decrease in the unit energy requirement as a
result of the new Alcoa Smelting Process.

The Alcoa Smelting Process requires 30% less energy than the most efficient
Hall-Heroult cells presently used worldwide. The process is diagramed in
Fig. 2. Essentially, aluminum oxide is converted into aluminum chloride
which is electrolyzed to produce aluminum and chlorine. The chlorine
recirculates for aluminum chloride production.

Ground was broken for the first commercial plant utilizing this process
at Anderson County, Texas, in September, 1973. One unit is scheduled to
come on stream in 1975.

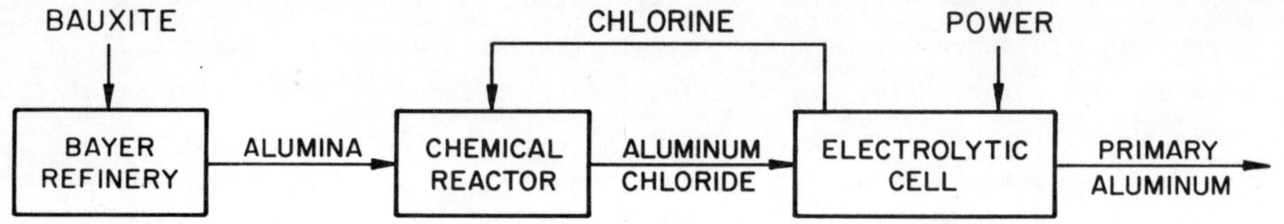

FIGURE 2 — ALCOA SMELTING PROCESS

SUMMARY

Here are the nine major points about energy conservation developed in this paper:

1. Using data from the aluminum industry, a more rigorous method for calculating energy requirements has been illustrated. Just under 10,000 BTU were required to produce one cubic inch of aluminum mill product in 1972. This is about the energy equivalent of a 2.5 in. lump of coal. This includes the energy of each processing step and of all consumed materials after conversion of various energy sources to a common base and allowance for the large scrap recycle.

2. About 5-6% of total U.S. energy is consumed in steel production, 1.4% for aluminum and less than 1% for all others - copper, zinc, magnesium, titanium, etc. The commonly used metals do not differ greatly in energy requirement per unit volume of mill fabricated product. There are promising opportunities to decrease the nation's total energy requirements by proper use of these metals.

3. Metal producers are emphasizing practices to save energy reserves by careful attention to operating procedures, avoidance of excessive peak levels and providing for alternate energy sources.

4. One of the greatest opportunities for energy conservation is in scrap recycle. Less than 5% as much energy is required to remelt scrap aluminum as to derive it from its ores.

5. Many opportunities exist for heat recovery from the waste gases of metallurgical operations. This heat is useful for steam generation, space heating and for recovery of useful chemicals from process waste solutions.

6. Improved blast furnace operation, continuous casting and basic oxygen steelmaking reduce the energy for steelmaking.

7. Lower energy processes are said to be under development for copper, magnesium and zinc.

8. Fluid calcination of aluminum oxide uses 30% less heat than conventional kiln calcination. This saving equals one percent of the total energy used for aluminum.

9. The Alcoa Smelting Process, due to start commercial production in 1975, uses 30% less power than the most efficient present-day cells employed to reduce the purified oxide to the metal.

CONCLUSION

A combination of short range "turn out the lights" actions already under way, recycle and waste heat recovery increasing over the next decade, and long-range, new energy saving processes give promise of significant energy reductions in primary metal production.

REFERENCES

1. Aluminum Statistical Review 1972, The Aluminum Association, New York City

2. Anon, Metals Progress, p. 13, January, 1974

3. Battelle Memorial Institute, Opportunities for Increased Solid Waste Utilization, Vol. II, June 1972; L.C. Blayden, AIME, February 27, 1974

4. R. W. Bouman Some Implications of National Energy Problems on Future Steel Industry Processes and Fuel Sources, C. C. Furnas Memorial Conference, State University at Buffalo, November 12 (1973)

5. J. C. Bravard, H. B. Flora, and C. Parks, Energy Expenditures Associated with the Production and Recycle of Metals, Oak Ridge National Labs., ORNL-NSF-EP-24 November (1972)

6. J. F. Elliott, Uses of Energy in the Production of Steel, C. C. Furnas Memorial Conference, State University at Buffalo, November 12 (1973)

7. H. H. Kellogg and J. Tien, Energy Considerations in Metal Production, Selection and Utilization, TMS-AIME, Chicago, October 2 (1973)

8. F. W. Luerssen, Perspectives - Energy and the U.S. Steel Industry, C. C. Furnas Memorial Conference, State University at Buffalo, November 12 (1973)

9. H. E. Miller, Steel Industry Report, Industrial/Non-Energy Task Force, National Petroleum Council (1974)

10. S. W. K. Morgan and S. E. Woods, Review 156, 161-74, Met. Rev., November (1971)

11. National Commission on Materials Policy, Final Report, Material Needs and the Environment Today and Tomorrow, June, 1973, GPO 5203-00005

12. M. Tenenbaum and F. Luerssen, Energy and the Steel Industry of United States, Fifth Annual Conference, International Iron and Steel Institute, Toronto, October 1971, Quoted in J. Metals, p. 18-20 December (1971)

13. Year Book of the American Bureau of Metal Statistics, 52nd Issue, American Bureau Metal Statistics, New York, June (1973); Metal Statistics, (1973), 66th Ed., American Metal Market, New York.

ENERGY CONSERVATION IN THE CEMENT INDUSTRY

Hoke M. Garrett[*]
James A. Murray[*]

The paper presents the U. S. Cement Industry's potentials for conserving energy and identifies some capital and time constraints which affect this objective. Reductions in per capita cement consumption are unlikely and the rate of increase in demand may even accelerate because of overall energy shortages. The energy requirements of the cement industry are discussed in detail. The kiln dominates a cement plant's demand for energy. Therefore, thoughts on how to improve kiln thermal efficiency in the short term are presented. Various cement making processes are discussed which, in the long term, should reduce total energy consumption to less than 60% of today's industry average of 7,000,000 Btu per ton of cement.

I. INTRODUCTION

"Take two cups of crushed limestone; add one-half cup of clay or pulverized shale (plus perhaps some sandstone or iron ore, depending); mix thoroughly and grind up fine. Bake in white hot oven at about 2,600 F. Cool. Add a tablespoon of gypsum and, again, grind very fine.

---

* Kaiser Engineers, Oakland, California

"If you can do this for about a penny a pound, you can almost compete. "

Dr. R. L. Handy of Iowa State University's Engineering Experiment Station came up with that humorous recipe to demonstrate the complexity and economics of the cement industry.

A not-so-humorous and very crucial impact on today's penny-a-pound cement industry is the supply of fuel needed to keep the kilns burning. To a domestic industry that demands about 550-trillion Btu annually, reduced fuel use has become a necessity in a time of energy shortage.

In terms of the overall power picture, that 550-trillion Btu, which includes both fossil and electrical energy, means about 0.8% of all energy consumed in the U. S. The various factors influencing reduction of this energy consumption are the subject of this paper.

Nationally, the energy shortage is being attacked on two fronts--increasing supply and reducing demand. Making cement consumes energy; therefore, the industry cannot join the effort to increase supply except indirectly as noted later.

Demand can also be attacked on two fronts--by reducing cement consumption or by reducing the energy required to make cement. The following discussions on reducing cement industry energy consumption center briefly on the prospect of reducing cement consumption followed by a broad evaluation of the potentials for improving cement making thermal efficiency.

II. <u>REDUCING CEMENT CONSUMPTION?--NOT LIKELY</u>

The approach of reducing energy demand by reducing cement consumption introduces both philosophical and practical questions. Theoretically, energy reduction can be achieved through reduction of economic and social activity or by zero or even negative population growth. Such considerations, short of disasters such as a world economic stagnation or food shortage, do not seem pertinent to the immediate problem.

These philosophical questions and their answers are in the province of others.

The possibility of using less cement still remains. But, the question arises, "Is it an essential product and are there alternates to its use?" First, cement is the basic ingredient for binding aggregate and sand into concrete and mortar. Although lime, gypsum and until recently, epoxies and polymers are used modestly as specialized binders, there are no known or foreseeable alternates to the massive use of cement for this purpose.

The essential nature of concrete is best answered by looking at the distribution of cement usage shown in Fig. 1 (Ref. 1). These data indicate virtually all categories are essential to our society as presently structured. The effect of the energy shortage on cement consumption for each category is difficult to predict at this time. For example, we might expect reduced consumption for highway construction simply on the thesis that highway programs will be stretched, modified or curtailed. However, this might be offset by more use of concrete highways if the asphalt industry is affected by the petroleum shortage to a greater extent than the cement industry. Also, the probable reduction of highway construction could be offset by increased rapid transit, railroad, peoplemover, and other mass transportation systems. In such systems, wood or concrete railroad ties can be used. Bridges may be constructed of concrete--or steel. Thus the effect of the energy shortage on cement is interrelated with effects on other construction materials. Furthermore, the cement industry will probably accelerate its ongoing efforts to reduce the quantity of cement and concrete required to perform a specific job. It must do this to remain competitive with other construction materials.

The ability to forecast such interrelated effects is typically based on reviewing historical trends. But the U. S. has not experienced a real energy shortage, hence, there is no apparent basis for forecasting the

effect of such shortages--or is there? Fig. 2 (Ref. 1) presents the
U. S. per capita consumption of cement since 1913. Also shown is a
straight trend line fitted to historical increases and projected accordingly to the year 2000. The actual consumption curve clearly shows
the effects of the major depressions, wars, recessions and other
phenomena--except a bonafide energy shortage. Western Europe,
however, has a long history of energy shortages and consequently high
energy costs. Its economy, climate, construction methods and many
other factors are similar to the U. S. In general, it has been where
we seem to be going--that is, we now accept many aspects of its
economy already influenced by small cars, mass transit systems,
bicycles, higher cost materials relative to labor, and so forth. Thus,
the effect of the energy shortage on the U. S. might be disclosed by
looking at the overall experience of Western Europe. Fig. 2 has been
supplemented to include such data (Ref. 1, 2, 3, 4). It shows that
Western Europe per capita consumption of cement was once well below
but has now rapidly surpassed U. S. consumption and was also once
below but has surpassed West European steel per capita consumption
at a slow but steady rate. Conversely, U. S. per capita consumption
of cement has always been below and is increasingly falling behind the
U. S. per capita consumption of steel. Relative to West European
experience, this is probably attributable to relatively low energy costs
and relatively high labor costs. This broad overview suggests that
the projected increase of U. S. per capita consumption of cement will
not be reduced due to the energy shortage. Instead, it appears that
the rate of per capita consumption will most likely increase as in
Western Europe history. The adjustment of the Portland Cement Association forecast, as shown in Fig. 2, is based on the foregoing discussions and the following suggestions:

● The adjusted forecast represents a shotgun pattern aimed above the
  basic PCA projection developed prior to the energy shortage.

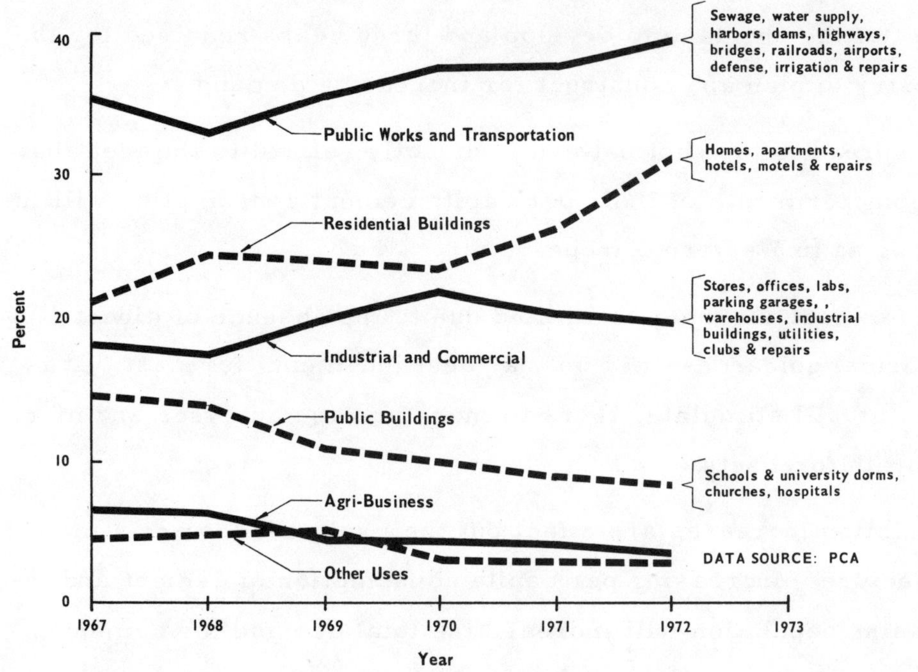

Fig. 1 U.S. Cement Distribution by Use Category

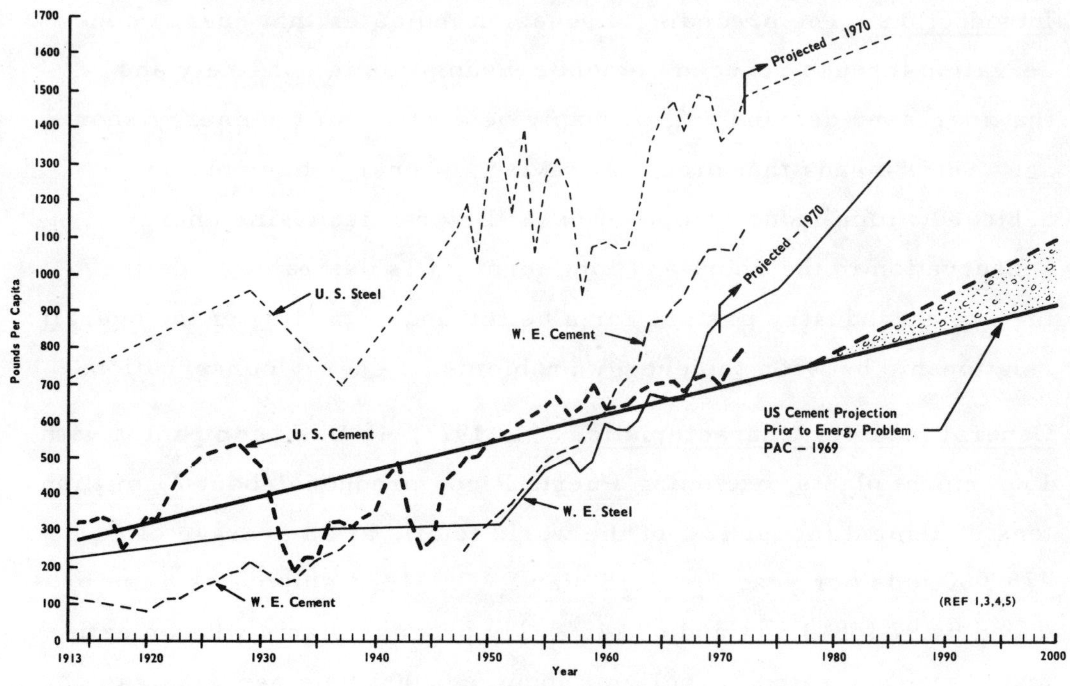

Fig. 2 U.S. Per Capita Consumption –
Historical and Projected

- The shotgun trigger is pulled in 1978, which allows time for the effect of the shortage to develop and three years required for the industry to plan and construct for increasing demand.

- The spread of the shot pattern is directly related to the idea that the long term rate of U. S. per capita cement consumption will increase as in Western Europe.

- The forecast accuracy is limited due to the absence of direct historical guidance--it is only a "best judgment" forecast. It is hoped it will stimulate others to more precise analyses and more accurate forecasts.

- Population increases are a fact but their magnitudes are a matter of conjecture. Increasing per capita consumption of cement and increasing population will increase the total demand for cement.

III. INDUSTRY POSTURE RELATIVE TO THE ENERGY PROBLEM

A. <u>Introduction</u>. The preceding discussion indicates that energy conservation through reduced cement consumption is not likely and that increased demand will probably be a result of the energy shortage. This means that practical savings of energy can only be achieved within industry operations. Before discussing energy conservation in the short and long term, it is desirable to define the cement industry posture for a better understanding of the overall relationship between the energy problem and energy conservation.

B. <u>General Industry Characteristics</u>. In 1972, 46 U. S. companies with 166 cement plants, excluding Puerto Rico, produced about 79 million tons of cement (about 12% of the world total), at an average of 475,000 tons per year for each plant. The 1972 shipments were produced by 97 (58%) wet and 69 (42%) dry process plants with an estimated average rated capacity of about 530,000 tons per year, as shown along with other industry data in Fig. 3.

In the 97 wet-process plants, grinding and blending take place while
raw materials are in a water slurry. The slurry is then fed to
rotary kilns, where the water is evaporated, the raw materials are
calcined, and cement clinker is formed.

In the 69 dry-process plants, moisture is removed prior to and/or
during grinding, resulting in blending of dry meal which is then fed
to the kiln. Six of the dry-process plants are equipped with a total
of twelve suspension preheater kilns. In this process the dry meal
is fed to a preheater tower (series of cyclones as described later),
where it is partially calcined before being discharged into a rotary
kiln.

The 1972 production was accomplished in about 454 kilns: 250 (55%)
wet-process and 204 (45%) dry-process kilns as shown in Fig. 4.
This figure also shows the age categories for all U. S. cement kilns,
as well as other kiln data. The average number of kilns per plant
is about 2. 7 and the average kiln rated capacity is about 194,000
tons per year. The average kiln age is about 25 years, with 27%
of the wet-process and 21% of the dry-process kilns, or about 110
kilns, being over 40 years old.

C. <u>Industry Energy Consumption</u>. For energy of all types, the cement
industry in the U. S. demands about 550,000,000,000,000 Btu annually or almost 7,000,000 Btu average per ton of cement. This demand includes the direct purchase of fossil and electrical energy
and the transmission and generation losses incurred after purchase.
This amounts to about 2. 0% of all the energy consumed by U. S.
industry or a little more than 0. 8% of energy consumed in the U. S.
by all activities. Therefore, the U. S. cement industry's efforts
to reduce its energy consumption will not have a significant impact
on the nation's energy shortage; however, it can significantly affect
the total amount of cement produced within limited fuel supplies and
it can also significantly affect costs and profits.

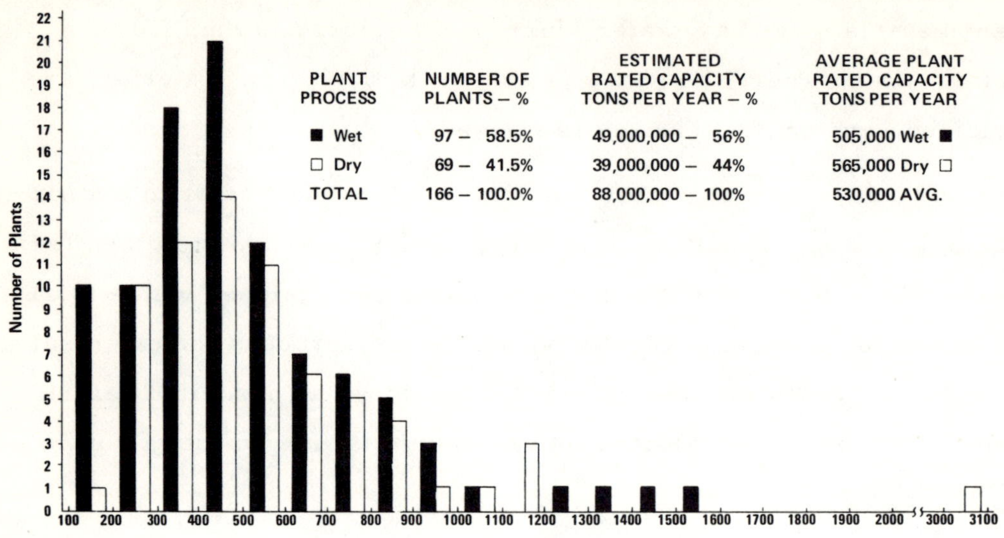

Fig. 3 U.S. Cement Industry - Plant Data

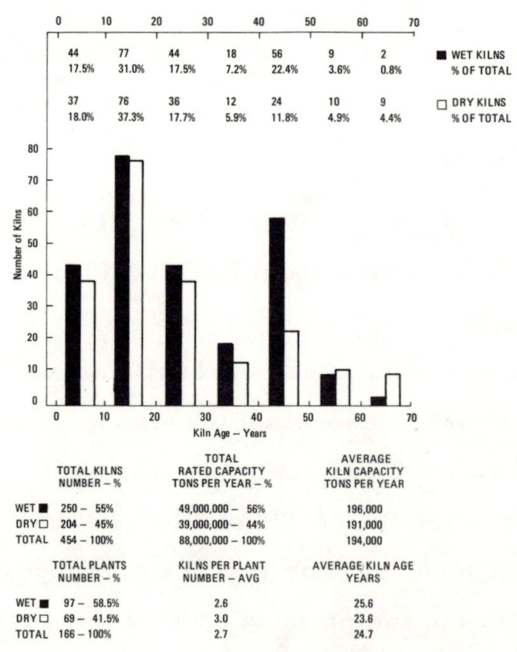

Fig. 4 U.S. Cement Industry - Kiln Data

The energy consumption by energy type for typical, relatively new
and efficient wet, dry and suspension preheater plants is shown in
Fig. 5. The plants are more efficient than the industry-wide
average, but not as efficient as the most modern plants. In all
three processes, the dominant energy requirement (about 90%) is
fuel, which includes both kiln burning and motor fuels. As indicated
in the figure, electrical and other energy sources (explosives and
manpower) are small. It is apparent that to conserve important
amounts of energy, kiln fuel consumption must be reduced. Of
course, any other energy savings, especially grinding which consumes most of the electrical energy shown, would be desirable.

D. <u>Impact of the Energy Problem</u>. The preceding discussions on the
general character and energy consumption of the cement industry
outline the challenge presented to the industry by the rapid escalation of the energy problem. But the industry has faced and solved
a similar problem--the labor cost push. Over the past 25 years,
the cement industry has been able to reduce the amount of labor to
produce a ton of cement by a factor of about three, while labor
costs increased by a factor approaching six (a net 200% increase)
as shown in Fig. 6 (Ref. 1). Labor reduction is largely attributable
to modernized cement plants starting in the late 1950's when
industry capital expenditures and technological developments to
provide large production units, centralized control and automation
techniques helped to reduce labor requirements dramatically as
shown. Conversely, in Europe where fuel instead of labor cost was
the dominant factor, significant efforts were made to improve plant
thermal efficiency. There, the trend has been to the dry-process
suspension preheater and roller raw mill wherever practical to
reduce energy consumption.

The typical direct production costs in a U. S. cement plant in 1970,
as indicated in Fig. 7, show that these costs are still dominated

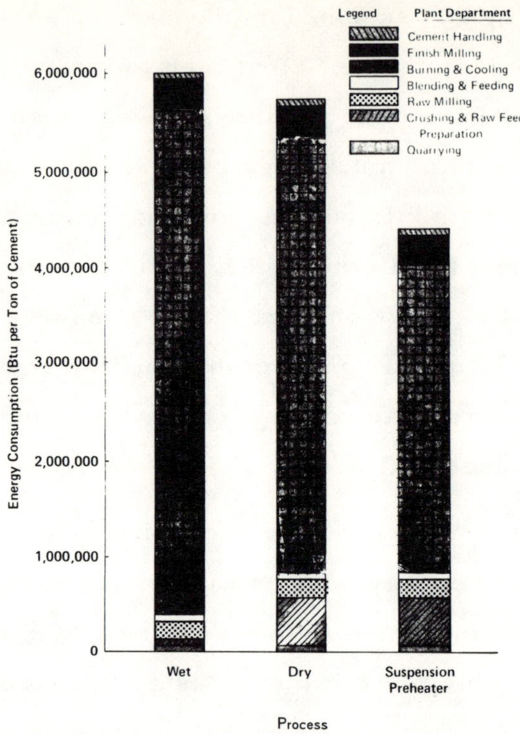

Fig. 5 U.S. Cement Industry - Energy
Consumption by Energy Type

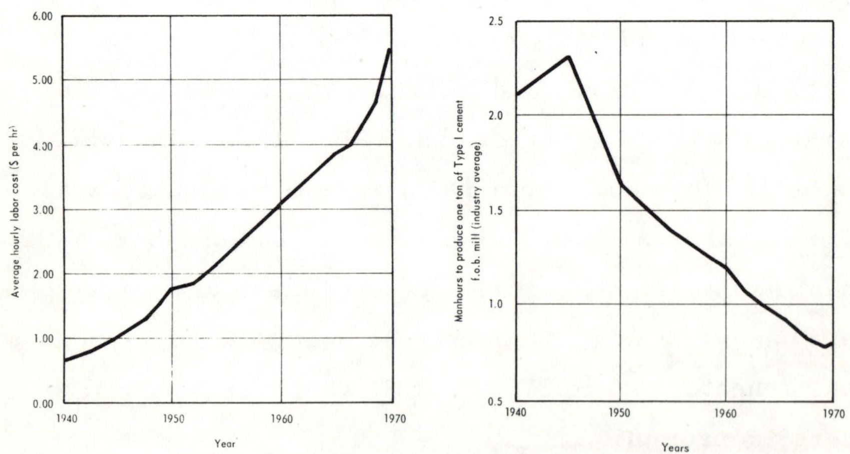

Fig. 6 U.S. Cement Industry Labor
Consumption and Costs

148

by labor at about 40%.   Labor is followed by fuel at about 25% and power at about 15%.   The figure also shows the history of these cost elements since 1945.   Here again it clearly indicates that the U. S. industry concentrated on the historically major cost element--labor.   Less effort was required to limit the net cost increase of fuel also to about 200%.

Also shown in Fig. 7 is the projected minimum plateaus anticipated for labor, fuel and electrical consumption which might be achieved with present technology within the next 10-25 years.   These projections are based on the following:

<u>Labor</u>--Many modern automated plants with central control facilities already require only 0. 35-0. 75 manhours of labor per ton of cement.   The larger plants are below 0. 40 manhours per ton.   The projection indicates a minimum of 0. 40 manhours per ton, which is about 50% of 1970 and 17% of 1945 average requirements.

<u>Fuel</u>--The modern well-maintained cement plant with suspension--preheater kiln should achieve a total fuel consumption somewhat below 3, 500, 000 Btu per ton.   In the future, the industry-wide average could reach a total fuel requirement of less than 4, 000, 000 Btu per ton allowing for the continued use of some long dry and wet-process kilns.   The 4, 000, 000 Btu per ton rate is about 50% of the 1945 average requirements.

<u>Electricity</u>--The history of electrical consumption of about 130 kwh per ton of cement shows only a small increase over the past 25 years.   There is no significant basis for altering this consumption rate in the projection.

Thus, the cement industry can save substantial energy and therefore mitigate shortages and higher costs, but to do so it must modernize existing equipment and wherever practical convert to the dry process using suspension preheater or superior processes.

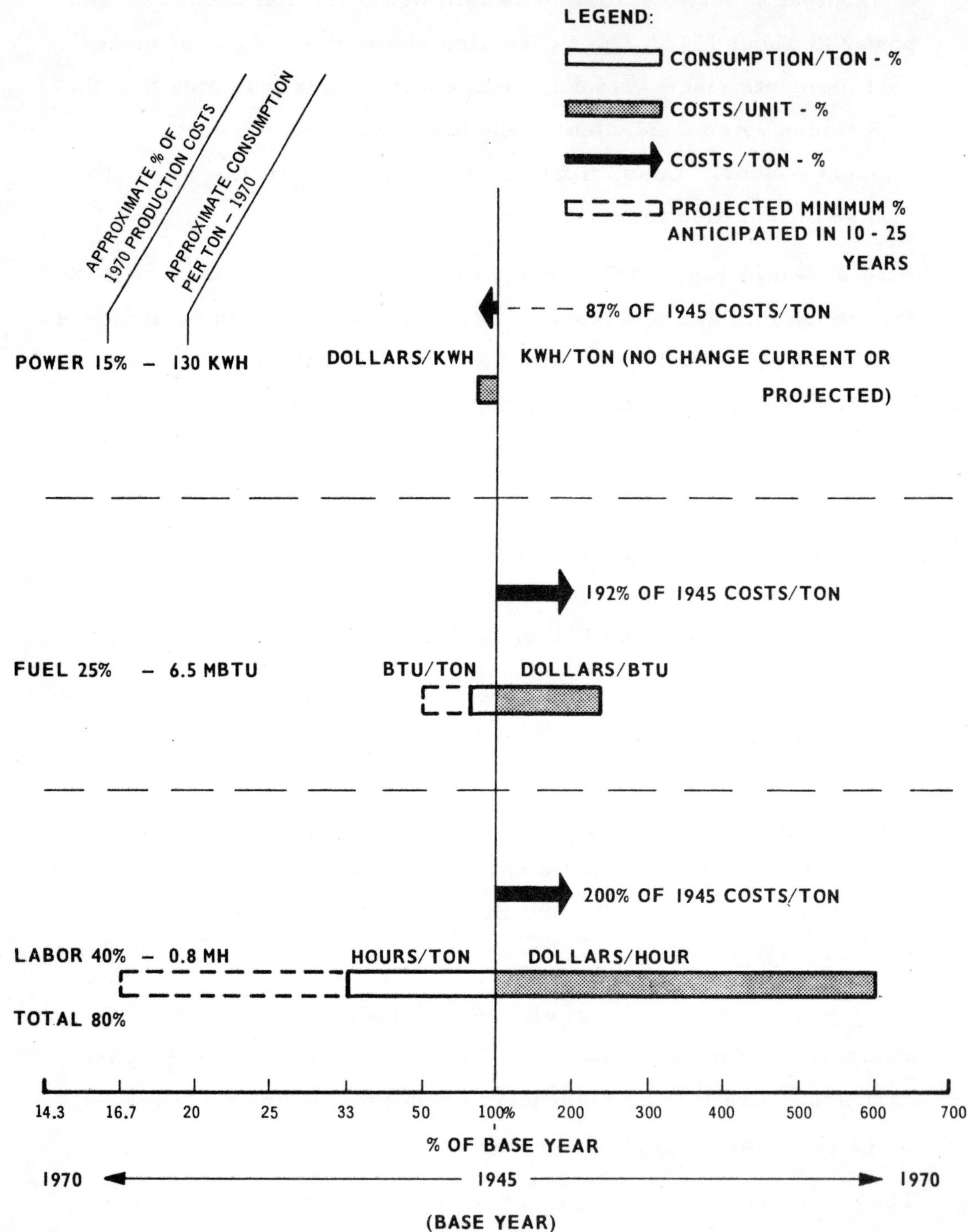

Fig. 7 U.S. Cement Industry - Labor, Fuel, and Electrical
Consumption and Cost Relationship (1945-1970)

150

It is believed that given sufficient capital the industry can achieve significant reductions rather quickly--certainly faster than the 25-year effort to reduce labor requirements.

This is easily stated but difficult to accomplish in view of a decade of industry economic doldrums. The selling price of cement, shown in Fig. 8 (Ref. 5, 6) has increased about 200% since 1945. More significant is the price erosion indicated during the 60's due to industry overcapacity. This situation developed as a result of substantial profitability established during the mid-50's coupled with, as it later became evident, overly optimistic projections of demand growth. Then the highly fragmented industry responded with a plant construction boom in the mid-50's, only to have it's bubble burst and thus sinking into price wars and reduced profitability.

The mid-50's boom and the mid-60's bust as measured by "Return on Net Worth" is shown in Fig. 9 (Ref. 7, 8). The data shown are not an absolute measure of the cement industry's profitability since the data are derived from an industry that has become increasingly diversified primarily because of low profits. However, the data do establish broad trends relative to cement history and the history of other industries. Also shown is a nominal improvement in profitability in the last few years; however, this is not commensurate with production capacity utilization relative to price and return on net worth shown in Figs. 8 and 9. The failure of profits to increase to 1950 levels is primarily due to recent government price controls which caught the industry at the lowest point of profitability in 20 years. Now that price controls have been lifted in recognition of cement shortages developing in localized areas throughout the U. S., one might hope that lifting of price controls will result in rapid expansion and modernization of the industry. However, the industry has been increasingly using what little capital is available after price wars and price controls to conform to mandatory and stringent air pollution requirements.

151

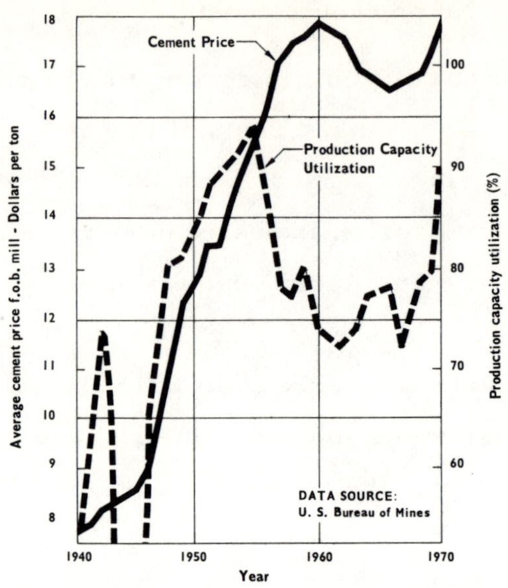

Fig. 8 U.S. Cement Price and Production Data

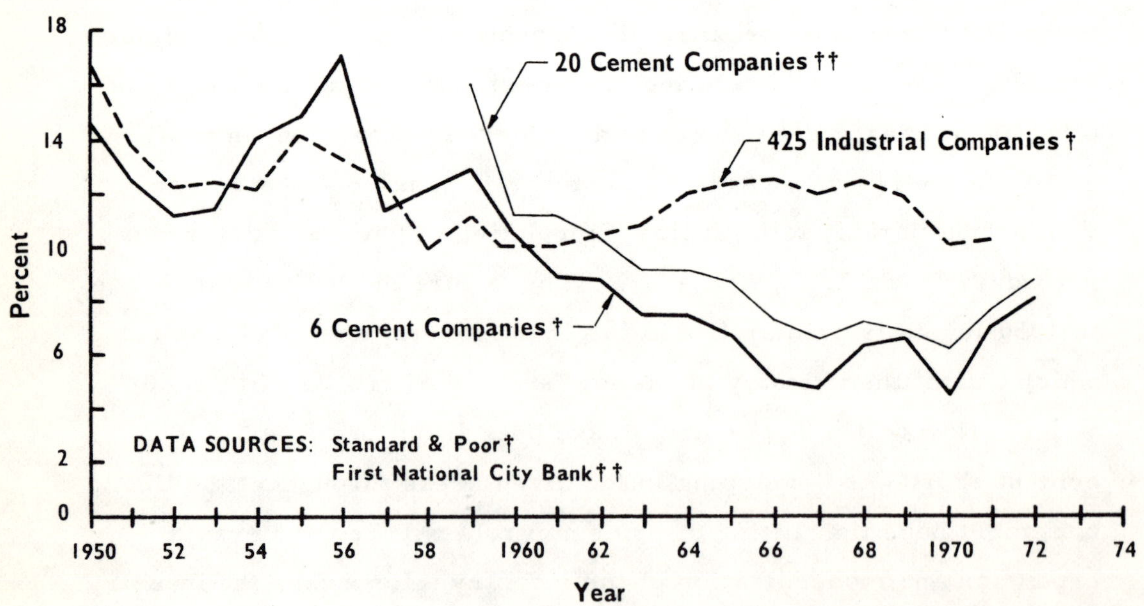

Fig. 9 Return on Net Worth of Selected
U.S. Cement Industry Plants

The preceding discussions establish the general posture of the industry relative to the energy problem. This posture and the challenge it presents to the industry may be summarized as follows:

- About 90% of all energy is required for kiln fuel.

- Almost 60% of the 166 plants are wet process, which requires more energy than other cement making processes.

- The typical plant is small and equipped with an average of 2.7 kilns each averaging under 200,000 tpy--also small.

- The average kiln age is about 25 years and about 25% of the kilns are over 40-years-old.

- The industry has a record of meeting economic challenges, but currently, as a whole, the industry is drained of capital.

- The energy consumption of the industry can be reduced in the long term by kiln replacement with suspension preheater or superior systems.

IV. <u>ENERGY CONSERVATION IN THE SHORT TERM</u>

A. <u>Introduction</u>. The preceding disucssion focuses attention on the cement industry kiln as the major area where significant energy savings can be realized. It is also clear that the present industry posture will not allow replacement and modernization of existing kilns on an "overnight" basis. Until capital is generated, the logical question arises, "What can be done to improve thermal efficiency of existing kilns?" The following discussion presents a number of thoughts on improving existing kiln thermal efficiency.

Each energy saving thought is identified relative to the typical rotary kiln process in Fig. 10. Should the industry implement all concepts to the maximum extent possible and continue the emphasis on dry-process plants, it is reasonable to expect that the present industry fuel consumption of 6,300,000 Btu per ton might be reduced

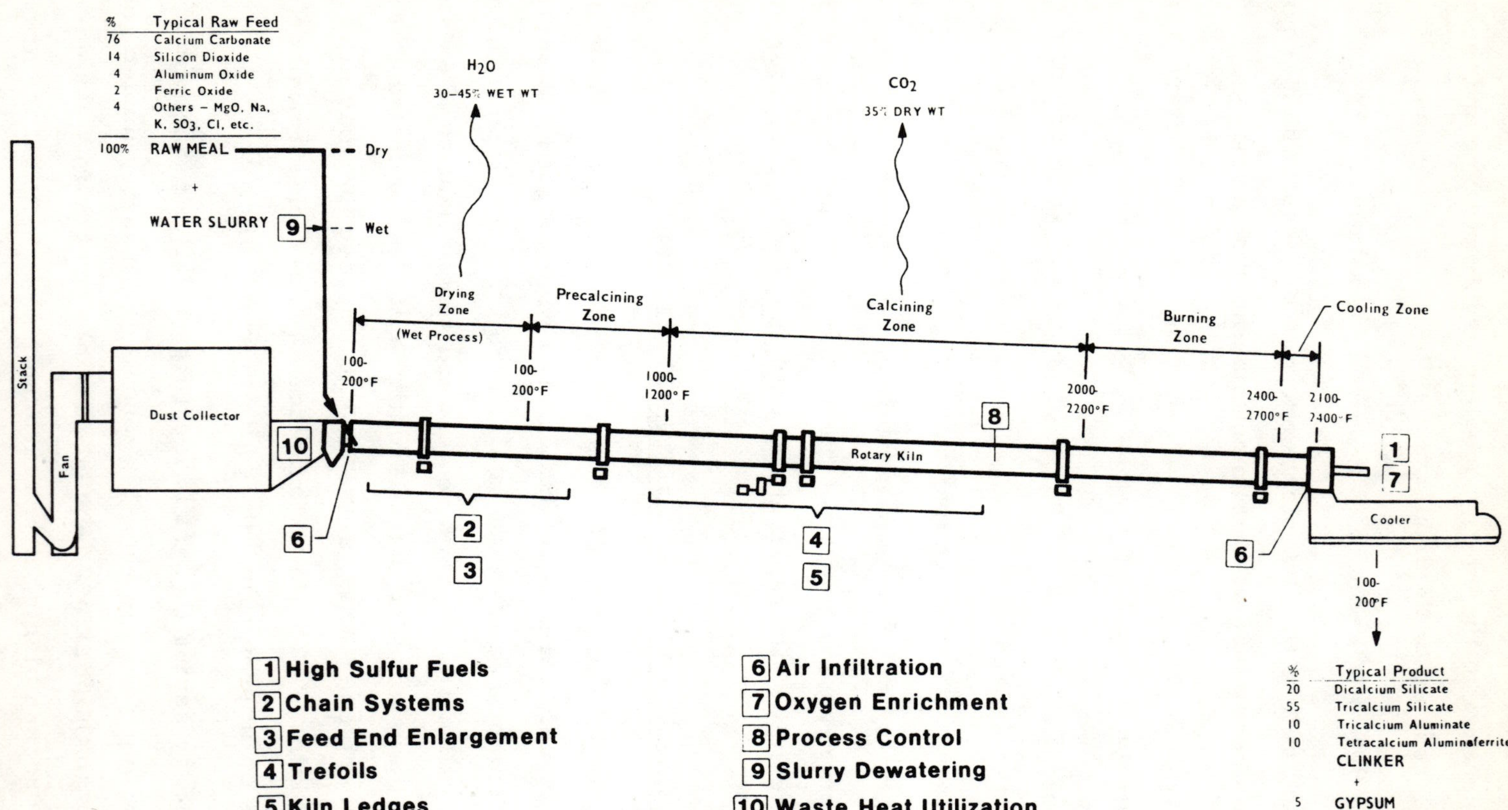

Fig. 10 U.S. Cement Industry - Energy Saving
Concepts in the Short Term

to about 4,700,000 Btu per ton or about 75% of today's average rate.
Also, such improved thermal efficiency will, in general, assist the
industry to generate needed capital, alleviate air pollution problems,
help toward meeting today's need for increased production.

B. <u>High Sulfur Fuels</u> (Fig. 10, Item 1). As briefly noted previously,
the industry can indirectly contribute to increasing energy supply.
From an overall energy standpoint, if an industry can use a lower
grade fuel, the nation will not require as much energy to produce or
refine higher grade fuels. The planned use of high sulfur fuels
should reduce industrial energy costs by reducing demand for premium low sulfur fuels. This concept is applicable to the cement
industry since about 75% of the raw feed is converted to CaO which
reacts with SO_2. Probably even more reactive with SO_2 are the Na
and K present in raw material compounds at many plants. Unfortunately, the variable chemistry and operating conditions in the 166
U. S. cement plants affect the amount of SO_2 entrapment and, in some
cases, the quality of the product. Entrapment of sulfur contained in
fuel and/or raw material feed seems to center at 75% in the plants
for which preliminary data are available (Ref. 1).

The cement industry's ability to use high sulfur coal and oil is contrary to the long term swing away from coal toward natural gas as
a dominant fuel, as shown in Fig. 11 (Ref. 1, 5, 6). This figure
also shows a rapid acceleration of this trend since 1969, probably
due to changing availability and cost relationships among coal, gas
and oil. Now the energy problem introduces a counter trend incentive to reduce gas and oil consumption and increase coal use. The
cement industry can participate in the effort to revert to coal as
the dominant fuel providing it is available and cost incentives are
reestablished. Further, the ash content of coal, consisting primarily of silica, alumina and iron, composes a part of the raw materials
required for making cement clinker. Also, sulfur can be consumed

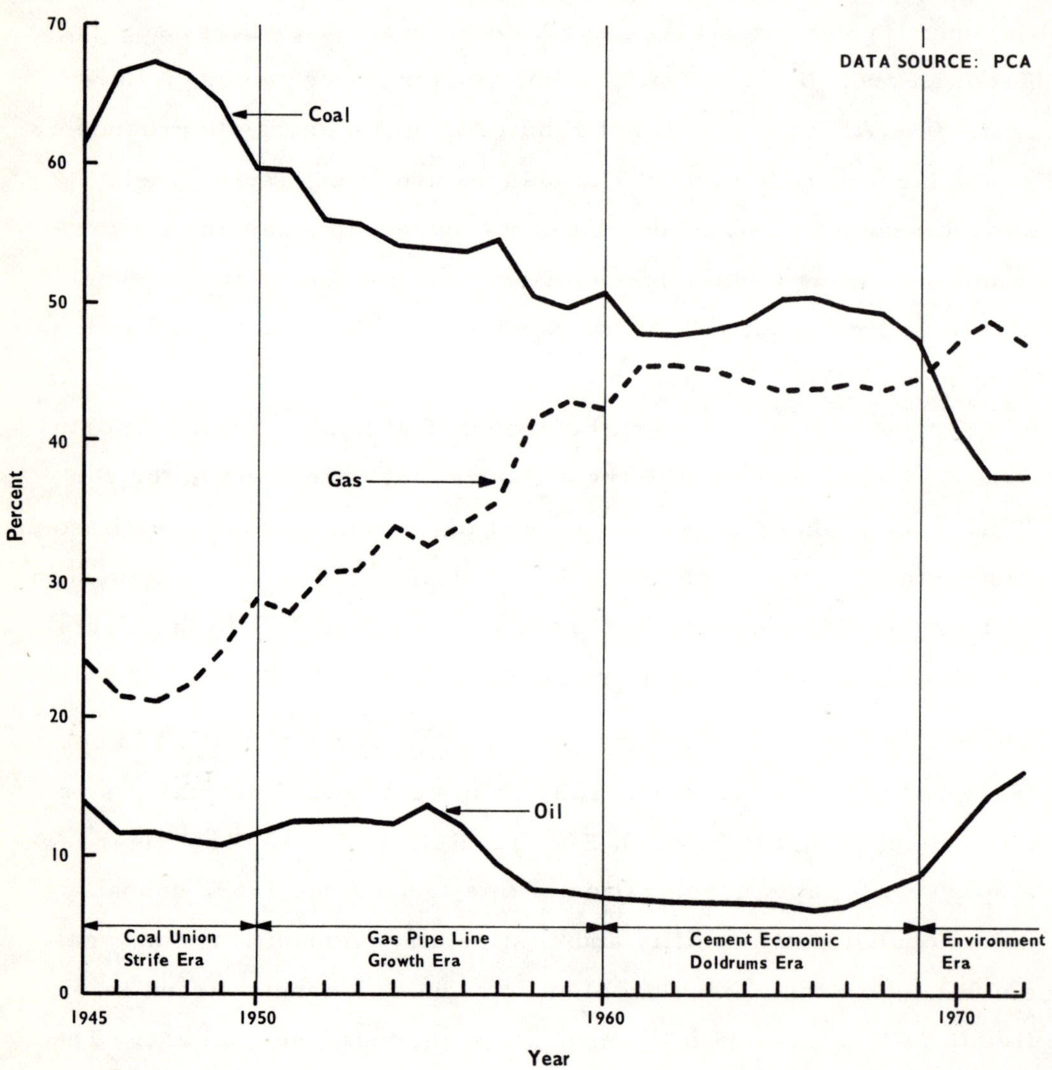

Fig. 11 U.S. Cement Industry - Fuel Usage

156

to varying degrees as a part of the product. Based upon limited
and variable data, it appears that increased SO_2 levels would not
violate current emission standards at most plants.

C. <u>Heat Transfer Efficiency Improvements</u>. The rotary kiln is a heat
transfer device where the hot gases flow counter-current with the
material. The material first undergoes the endothermic processes
of drying, preheating and calcining and then the exothermic processes
of cement sintering and finally cooling. Any improvement in the heat
transfer efficiency in the kiln will result in less waste heat.

Probably the most commonly used technique for improving heat
transfer characteristics is kiln feed end chain systems (Fig. 10,
Item 2). The basic principle of this well developed concept is to
absorb heat from the gas stream for transfer to the raw mix as the
chains revolve with the kiln into and under the material bed. Frequently chains are hung in simple curtain fashion; however, primarily in wet kilns, chain systems are hung in zones with the mid-
zone arranged in festooned garlands to facilitate conveying raw
feed during the plastic stage of slurry drying. A case history of
one wet-process chain installation is presented below to characterize the effect of chain systems on thermal efficiency and
productivity.

Chain Installation	Chain Weight Total Tons	Production Tons/Day	Fuel Consumption Btu/Ton
Initial	110	1,350	6,500,000
Revision 1	146	1,500	5,300,000
Revision 2	172	1,550	4,900,000

In some cases the kiln feed end can be enlarged to improve efficiency
and capacity (Fig 10, Item 3). This is an important consideration
since about 75% of the 454 total U. S. installations are single diameter kilns. Feed end enlargement will increase production which

also increases thermal efficiency by reducing unit radiation losses. The enlarged feed end will facilitate installation of a greater quantity of chains or other heat recuperative devices to achieve increased thermal efficiency. Thus increased production and thermal efficiency can be achieved simultaneously.

One cement company has modified six dry-process kilns by enlarging 100 feet of 400-foot kilns from 12-feet to 14-feet 6-inches diameter and by installing high density chain systems. This modification increased production by about 30% and reduced specific fuel consumption by about 20% to a consumption level approaching 3,900,000 Btu per ton. Similar production increases should be achieved for wet-process kilns. In most cases, thermal efficiency improvement should be nominal (reduction in unit radiation losses) since many kilns are already equipped with sufficient chains to reduce exit gas temperature to the maximum extent practical (about 400 F for the wet process).

Due to steel temperature limitations of about 1,700 F, refractory devices must be used in hotter regions of the kiln to improve heat transfer efficiency. The Trefoil system (Fig. 10, Item 4) consists of a series of refractory arches which divide a portion of the rotary kiln zone into three segments. This divides the material into three streams, threby exposing more material and refractory surface area to the hot gas stream. The system is used in about 65% of all lime industry kilns with production increases reported to range from 5%-15% and thermal efficiency improvements from 5%-20%. However, recent Trefoil installations at two cement plants improvements have been difficult to measure. The general conclusion is that a nominal improvement in fuel consumption has been realized. Certainly more development work is needed to transfer considerable success in the lime industry to the cement industry.

Thermal efficiency is improved if the material tumbles as the kiln rotates. This tumbling action exposes more fresh surfaces to be heated. Kiln "tumbling ledges" (Fig. 10, Item 5) are a means of creating positive tumbling action in a rotary pyro-processing vessel. A ledge installation consists of multiple rows of 2-inch to 10-inch high curbs or similar discontinuities installed as part of the refractory lining along the kiln axis. The primary function of ledges is to assure material tumbling in lieu of material sliding on the shell refractory lining. Material sliding limits heating to the peripheral surface of the bed exposed to the gas stream and hot refractory. Further, the center or "core" of the material bed is not exposed to radiant heat above the bed or contact heating from the refractory under the bed. The core must be heated by inefficient conductive heat transfer from the relatively stationary hot surface materials. Thus a sliding bed results in a "cold core". The use of ledges has been very effective in rectifying material sliding action in rotary vessels processing coarse nonfluidized material. However, the results in the cement kiln calcining zone have been mixed. In any event, efforts should continue to develop better methods to assure tumbling of fluidized material in the kiln calcining zone, where most of the kiln heat is required.

D. <u>Waste Heat Reductions</u>. Kiln exhaust gases are the major source of waste heat. Reducing excess air and air infiltration (Fig. 10, Item 6) will reduce waste gas volume. Air infiltration at the burner hood causes preheated secondary combustion air from the clinker cooler to be displaced by cold ambient air. Thus, more fuel is required to achieve necessary kiln gas temperature. Depending on kiln capacity and fuel consumption, hood leakage will account for 6%-15% of the combustion air, or about 50,000 to 200,000 Btu per ton. This can be reduced by minimizing the seal gap and/or hood draft. At the feed end of the kiln, the wet-process plant also pays a fuel penalty for any air infiltration. The temperature of exhaust gases from the wet-process kiln at design conditions should be as low as possible while still being consistent with sound dust collection and corrosion condensation prevention techniques. Below that temperature, many problems may be anticipated and additional fuel will have to be added to heat any air leakage at the feed end to the required exit temperature. Here again, seal gap must be minimized.

Another method for reducing the quantities of waste gases is to enrich the secondary combustion air with O_2 (Fig. 10, Item 7) and thereby reduce the proportion of N_2 in waste gases. The primary purpose of oxygen enrichment is to increase kiln production capacity and achieve attendant reduction in specific fuel consumption. Recently a cement company commenced oxygen enrichment on a gas fired dry kiln rated 700 tpd. Using one ton per hour of oxygen the company has achieved a 20% production increase during a recent month (840 tpd average). It now plans to modify kiln auxiliaries in the future, when it is estimated that the present fuel consumption of about 4,200,000 Btu per ton can be reduced to 3,700,000 Btu per ton--a 12% reduction.

Waste heat can also be reduced through better process control (Fig. 10, Item 8). Starting and stopping a kiln wastes large amounts of heat because of the system's vast thermal inertia created by 2-3 hours of material residence time. Fuel consumption can be reduced up to 5% whenever the process is operated at steady state conditions. During the 1963-72 period, approximately 20 computers were installed in the cement industry. Initial kiln control efforts were based upon development of complex process mathematical models. Process complexity and the inability to develop reliable remote sensors reverted computerized control efforts back to the time honored empirical methods. Although computer control has been effective in other cement plant areas, reliable and continuous control of the kiln process has eluded the industry.

The most promising way of reducing a major portion of the energy requirements for the wet-process cement plant is to dewater the slurry fed to the kiln and thereby reduce the latent heat in waste gases (Fig. 10, Item 9). About 20%-25% of the process heat required is used to evaporate water in the feed. If the slurry could be dewatered, a proportionate decrease in the amount of heat required would result. From a total energy basis, this dewatering must be accomplished by a process which requires less energy than evaporation and permits gas waste heat to be recovered. In the past, drum filters and the like were used to dewater slurry. These filters, constructed of wood, steel and cloth, required considerable operating and maintenance labor. Such labor requirements at a time when labor costs were rising far more rapidly than fuel costs resulted in abandonment of existing filter systems during the 1940-1960 era. However, the current increase in fuel costs suggests reevaluation of various available filtering methods of improved design for dewatering slurry.

In the cement making process about 25% of wet or dry-process kiln energy is wasted in exhaust gases. In the dry process, exhaust gases offer a source of recoverable heat (Fig. 10, Item 10). Unfortunately, waste gases from efficient well maintained wet-process kilns are too low in temperature and too high in moisture content to permit practical waste heat recovery. Two methods are common for waste heat recovery from dry and preheater-process kiln exhaust gases--waste heat boilers and raw material drying. The long term demise of waste heat boilers on U. S. kilns stems from improving chain systems which allow direct heating of materials in the kiln. Raw material drying by waste heat from suspension-preheater kilns is effective since there is enough heat to dry raw

materials from about 7%-8% moisture down to about 1/2%. Typical
fuel saving is about 500,000 Btu per ton of cement. There is more
waste heat available in the exhaust gases of the long dry kiln than
for the preheater; both the quantity and temperature of the gases
are greater. However, the gases are above the inlet temperature
limits for drying in some roller mills. For rotary dryers, the
high volumetric flow rate from long dry kilns requires larger diameter shells to keep the dryer gas velocity sufficiently low to prevent excessive dust entrainment. These drawbacks have limited
waste heat use from long dry kiln. In view of the energy problem,
a fresh look should be given to utilizing waste heat from long dry
kilns, and perhaps even clinker coolers.

V. ENERGY CONSERVATION IN THE LONG TERM

A. _Introduction._ The preceding discussions clearly demonstrate that
the energy crunch dictates the use of dry-process cement kilns
wherever practical. In summary, the typical kiln fuel consumption
and exit gas volumes, excluding infiltration and raw material drying, per ton of clinker for the three basic processes are
approximately:

Kiln Type	Wet	Long Dry	Suspension Preheater
Btu/Ton	5,150,000	4,200,000	3,000,000
% Relationship	100%	82%	58%
SCF/Ton	100,000	70,000	50,000
% Relationship	100%	70%	50%

The above relationships are admittedly generalized since the many
variables associated with each plant affect fuel consumption and
exit gas volume. The shown fuel consumption rates are lower than
typical industry averages for the industry energy analysis in Section
III. The thrust is to emphasize the well-known fact that the Suspension Preheater (SP) or superior type kiln is the most advantageous
for long term fuel and pollution control problems.

The following presents the basic characteristics of the SP kiln and explains the reasons for the delay in widespread U. S. industry acceptance since SP development in 1950. Also included is a brief discussion on the Secondary Furnace kiln (a refinement of the SP Process) and several other kiln concepts.

B. <u>The Suspension Preheater (SP) Kilns</u>. The Humboldt four-stage preheater shown in Fig. 12 was first to be developed. Its operation involves four-parallel, flow cyclone heat exchangers. Each cyclone includes a long riser where raw feed is introduced and conveyed upward into the cyclone in parallel flow with the hot gas from the preceding cyclone. The intimate contact of individual raw feed particles with hot gases in the riser and cyclone results in rapid heat exchange. Each cyclone serves to spin out the feed for gravity discharge into the riser of the next lower cyclone for repeated parallel flow heat exchange with successively hotter gases. This process results in increasing raw feed temperature from about 160 F to 1,450 F in about 30 seconds and achieves an apparent calcination of about 45%. The preheater product is then discharged into a conventional but foreshortened rotary kiln where calcining is completed and material converted to cement clinker at about 2,600 F.

Major western competitive systems, similar to the original Humboldt, and year of their development include the Danish firm, F. L. Smidth--1955 and the German firms of Polysius--1958, Wedag--1962, Krupp--1964 and Miag--1968. These firms are selfrepresented in the U. S. except for Humboldt until recently represented by Fuller and Miag represented by Allis-Chalmers. Today there are approximately 267 Humbolt, 75 F. L. Smidth, 132 Polysius, 15 Wedag, 26 Krupp and 5 Miag preheaters operating worldwide (over 500 SP Systems total). These figures indicate a dominant sales position by Humboldt, the first commercial developer in 1950: 17 years after the original Czechoslovakian 1933 patent. The

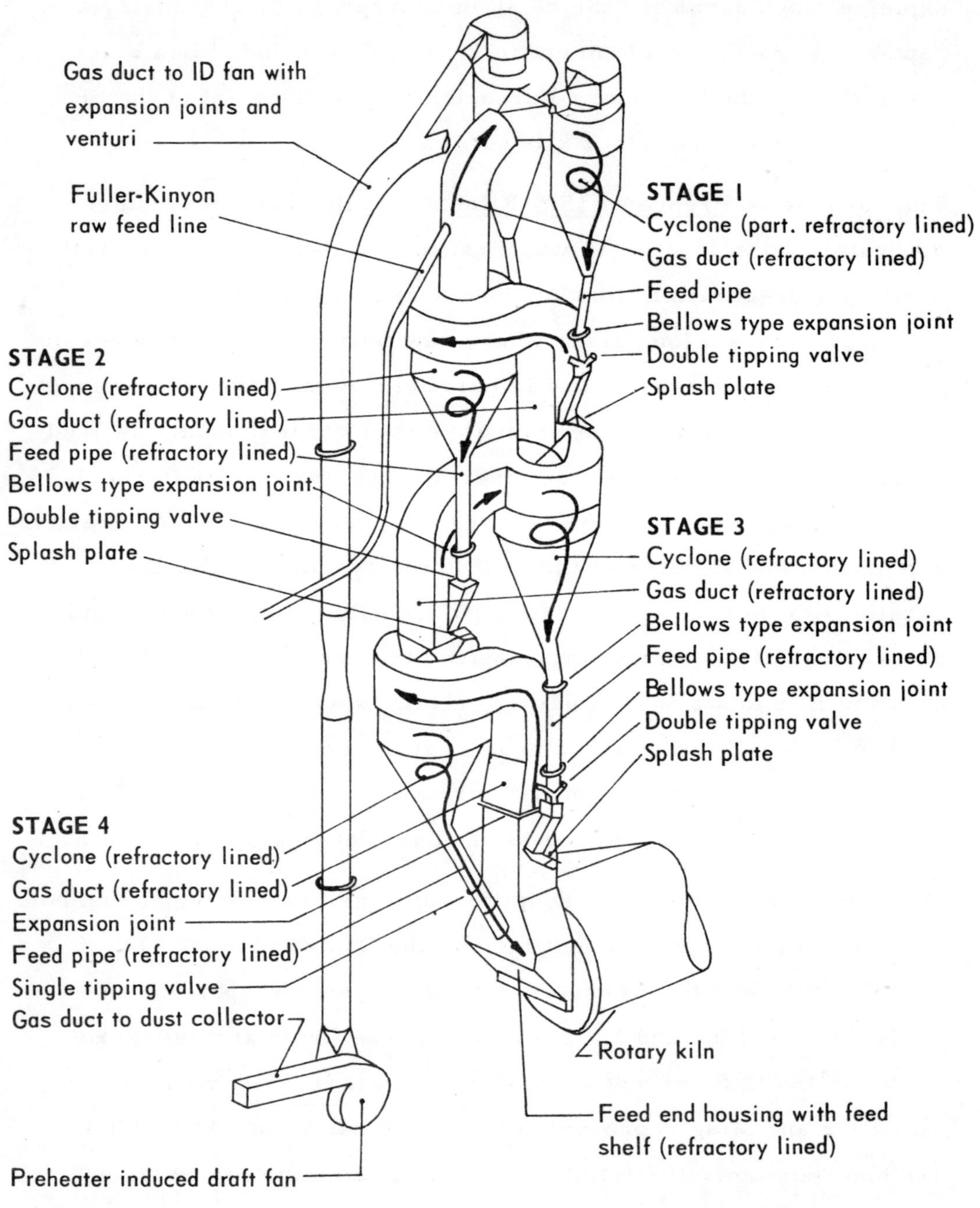

Fig. 12 Suspension Preheater Kiln 4-Stage

figures also indicate a subsequent 18-year spread in the development of competitive systems. Significantly, only 22 units or 4% were sold through 1971 in the U.S. where about 12% of the world's cement is produced.

The history of SP sales in the U.S. is summarized in Fig. 13. Sale of the first Humboldt unit was made in 1953, three years after the unit's initial development in Germany. This was the fourth unit in the world, preceded only by three units in Germany. The unit's immediate acceptance was followed by rapid sale of 12 additional Humboldt units by the end of 1955. These plants (13 in number) came on-stream during the period 1955-1958. Unfortunately, the alkali problems at many of the plants and the problem of kerogens (combustibles in raw materials) at two plants resulted in virtual disaster for the newly introduced process. The alkalis in raw material evaporate near burning zone temperatures and condense primarily as sulphates and chlorides in the lower portions of the preheater. Continual condensation leads to plugging which often results in frequent kiln shutdown. Kerogens typically start burning in the preheater at 500 F, which produces high temperatures and waste heat which can literally make the system inoperable. As indicated, about half of these 13 original installations are now shut down. This situation stemmed from inadequate knowledge regarding the basic process and the effects of raw materials on SP System operation. In addition, the first installations were essentially prototype units without benefit of design refinements resulting from experience in practical operations.

Thus the stigma was born that the SP System with its many early chemical and design problems was unacceptable considering U.S. requirements for continuous operation without shutdown during a time when attention was being focused on reducing manpower. Therefore, SP kiln sales were limited to two units during the

14-year period between 1955-1970. Today, the one-two punch of
increasing fuel costs and air pollution control requirements has
spurred the sale of SP Systems in the U. S. SP System sales have
increased in the last three years from zero to twenty units. Today's
acceptance of the SP System by the U. S. cement industry is based
upon a reasonably thorough understanding of the chemical requirements in raw materials, a means of handling high-alkali raw materials and producing low-alkali clinker, and refinements in design
developed over the years (primarily in Europe). Current SP System
success sets the stage for acceptance of the Secondary Furnace
System (SF) which is a refinement of the SP System.

C. <u>The Secondary Furnace (SF) Kilns</u>. The conception and development
of the SF System originated in Japan. The Ishikawajima Harima
Heavy Industries (IHI) Suspension Flash Preheater (SFP) was developed in 1966 and is the offspring of the parent Humboldt SP for
which license rights were acquired by the Japanese firm in 1963.
Parallel or subsequent SF Systems developed by other Japanese
firms are offsprings of other German SP Systems.

The basic concept of all SF-type kilns is the addition of a separate
combustion chamber at the base of the conventional SP preheater--
just ahead of the rotary kiln. The IHI-SFP is shown in Fig. 14.
The preheater is fed raw meal which progresses through the cyclones
as described in the preceding section on SP Kilns. However, the
third-stage product is discharged into a flash furnace (FF) which
contains several roof mounted burners. Combustion in the FF takes
place in intimate contact with suspended raw meal particles preheated by the combined exit gases from the kiln and FF in the preceding cyclones. The FF uses about 60% of the total kiln fuel demand and reduces the rotary kiln burner demand to about 40%. This
dramatic shift of firing location is possible since the primary demand for heat in the cement making process is the highly endothermic

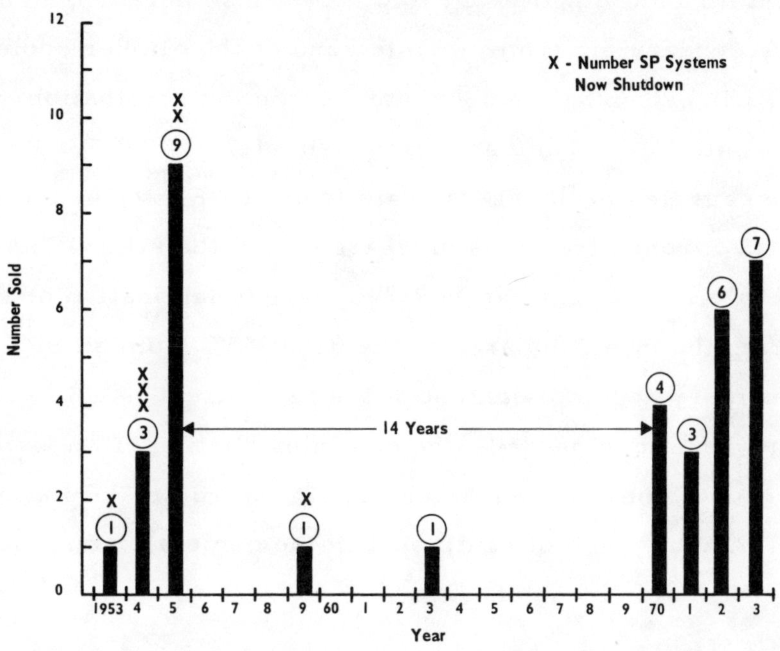

Fig. 13 U.S. History of 4-Stage Suspension Preheater Sales

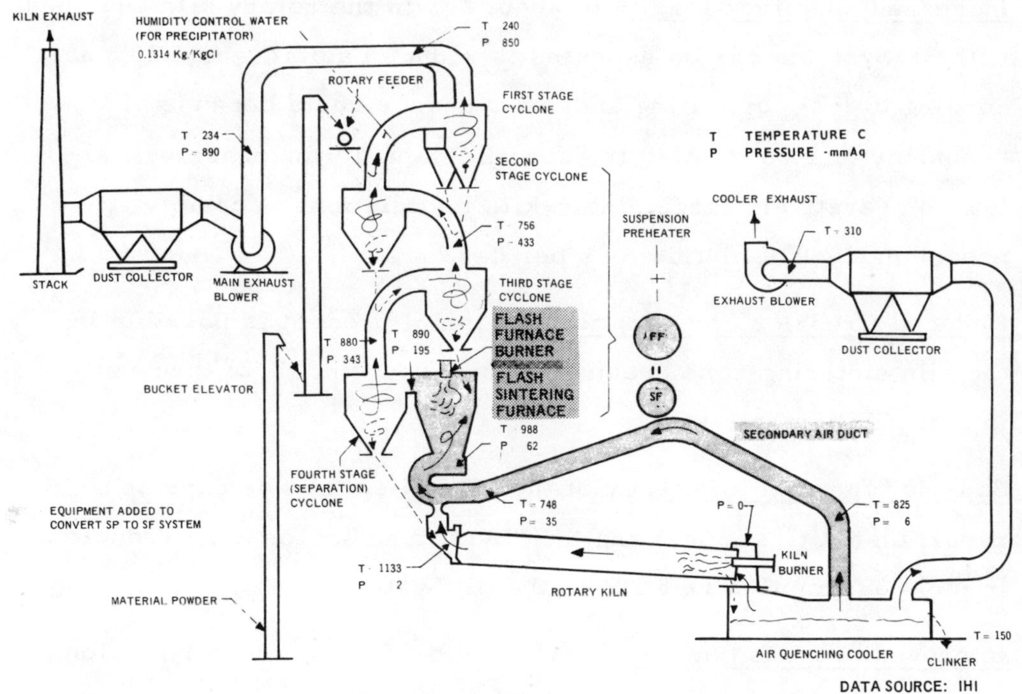

Fig. 14 Flow Sheet of Rotary Kiln with
Suspension Flash Preheater

requirement for calcining $CaCO_3$ to CaO. These burners, using
preheated secondary air from the mid zone of the clinker cooler
mixed with kiln exhaust gases, create a turbulent calcination
atmosphere within the chamber at approximately 2,000 F. Raw
meal is then carried in the gas stream from the FF to the fourth-
stage cyclone, from which it is discharged into the kiln. This
process results in approximately 90% apparent calcination of raw
meal entering the kiln compared to the usual 45% from an SP.
Thus the primary accomplishment of the SF concept is, "to provide
direct heat where it is needed--the calcining portion of the cement
process." Also, the reduced firing rate in the rotary kiln allows
the largest moving piece of equipment in the cement making process
to be reduced in size by about 50%.

The basic characteristics of the SF System can be expected to achieve
the following benefits compared to the parent SP System:

- Increased specific capacity of about 2.2 in the rotary kiln equipped
  with SF Systems can be expected to reduce capital (about 20% according to IHI), operating and maintenance costs because of
  "economy of size." Also the specific capacity increase will allow comparatively large single-kiln installations, possibly approaching 10,000 metric tons per day.

- Reduced firebrick consumption per ton of product is possible in
  the kiln sintering zone because of over doubling (2.2) thru-put
  in a small diameter kiln.

- Superior thermal efficiency of the SF System can be expected. A
  report on the first operating SFP System indicates a 5% reduction
  in fuel consumption relative to the SP System.

- Improved process control is expected because a greater portion
  of material processing is accomplished in stationary units as
  opposed to the rotary vessel. This will allow manual, automatic,

or computerized control of the stationary portion of the system
and probably even the rotary kiln since its role is virtually reduced to the single process of clinkering. Also, improved process control with attendant steady state operation should result
in greater fuel savings than the 5% achieved in the one operating
SFP unit.

- <u>Markedly increased alkali control</u> should be possible in an SF-
  type system. This possibility exists because less than one-half
  of the combustion air is needed in the rotary kiln, and the alkalis
  are evolved into the reduced volume of exhaust gases. This
  would more than double the concentration of alkali fume exiting
  the rotary kiln. This should increase the efficiency and reduce
  the cost of conventional SP alkali bypass systems.

The foregoing indicates the SF and comparable systems have the
potential of becoming the dry-process system of the future.
Japanese acceptance of the concept is substantiated by current
orders for 12 of the SFP Systems by IHI including several rated
7,200 metric tons per day. During this period IHI has not sold
any SP units.

D. <u>Other Kiln Concepts</u>. The basic characteristics of several U. S.
developed concepts and current status of development are briefly
summarized below. They are similar in many ways to the SF kiln
wherein direct heat can be supplied for the highly endothermic
calcining zone.

<u>Trough Kiln</u>. Thermo Electron Corporation (TECO) studies in 1968
led to development of a special burner which could be operated while
submerged in the cement mix and would continue to operate as it
was covered and uncovered by the mix as the kiln revolved. A
stationary "trough kiln," shown in Fig. 15, was then developed to
utilize the submerged combustion technology.

The first zone of the "trough kiln" is a submerged preheater section utilizing waste heat gases from the calcining and clinker cooling processes. These gases are introduced under fan pressure through a high temperature porous media thereby fluidizing the mix for conveying through the sloped stationary kiln into the calcining zone where the mix is fluidized and calcined by the special submerged burners. The burners utilize preheated air from the clinker cooler which is cleaned by high-temperature cyclones. The fully calcined mix is then discharged into a short rotary kiln where clinkering takes place.

Based on laboratory kiln batch experiments and the use of available energy conservation techniques, the heat requirements of a "trough kiln" are predicted to be 2,700,000 Btu per ton, which is comparable to Japanese SF Systems. As in the SF Systems a smaller rotary-kiln size will be required due to the increased heat transfer rates that can be achieved. Also as in SF Systems, and probably most important to the U.S. cement industry, is the ability to instrument the stationary kiln for process control and to facilitate bypass control if required to reduce alkali concentrations.

HOC Kiln. A new stationary type calciner concept now in use in the catalytic industry (three units sold) has recently been successfully tested for Portland and specialty cement production. The patented furnace was developed by Harrop Ceramic Service Company of Columbus, Ohio in 1965.

The Harrop Osciplate Calciner (HOC) operates in a manner not unlike the clinker grate cooler, with the material bed moving over a series of stationary, overlapping hearths. The hearths, refractory lined for high temperature zones, are arranged in a stairstep array as depicted in Fig. 16. The material is transported over the stationary hearths by oscillating pusher plates, one for each hearth, located in a cooled chamber under each preceding hearth. The flow

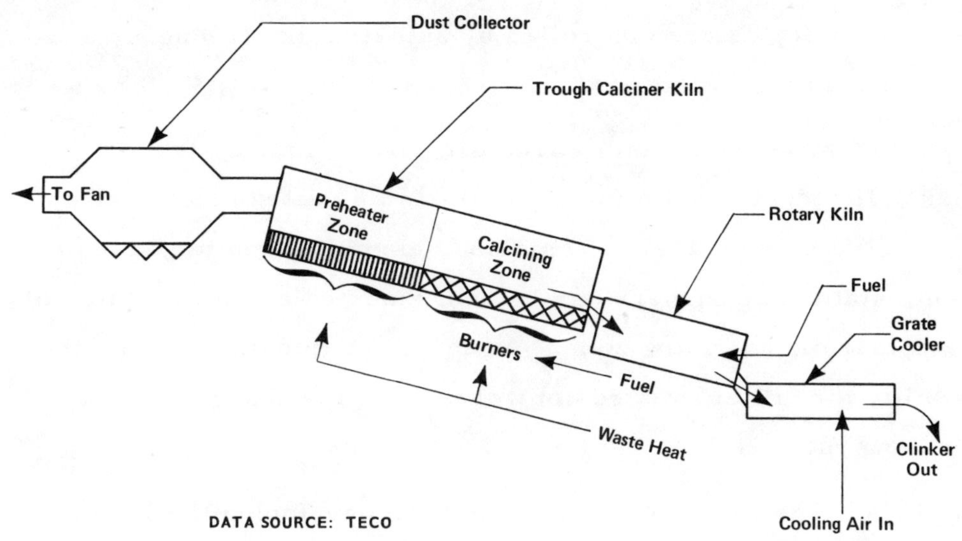

Fig. 15 Schematic Diagram of Teco Stationary Trough Kiln

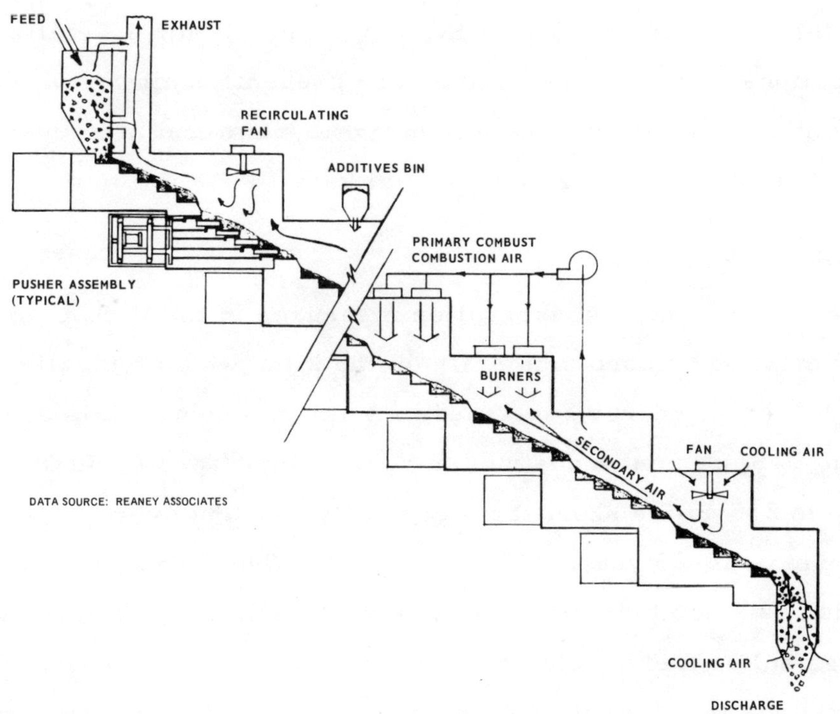

Fig. 16 Schematic Diagram of Hoc Calciner

rate and bed depths are controlled by adjusting the frequency and stroke of the pusher plates.

The stationary furnace can be fired directly or indirectly with gaseous, liquid, and solid fuels, or it can be heated electrically with metallic or nonmetallic resistance elements. In the case of cement, stationary burners can be roof mounted and located to suit the endothermic calcining zone. As in any stationary system, the possibility for full automated control is greatly enhanced compared to the long rotary kiln.

<u>Pyzel Kiln</u>. The Fuller Company has been developing the Pyzel fluid bed-type kiln since 1955. This effort included extensive pilot-plant work and funding in excess of $5,000,000 over a 16-year period. Unfortunately, acceptable thermal efficiency was not attained primarily because of high-temperature waste gases. Consequently, Fuller is offering the Pyzel System primarily for specialized applications only. However, they are presently being rewarded with a dominant sales position on their SP System and are developing an SF-type system for commercial sale.

VI. <u>CONCLUSIONS</u>

In conclusion, energy conservation measures in the U. S. cement industry must be focused primarily on the kiln, which typically uses about 90% of all energy consumed in making cement. It is anticipated that the U. S. cement industry can reduce plant specific fuel consumption up to 25% in the short term of 5-10 years and over 40% in the long term of 10-25 years or by year 2000. The U. S. fuel supply situation can also be indirectly relieved by allowing the industry to use high sulfur fuels. This can be done without major environmental impact since the cement making process automatically traps sulfur to varying degrees.

In the short term, energy conservation measures range from various
kiln heat efficiency and recovery concepts to better operating and
maintenance procedures and finally, shutdown of inefficient operations.
The rapidity of fuel cost increases indicates such concepts should be
implemented promptly--certainly more prompt than the 25-year
industry effort to reduce labor requirements.

Although technical and practical problems nullified the initial rapid
acceptance of the SP System in the early 50's, it is now making a U. S.
comeback.   The "SF" kiln offers even greater thermal efficiency possibilities through better process and alkali control characteristics;
also, several new U. S. kiln developments might become technically
and commercially viable.

Historically, the time frame for successful (and unsuccessful, as
noted in one case) development of a new process seems to span about
15 years.   However, the urgency of the energy problem and the possibility of government incentives through a national program may accelerate development and commercial implementation of promising
new processes.

Thus, the means for reducing demand through energy conservation
methods are clearly visible.   The optimistic outlook must be tempered
by the time requirements to fund and implement available and developing technology.   It also must be tempered by the realities of the
market, overall economics and the impact of government legislation
and regulations.   Nevertheless, the present industry outlook is long-
term optimism based upon the following projection of annual U. S.
cement industry energy consumption by year 2000:

- Cement Consumption per Person-Tons   0. 5 Ton
  (see Figure 2)
- Population Year 2000   300, 000, 000

(Portland Cement Association mid-range
projections)

- Energy Required per ton of Cement-Btu   4, 000, 000

or 600 trillion Btu per year

The population increase of almost 50% and the cement consumption
increase of about 20% is largely offset by a reduction in energy required
to produce a ton of cement.   Thus, by year 2000 it is possible that
total industry energy consumption will exceed today's rate of 550 trillion Btu per year by only 10%--a solid basis for long-term optimism.

VII. <u>ACKNOWLEDGEMENTS</u>

Much of the data was contributed by industry members as well as by
organizations, engineers and equipment manufacturers associated with
the industry.   The high degree of interest, assistance and cooperation
from contributors are gratefully acknowledged.

It is hoped that this paper will further foster the open exchange of new
ideas and data to accelerate conservation of energy in the cement
industry.

VIII. <u>REFERENCES</u>

This paper is, for the most part, a supplement to the paper "Thoughts on Cement Kilns Relative to the Energy Crisis" presented by the author at the 9th International Cement Industry Seminar under the auspices of Rock Products. Additional reference material included under Sections II, III, and IV-s preceding are:

1. Portland Cement Association--Private Communications.

2. Committee on Economic Studies, International Iron Steel Institute "Projection 85," March 1972.

3. Secratariat of the Ecomonaic Commission for Europe, "Long Term Trends and Problems of European Steel Industry," Geneva 1959.

4. Staff Study of the Committee on Finance, "Steel Imports," Dec. 19, 1967, United States Senate.

5. "Cement in 1970, 1971 and 1972, Mineral Industry Surveys," Bureau of Mines, U.S. Department of Interior, Washington, D.C.

6. "Cement," <u>Minerals Yearbook</u> Vol. I, U.S. Bureau of Mines.

7. Analysis Handbook, Standard and Poors, 1971 edition and Monthly Supplements, April 1972 and 1973.

8. Monthly Economic Letter, First National City Bank, New York, April issues 1961-1973.

SESSION IV

CHANGES IN DEMAND: CULTURAL AND SOCIETAL
ASPECTS

Chairman: Richard N. Adams
           Professor of Anthropology,
           University of Texas, Austin, Texas

ENERGY AND THE STRUCTURE OF ADAPTATION

Roy A. Rappaport[*]

    Issue is taken with the assumption that increases
in the amount of energy harnessed by a society improves
either its adaptation or its adaptiveness. A conception of adaptation that subsumes both "reversible"
"function" or "systematic" changes and irreversible
evolutionary changes is advocated, and the cybernetic
and hierarchical structure that it implies is outlined. In light of this discussion a conception of
mal-adaptation as structural anomoly is proposed,
and some suggestions toward a taxonomy of mal-adapta-
tions and their development are made. A number of
ways in which increases in energy flux may encourage
and accelerate mal-adaptive trends are also suggested.

---

* Dept. of Anthropology, University of Michigan

ENERGY AND POLITICS

Robert Engler[*]

The headline issue of energy provides a rich opportunity for scientists, engineers, and social scientists to share in the raising of fundamental questions about the premises and directions of the modern world. Energy is the mainspring of civilization. And, the immediate crisis is not a technical one of resource supply. It is rather a social one of organization. The critical shortage is not of energy but of public perspective and responsible policy.

Scientists and engineers have helped all of us to learn to "think in terms of the system". As Harrison Brown has written, the industrial world, like the natural world, is an elaborate network of cause and effect which leaves no area of the globe uninvolved. Both worlds are vulnerable to disturbance or breakdown. Both require stability. One achieves order through natural laws, the other seeks it through man-made institutions.

Toughminded social science also demands "thinking in terms of the system" if one is to penetrate the underlying principles and controls shaping events. The alternative is a trained innocence which treats related issues in isolation from one another and which perpetuates reliance upon unverified data and official positions of private corporations and public agencies with stakes in the going arrangements.

Scientists and social scientists need to understand the links between their worlds. Both may have to ask uncomfortable questions about the going structures of political power and be involved in the search for new options. Where they abdicate, public policies are likely to

---

[*] Professor of Political Science, City University of New York

remain fragmented responses to the cohesive forces behind problems labeled "emergency". The issues are then co-opted, and formal administration is placed in the hands of professional crisis managers. Thoroughgoing solutions are ruled out and the structure of power is preserved. Citizens are left with their confusions, the social costs and the recognition that the political process and the intellectual world offer more rhetoric than help for coping with the giant pressures of their daily lives.

Four central considerations shape this paper:

1. The nature of the private controls over energy
2. The limits of public government
3. The need for responsible planning for longrun public energy policy
4. The need for a reappraisal of the American political economy and for the development of alternative patterns which might make such planning meaningful in the lives of the people and the directions of the country.

1. NATURE OF PRIVATE CONTROLS

There is a much circulated thesis that the rules of a free market have governed the production and distribution of energy - save when interfered with by the clumsy hand of public government - and that were we to remove the public impediments, the private sector could again serve us well. But the oil industry has been a private planning system which operates on an essentially global scale. It commands extraordinary financial and physical resources greater than those of most industries and many nation states. The major integrated companies have sought effective control of the flow of oil from well to pump. The ability to shift profits from one stage to another has been vital in driving out weaker competitors and avoiding or taking advantage of tax laws. This ability is pertinent to an understanding of the so-called "energy crisis". Faced with greater demands from the increasingly sophisticated producing nations, the international oil companies have attempted to increase their profits downstream at the refinery and marketing stages.

Despite the great number of presumably independent entrepeneurs
involved and the competition for new markets and reserves, in important respects the industry functions as a cartel in setting prices, protecting profits, and minimizing competition from intruders or regional
independents. The twenty majors account for approximately 94 percent of domestic reserves, 70 percent of domestic oil production, 72
percent of natural gas production and reserve ownership, 86 percent
of U.S. refinery capacity, 79 percent of gasoline marketing. The top
four corporations account for over 30 percent of these operations. The
majors also own or have interest in 70 percent of interstate pipelines.
They own or lease half the world's oil tanker tonnage. The seven
worldwide giants of oil have assets whose combined worth, at a minimum, is $85 billion. Their community of interest is guarded through
patents, bidding arrangements for public lands, joint ventures in
drilling and pipelines, banking ties, regional price leadership and
recognized territorial prerogatives. Brand distinctions are an advertising fiction; gasolines are a standardized commodity which often flow
through the same pipelines, refineries, and tanks.

Restricting new refinery and pipeline construction, curtailing refinery
runs, driving out independent refiners and marketers by cutting off
their supplies, and generally doing everything to forestall the night-
mare of competition have been central factors in the current shortage.
The mechanisms they have created for an "orderly flow", that is, for
limiting production to effective market demand, have been inadequately
responsive to increases in or miscalculations of demand.

Meanwhile, the international oil companies have long anticipated and
prepared for the rise of bargaining power by the Middle Eastern producing countries. Indeed, there has been little protest or resistance
to the upward spiralling of prices by the Arab nations. All this has
increased the worth of their holdings at home and their profits.

They have been mindful of the ultimate limits of non-renewable fossil
fuels and have broadened their operations in coal, gas, tar sands,

shale, uranium as well as in the so-called non-conventional fuels. As
energy corporations, they now control over 30 percent of domestic
coal reserves and 20 percent of domestic coal production, over 50 percent of uranium reserves and 25 percent of uranium milling capacity.
They have helped to push up the price of coal. The insistence that
federal price regulation of natural gas must be ended seems related
not to any present unprofitability but to a desire to make gas prices
"competitive" with, that is, up to, coal and oil prices.

In all of these operations, the corporations who dominate energy behave
as highly accomplished profit-gatherers, roaming the earth in search
of reserves, markets, and profits. They are also prominent in rubber,
petrochemicals, synthetics, fertilizers, land developments, leisure,
and many other activities. Gulf's recent move to take over Ringling
Brothers Barnum and Bailey Circus suggests an unconscious identification with rulers of an earlier empire. Third quarter profits in 1973
increased for the industry as a whole 63 percent over 1972: 81 percent
for Exxon, 91 percent for Gulf, 274 percent for Royal Dutch Shell, and
483 percent for British Petroleum. Meanwhile, they shift risk to tax-
payers, costs to consumers, and now shift blame to environmentalists.

2.   LIMITS OF PUBLIC GOVERNMENT

Much of what is now called public energy policy is supportive of or
subservient to private industry objectives. Prorationing and so-called
conservation laws have protected prices when the key production was
domestic. Once the supply centers shifted abroad, federal import
controls were added without congressional review. This governmental
intervention, pushed for by the industry and maintained until last
spring, was a price stabilization subsidy which kept oil from abroad
in check, while protecting domestic pricing and allocation patterns
within the industry. The cost to consumer and taxpayer has been
estimated at a minimum of $5 billion a year. It is scarcely in accord
with the tenets of a free market economy.

The federal government has provided private industry with tremendous
financial incentives through the depletion allowance, depreciation of
intangible drilling costs, overseas depletion allowances, and tax
credits for taxes and royalty payments to foreign governments. When
crude oil prices go up, so do depletion allowance benefits (and corporate payments to themselves). All of these help to explain the minuscule
federal income taxes paid. Yet, there is little evidence that these tax
privileges have stimulated domestic production.

In addition, the corporations have been able to generate vast capital
for reinvestment by charging customers, not simply for the cost of
production but for its replacement. Thus, consumers have become
investors without rights. And, as they are now discovering, with
limited guarantees of future supply. The resulting reinvestments are
increasingly in overseas refineries and pipelines whose products are
intended for the burgeoning markets and retail networks being developed
in Europe and Asia. The shiny piece of glass or trinket motorists,
until so recently, received at the filling station here served to
emphasize their status as natives.

The federal government has been most responsive to industry definitions of "technological readiness" of alternative energy sources. The
current energy scare and the talk of technological crash programs rip
out of historical context the fact that we did have federally run experiments and plants as far back as 50 years ago. Coal-oil laboratories
and plants were set up by the Bureau of Mines during World War II.
The shale oil pilot plant at Rifle, Colorado, and the Laramie, Wyoming,
laboratory, along with coal-to-gas plants, were closed down under the
Eisenhower administration, not because of technological failures, but
because the industry insisted that the next steps - introducing such
fuels onto the market - should be a private determination, once the
price was right.

The energy industry has thus maintained vigorous surveillance and
veto power over all forms and sources of energy innovation whether
offshore oil, shale, gasification, tar sands which might upset their

control over the market. Yet, it endlessly suggests that it is on the
frontiers of research and development. Its laboratories are important
centers. But cases such as that of synthetic rubber offer revealing
documentation of higher loyalties to international corporate relationships
and to profits. And, a breakdown of industry budgets suggests that much
of the self-celebrated commitment to knowledge more properly belongs
under the heading of public relations and advertising.

New loans and grants to stimulate research in the private sector,
offered in a host of bills now circulating as approaches to energy
shortages, are not an answer. We are not in the world of Adam Smith
but of Exxon and Mobil. If the justification is to support private initia-
tive, where are the assurances of the possibility of entry by newcomers
and small businesses who might conceivably introduce competitive
fuels, technologies, and pricing? Why have the belated coal research
contracts gone to giant oil rather than to independent coal companies?

The industry has been aided by a sympathetic, innocent or captive
bureaucracy. Its members have come directly from industry and share
its premises. Or they have never been offered nor been encouraged to
think about guidelines for a genuinely public policy. Many current
proposals call for the consolidation of now scattered (in some 60
agencies) and half-hearted federal efforts. This could be a step
toward greater public accountability. But if the present Interior
Department is already carved out by coal and gas interests, if its
experts are generally from industry and industry-related universities,
if regulatory agencies such as the Federal Power Commission are
staffed by so many who identify with private industry and expect to be
on corporate payrolls, and if the White House, State, and Defense
Department decision-makers involved in energy planning are industry-
oriented or retain industry ties, what assurances does the citizen have
that any new agency will have more integrity and public responsiveness?

The federal government is also honeycombed with a network of advisory
bodies representing the leaders of the major energy corporations and
trade associations, along with some "independents" who in effect are

making public policy in the Interior Department, the Executive offices and elsewhere, while concealed from public view or review. The National Petroleum Council offers one classic example; the Emergency Petroleum Supply Committee is another.

The federal government has also provided immunity from antitrust actions as well as extensive diplomatic, military, and espionage support for the international operations of the oil industry. The assumption has been a mutuality of interests. The specific activities have generally been kept secret from the American people, in the name, of course, of national security.

Ignored in most discussions of energy policy is the political power of the industry. Its control over basic resources has created centers of wealth which have permeated the entire body politic. Oil "investments" in politics, although not unique among industries, are unusually substantial in elections. Watergate provides the most recent reminder. The Committee to Re-elect the President solicited and received several million dollars from oil corporations and individuals, including illegal cash contributions of $100,000 each from Gulf, Ashland, and Phillips Petroleum. Other funds came through Pennzoil, Occidental, Texas Eastern Transmission, and Brown and Root. A Greek national gave $25,000 and shortly after the election his oil company in Greece was awarded a $4.7 million contract for refueling the U.S. Sixth Fleet.

Such behavior offers one more illustration as to why the current administration has forfeited any moral right to govern. But one hastens to add that, in the case of oil, this corruption has a long bi-partisan history at all levels of government. As one notable critic concluded before the turn of the century, that "Standard has done everything with the Pennsylvania legislature except refine it".

The oil industry's leasing of one fourth of the land area of the United States and the identification of so many citizens and politicians with its privileges are also corrosive factors. We continue to see otherwise articulate legislators toady to oil's wishes, all presumably

explained by "regional loyalty". Extensive lobbying and public relations
activities, along with ready access to the mass media and manipulation
of such symbols of the society as "progress" and "private enterprise",
in the past have served to gain public support. Confronted by embar-
rassing profit figures just when consumers are fearing the cold and
politicians are feeling the heat, Exxon's chairman has announced that
"we are embarking on the oil industry's version of Project Candor".

Much of the official rhetoric from Washington presumes the energy
corporations are eager actors in search of a script. Their past record
and present performance suggest they already know their roles - extending operations and maintaining profits and power.

The issue is not one of conspiracy; that is, whether the present energy
squeeze can be traced to one specific action or meeting place. Rather
it is the assumption from the time of the first oil well that profit considerations could wisely and justly determine when, where, and how
natural resources are to be developed, distributed, or conserved.
Within that framework government has been assigned the primary role
of keeping supply down and prices up. It also serves as a relief agency
when the social costs become too glaring. Exhortations for increased
corporate responsibility and pledges of new governmental vigor are
illusions in such a setting, unless there is also to be a change in the
basic rules of the game.

Private companies did not put these resources in the ground. Nor did
they create their value. Energy planning is too important to be entrusted
to the international private governments who dominate so much of the
marketplace and who show such willingness to abandon the American
scene. No private interest should be permitted to negotiate with a
foreign government about matters so basic to the political economy.
When a Shell subsidiary replies to a state attorney general's charge
that it contrived to create an artificial shortage by keeping home heating
fuels off the market until prices went up with the explanation that such
action is a "normal customary practice", it is telling it as it is.

The current energy scare is a dramatic public case study of the failure
of this industry to give its first loyalty to the citizenry. It is also a
case study of the incapacity of government to govern for the people in
such an environment.

3.   PLANNING FOR LONG-RUN PUBLIC POLICY

The need is for long-run energy planning in harmony with some larger
public goals of development, responsibility and participation.

To this end, the following proposals are offered:

a.   An independent natural resources commission, perhaps with a
     maximum two-year life span, should be appointed with the man-
     date to appraise the desirability and specific requirements for
     the public ownership of all energy resources in the United States.

b.   The commission would also be asked to review the experience
     with and prospects of intermediate steps for accountability--
     centralized data, regulation, antitrust, federal chartering,
     breaking up of the integrated corporations, divestiture of control
     of competing energy sources. Examining the case for the inte-
     grated operation--private or public--would be valuable. The
     argument has generally been that only big systems can serve big
     publics with big technology. Perhaps this is no longer so. A
     fresh look, not at energy theory or public relations, but at the
     technology and economic practices of the industry as set against
     social goals and costs would be instructive.

c.   All special privileges and subsidies should be inventoried and
     reviewed. Who gains? With what public justification?

d.   The commission should also be asked to come up with proposals
     for stimulating local, regional, national, and international
     planning bodies who in turn might seek to place energy in the
     context of communal priorities. It would encourage projects
     which then receive federal support. These might include:

     (1) Regional development of small-scale alternative technologies
     (2) Experimental energy systems with emphasis on renewable
          rather than non-renewable sources
     (3) Local producer and marketing energy cooperatives
     (4) Land use planning
     (5) Natural farming, with less dependence upon chemicals
     (6) Local mass transit experiments
     (7) New building patterns

(8) New towns

(9) A national power grid system.

The commission could be asking what is to be learned from the
TVA, the REA, and from other countries.

The commission hopefully should be wary of going overboard
against what it is now fashionable to deride as "cheap energy".
The TVA, for example, with all its limitations, stimulated
development in a depressed area. We should be asking where
we want low cost energy and where we do not.

e. A technological Review Board should be planned for, not just for
energy technology but for all technological innovations. We may
not wish to be Luddites. But, we must learn how and when we
can anticipate social and environmental consequences before we
are confronted with them so that we may prepare for their impact
or develop the wisdom and the political will to say "no".

f. The control and development of all energy resources from federal
lands should immediately be placed within a public corporation.
This corporation, with access to perhaps half of our fossil fuel
(shale, offshore lands, coal, etc.) could serve as a yardstick for
conservation and development. (At present, the government sets
few standards and has minimum knowledge about corporate leasing.
Coal lands, for example, are being held rather than developed by
the oil industry.) It could also provide minimum protection against
an industry whose world power allows it to rig prices charged the
military during wartime, cut off supplies to the armed forces at
the behest of another nation, generate unemployment at home, and
control research and development on public lands.

g. As an interim stage until the natural resources commission has
reported, all energy operations should be given a public utility
status, so that the present shortage pressures can be minimized
and equalized. Rigorous controls should be placed on prices,
profits, and supply allocations. If there is to be rationing, it
should be a public function, not a byproduct of corporate decision-
making.

One basic purpose of such a commission and the related measures
would be to generate a great public debate. We need to place in the
public record the basic data about energy, natural resources, costs of
production, and consumption patterns. Much attention is now being
focused on corporate secrecy as to inventories and reserves. I hope
we will end the outrageous practices of basing public, domestic, and
foreign policies upon trade association figures, and of corporate and
government bureaucrats defending information about energy as proprietary and masking their moves behind "national security".

There are indeed questions about the "energy crisis" which need public airing. For example, what happened at Tehran in 1971 when oil company leaders met with the oil producing nations after receiving support for collective negotiations from the Justice Department? (A letter of inquiry which I had sent to the Antitrust Division produced the reply that in its judgment and in that of the State Department "release of this information at the present time would be contrary to the national interest".) Why does our State Department remain publicly so quiet about such negotiations between corporations and governments? What goes on in the meetings of the Emergency Petroleum Supply Committee and the Foreign Petroleum Supply Committee, composed of representatives of the international oil companies, who have been convened as industry advisory bodies within the Interior Department to help resolve the international oil crisis? And, how did the Cost of Living Council arrive at the special treatment it accords the oil industry under Phase IV regulations? A pass-through of increased costs at all levels on a once-a-month basis is allowed, as contrasted with yearly review in manufacturing. What are the figures on reserves and stocks held or controlled by each corporation? How does the industry arrive at cost figures?

But, we need not go overboard on the information issue. In a larger sense, the issue is not secrecy. We have had abundant warnings over the years of the consequences of our spiraling demand patterns and of the increasing costs of ultimately finite fossil fuels.

It is no secret to the people of Appalachia that their land has been ravaged in the quest for coal. It is no secret to Montanans who fear a similar fate. Nor is it a secret to the people of many oil-producing regions that the great wealth generated can be badly distributed. It is no secret to the people of Latin America and the Middle East that oil companies have enjoyed the political support of the American government. And, now millions of Americans are learning something about their own vulnerabilities.

Much of the knowledge about oil and oil policies is not that esoteric.
There is little excuse for the incompetence that characterizes most of
public energy leadership today. We have had and presumably still do
have experts in the government and outside who know the score. If we
were to create a climate where genuine public servants would no longer
be made outcasts for maintaining loyalty to public ideals, the knowledge
factor would not be so intimidating to legislators and citizens. I am
thinking of men like W. B. Watson Snyder who worked almost alone and
unsupported in the Antitrust Division for so many years; of economist
John Blair in the Federal Trade Commission and then the Senate Antitrust Subcommittee; of geologist-economist David Brooks whose research on imports, coal, shale, and helium caused such discomfort in
the Bureau of Mines. He found himself virtually blacklisted in the
United States federal government and moved on to direct energy research for the Canadian government.

In contrast, a head of Interior's Oil and Gas Division moves to the
American Petroleum Institute. And, the Assistant Secretary of the
Interior for Mineral Resources who defended corporate withholding of
data on reserves on public lands as "proprietary" information - "this
government of ours is a private enterprise government", he explained
to a Senate Committee - has moved on to head shale operations as a
vice president for Atlantic Richfield. Arco is the second largest
holder of coal leases on federal lands.

4.   DEVELOPMENT OF ALTERNATIVE PATTERNS

If such planning is to make for a more just society, it must be placed
within a broader frame which will counter many of our present patterns.

a.   There must be genuine respect for the environment. The use of
     the cry "energy crisis" to destroy the ecological gains of the past
     few years is tragic. We have a long way to travel if we are to
     create respect for nature and for ourselves as part of it. We
     must learn to say "no" to the development of resources rather than
     urging maximum efficient recovery when thinking in terms of the
     future may be more important than present definitions of need.

b.   Energy planning must include end-use planning. There is scan-
     dalous waste in our present industrial practices, whether in the

production of cars, cans, or of energy itself. We must create the
means for saying "no" to big cars and "yes" to mass transit.

c.   We must challenge the present technological and profit incentives
      for insatiable and inequitable consumption. We must develop
      visions of man more profound than that of robot producer lacking
      respect for his work and mindless consumer responding to the
      latest hard sell.

d.   We must ask about the inefficiences and immorality of an industrial
      system which unabashedly devours a third of the world's resources
      for its six percent of the population and whose corporate and polit-
      ical leaders make their pledge in the search for more and the "use
      of all we can get". The theme of "more", sustained by the magic
      of technology and the myth of limitless nature, too long has been
      used to evade the tougher questions about just distribution at home
      and abroad.

e.   We must ask about the relation of energy policies to our foreign
      relations. Is energy self-sufficiency by 1985 either a feasible or
      an appropriate target? Does banding together with other consumer
      nations while moving to keep imports out assume that there is
      little hope for a peaceful and cooperative world order, that the
      best we can do is play off one region against another? And, mean-
      while will we once more be protecting a high-cost domestic
      industry from the possibility of lowered foreign prices?

Will we be thus preparing fortress America while ruling out the possibility of world planning which would respect the needs and fellowship
of all mankind rather than the power of the more advantaged?

We must become sensitive to the likely impact of corporate-govern-
mental arrangements with the new producer cartels upon the truly
poor of the world, the great numbers to whom shortages or rigged
prices mean not fewer air-conditioned hours, but no kerosene for
warmth or fertilizer for survival. Where has our concern over energy
shortages been when the poorer countries have been pressured against
developing independent energy bases and when the high prices of the
purchased fuel have cancelled out foreign aid and painfully attained
growth gains. We have invested more in the search for oil in Latin
America than in aid for its people's development. And, we have taken
out in profits our investments many times over.

Nor can we ignore the heavy drain on energy by our imperial foreign
policy. An aggressive search for energy security, whether under

public or private auspices, could provide one more excuse for curtailing school bussing than for ending oil aid to the South Vietnamese
Army.

CONCLUSION

There is no magic in planning or in public ownership. We are all
aware of the curse of big bureaucracy, private or public, military or
civilian. Indeed, our economic system, like our political system, may
be toppling under the weight of big technology.

But, surely there is no gain in extending the power of the already
bloated and irresponsible private government of energy. And, we have
had ample experience with the failure of regulation when the regulators
are more powerful than the regulated.

There are many exciting frontier areas of energy research. One easy
temptation, however, is to continue to believe that science and technology will always come to the rescue with new techniques and resources.
Rather than reappraise or scale down our demands upon the environment, all that is needed, presumably, is a redoubling of our technological efforts. Yet, crash programs to maintain our present practices
have their own built-in dangers. Giant science, whether public or
private, can accelerate the growth of the giant institutions and their
irresponsible elites which already place so heavy burden on the human
spirit and make so difficult the task of political reconstruction.

The crisis is not of energy but of industrial organization and political
life. We need to be exploring not just alternative energy models but
alternative economic and political thinking which may yet help us to
attain the ideal of morally autonomous citizens sharing in the govern-
ance of peaceful communities. We need to stop assuming there is
always a technological answer when the questions may be moral and
social. And, we need to create a sense of the politically possible
which will liberate the energies within the people.

This, to me, is the real challenge of the current energy crisis.

REFERENCE

Robert Engler, the Politics of Oil: A Study of Private
Power and Democratic Directions, University of Chicago,
Phoenix edition, 1967.

ENERGY DECLINE - RAILROAD REVIVAL?

Fred Cottrell[*]

Twenty years ago I wrote a book called ENERGY AND SOCIETY.[1] I cite it primarily because the framework used there is that which I will use here. Its thesis is very simple. The energy available to man limits what he can do, and influences what he will do. Today it would be difficult to find anyone who would disagree with the first half of that statement. It is more difficult to get agreement on the extent to which and the specific ways in which man's choices are affected both by the amount and the sources of the energy available to him and the nature of the converters that he can use to achieve particular ends. But the fact that man is presently being forced to create new orders of choice as a consequence of changing access to natural gas, petroleum and nuclear power is attested to every day. The resultants of the struggle to realign their choices will show a great deal about the emergent value hierarchies and social structures of all societies, particularly those in the West.

Abundant evidence exists to show that there is a close correlation between the per capita consumption of energy in a given social system and the ability of that system to achieve a high place in most of the indices of social and personal well-being.[2] Evidence also shows that the rate of probable increase of available energy will not be equal to the probable growth of world population.[3] So the per capita consumption of energy must decline. The decline will not be evenly distributed among the continents, among nations, classes, raced and industries. Many systems will be forced into contraction. Such general contraction will reverse the expansion that has characterized the Western World for more than two hundred years. We have no significant body of social or economic doctrine that rests on the idea that tomorrow there will be less than today for men to share.

---

[*] Director of the Scripps Foundation for Research in Population Problems, Miami University, Oxford, Ohio

There is also an enormous social structure which depends for its effective functioning on continuing increases both in production and productivity. The only experience on which to base a new "dismal science" to explain and justify it will be that which is supplied by the examination of declining industries and societies. The most pervasive of such industries is the railroad which has recently been in relative decline in most parts of the world.

In ENERGY AND SOCIETY I emphasized that a crucial factor that made expansion of social systems possible was an increase in surplus energy (energy made available beyond that expended in making it available). Changes in the ratio of energy output to input which resulted from use of new sources of power; wind and water, coal and petroleum, gas and nuclear power, made it possible to do some things never before possible and to do others with the expenditure of far less human time and energy than had been required where plants were the source of fuel. But still other ends could be attained only by using as much human input as they previously required.

The model I follow in dealing with the way energy changes affects values is akin to that used by those economists who explain economic choice by opportunity cost--we know the value of a thing when we know what economic goods will be sacrificed to obtain it. To understand human behavior I enlarge the model to include all of the choices humans make, in terms of all that must be sacrificed to reach a given end or attain a given goal.

In this framework the influence of new energy flows, making some things less costly in terms of the human input required to obtain then, while leaving other ends just as costly as before, produces both overt and covert conflict resulting in a changed value hierarchy.

Energy flow affects and is affected by many variables, of which energy source and required technology are only two. Those who value highly the goods and services increased by flow in a given sector will attempt to increase that flow. Others who see it reducing what they value use whatever power is available to them to prevent it.

If railroad traffic is to increase, it will come only because of gains anticipated by some sets of people who have power sufficient to overcome the
resistance of those who value other things more highly.

Basically it was because the railroad made available to man an enormous increase in energy that it had such an influence in changing values and social
structure. The steam engine made possible the conversion of fossil fuel to
mechanical energy. The reciprocating piston engine was relatively inefficient in converting the potential energy of coal into mechanical energy. What
made it so valuable was the very low human cost of the fuel it could use. The
five or six hundred pounds of coal that a miner could daily dig out with pick
and shovel would yield as it burned hundreds of times as much heat as would
the average amount of food or feed that the farmers produced per day. Today some mines produce coal with a heat value hundreds of thousand times
the energy consumed in mining it.

Even when it functioned at less than one percent efficiency the steam locomotive was, in energy terms, twenty or thirty times as efficient as its primary competitor in transportation, the horse. The steam locomotive was
never more than about five percent efficient in terms of the power produced
at the drawbar as compared with the heat value of the coal put in the firebox.
Nevertheless, so long as there was no other locomotive more efficient in
converting coal into motion, those who controlled the railroads could make
enormous profit and still deliver goods at costs so low as to destroy their
competitors. And in many cases that is exactly what they did.

It is neither necessary nor possible here to recapitulate the early history of
railroads in the United States. But the work of the "Robber Barons" whose
slogan seemed to be "The Public Be Damned" produced countervailing effort.
The Granger movement and similar activities resulted in much state regulation, and eventually produced the Interstate Commerce Commission. New
factions arose to force reduction of railroad power. They still have much
to do with the legislation that affects shippers, passengers, railroad labor,
investors, and the communities through which railroads run. Bureaucracies

were built to serve the interests of these different sets of people. They are still able to prevent change that might permit the railroads more adequately to serve other segments of the population.

The development of high energy technology distorted the traditional methods of distributing goods and services. At first speculators were able to claim most of the gain made with tremendous amounts of surplus energy newly available. As time passed others forced them to disgorge part of it.

The railroad engineer became the aristocrat of labor. The rules he and his fellow workers managed to impose to protect their positions are still extremely resistant to change. Communities through which the railroad passed were often without other access to the product of new energy. They could tax railroad real estate, and thus secure the means to run schools and perform other necessary services. State public utilities commissions forced the railroads to make rates that favored those with political power in that state. Those with an interest in alternative means of transportation devised ways to prevent the railroads from effectively competing with them.[4]

A very important set of restraints came from the Federal government. Railroads were classified as common carriers, and forced to transport almost every kind of goods offered them. They are required to set a rate, win its approval, often before both the Federal and State agencies, and then publish it. These published rates cannot be changed without government permission. Today the common carrier, supposedly protected against competitors, confronts a tremendous array of them whose rates or services it is not legally permitted to match.

What we have interest in doing is to account for a great deal of legislation and the bureaucratic structure which presently controls the railroads. It was their inability and/or unwillingness to use pricing as a primary means of adjustment to changing technology that led to the situation we now face in which the less efficient means (in energy terms) have forced the railroads into contraction.

Even so, they probably could have maintained a dominant position had not their monopoly on the conversion of fossil fuel been destroyed. The railroads themselves not only ran on coal, but they also hauled the coal necessary for most other high energy enterprise. With that monopoly they remained in a strong position.

The development of the petroleum and natural gas industries changed it vitally. At first they did not seem to be a major threat. Transportation using the horseless carriage was expensive and uncertain. Its widespread use required the creation of an extremely expensive network of roads.

The use of petroleum in freight hauling was limited by size and power of the internal combusion engine. In early days those engines were small and seldom more than ten percent efficient. Eventually they reached a much higher level of efficiency than the reciprocating steam locomotive.

Potentially petroleum offered many advantages as a fuel. Gushers often produce in a day fuel with more heat value than all the energy expended in drilling and the other activities necessary to get the oil. Slowly it replaced coal in many applications. Finally many railroads shifted to petroleum as fuel for the steam locomotive, and eventually adopted the diesel. But by this time railroads faced competition by a complex set of forces which had grown up around petroleum and the internal combustion engine.

In terms of energy costs, the railroad retained many of its advantages. But other elements of cost such as time, convenience, public esteem, and particularly political power, moved it down in the hierarchy of values common among the people of the United States. The result was a rapid decline in the wealth, the power and the prestige of those attached to the railroad as compared with those producing, distributing and using automobiles, trucks, tractors, and buses.

This poses an interesting question. If in fact the railroad was from the beginning more efficient in energy terms than competing means of transport, and still is, how likely is it that a mere shortage of energy will bring it back into favor? this is the essence of our problem.

It is not difficult to show that solely in terms of energy the railroad is today far superior to its competitors in transporting goods and people.[5] Almost all of the locomotives now used by the railroads are diesel electric. The engines operate at about the efficiency of those found on diesel trucks, somewhere between 35 and 40 percent. This is about half again more efficient than the gasoline engines very widely used in private automobiles. Trains vary in their efficiency, but a 100 car freight delivers about three times as many ton miles per gallon of distillate as does a thirty ton truck. It delivers 90 times as many ton miles per fuel unit as air liners. There is no way accurately to estimate the energy costs of producing and maintaining highways and the facilities required for the operating of the airways, nor has it been done for the railroads.

But in a sense these figures prove too much. The advantage of railroads in terms of energy is probably no greater today, if as great, than it has been during most of the time railroads have been losing traffic to vehicles powered by the internal combustion engine. The reason is, of course, that passengers and shippers were influenced by other considerations than efficient use of energy.

First of all after the advent of petroleum, gas and hydropower, it was not possible in the United States for anybody to keep a monopoly on all kinds of fuel and force the price of energy up to what the traffic would bear. And as a matter of public policy low priced fuel was given priority over the conservation of what we now are forced to recognize as being limited natural resources.

Second, petroleum and natural gas can be delivered by pipeline at very low energy costs, and electricity can be transported by high tension lines at extremely low costs. Thus, the railroads were forced to carry coal and petroleum for their competitors at very low rates and/or lose a considerable part of their business.

Third, the policy of regulating utilities so that energy was sold at relatively low price provided energy users with new cheap alternatives to rail transportation. Changing methods of production reduced dependence on railroads. In many cases goods that were once produced at central points, adjacent to rail centers where cheap fuel could be obtained, came to be manufactured at points near their markets and where costs of land were low. So through the relocation of industry the railroads lost a lot of long haul shipping, in which their technology was most efficient, and were forced to compete for short haul business where they were at a disadvantage in competition with trucks, pipelines and the power grid.

The redistribution of population was also a factor. The industrialization of agriculture and the shift of population from North to South and to the West Coast left many communities with no economic base with which to hold their populations. Thousands of rural towns and villages lost out to megalopolis. As farm regions adopted the truck and tractor, the number of inhabitants of many rural communities fell to a point where they could not support schools and other necessary services. The railroad was in most cases the only kind of industrial asset local government could tax. Efforts by the railroads to abandon unprofitable lines were met by political power stemming from the voters in areas that should, from the point of view of corporate profit, have been abandoned long ago. Many of these areas have taken in taxes from the railroad more money than the total revenue they generated for the railroad by shipping or traveling on it. They remain adamantly opposed to abandonment that deprives them of this income.

Finally, probably the most significant factor forcing the railroads into contraction was the love affair between the American and his automobile. The cost of owning and driving an automobile are very high. It is, in energy terms, very inefficient. It is a primary source of pollution. It maims and kills more people in a year than were lost in that time in any war we have fought. On the other hand, the private automobile provides more freedom

of choice for its owner than anything else he can possess, makes available
services he otherwise would be deprived of, and in many other ways contributes to the enhancement of living.

Railroad management may be shown to have been guilty of complacency in
dealing with it, but it is highly unlikely that anything it could have done would
have stopped its widespread adoption so long as the exhaustion of natural resources and the dangers of pollution were blithely disregarded as they have
been in the American system of values. The automobile took over and the
passenger train ceased to be the backbone of travel.

What was more significant perhaps is the way the automobile owner subsidized the railroad's competitors for freight. In the earlier years, it was
considered normal to use taxes paid by the railroads to build streets and
farm-to-market roads. And since these roads were often used to haul to
and from a rail shipping point this kind of symbiosis was not only tolerated
but encouraged by railroad supporters. But the good roads movements produced a network of highways that were capable of destroying the near monopoly the railroads had held in long distance shipping. Once it became acceptable to use fuel taxes exclusively for road building an enormously costly substructure for road freight shipping could be created without any great
investment being made by the trucker.

The point is made, not to accuse the trucking system of parasitism but to
indicate how the capital required to make the present truck freight system
work came into being. Whether or not it is justified in terms of some theoretical cost-benefit analysis, it is there for use at a cost to the trucker
far below that which the investor in railroads was required to pay for his
tracks and right of way.

With a network of roads available, and an increasing number of vehicles
capable of carrying freight in the hands of small firms and individuals,
the system of rate control previously set up to protect the public became

every day more vulnerable. Railroads have hundreds of rates for carrying different goods, even on the same train. Because of social objectives and/ or political power, common carriers were required to carry some goods at low rates. Other goods of high value move at rates that are expected to subsidize these profitless activities. Increasingly, legal means were created to permit private haulers, not bound by the government-authorized-and-published rates that controlled common carriers, to skim the cream of freight traffic. We cannot here present the multitude of devices used nor the extent of the inroads on profitable lading that resulted. Again we are at once trying to show why the railroads were not able to use their greater mechanical efficiency to meet the tactics of their competitors, and also to indicate the stake which truck freight operators have in <u>continuing their operations under the protection they now have.</u>

The melancholy recital of the limits imposed on railroads would not be complete without reference to the subsidies given by government to support their competitors. Locks and dams were built with tax money to develop freight shipped on lakes and rivers. The railroads were even forced to pay a large part of the costs of building overpasses and relocate track for the benefit of their competitors. [6]

The power of labor unions succeeded in forcing the railroads to compensate workers for jobs lost in the effort to make railroads more efficient in the use of energy. This reduced the incentive to change. Most of the technological advance came as a result of the activities of outside suppliers, not subject to controls the railroads faced. They operate in a free system and profit from it. But those profits are not available to maintain track and other fixed structures.

The upshot of all these conditions was that railroads were unable to maintain their tracks, build or buy sufficient new equipment or attract more efficient management. The results were made manifest in inferior service, high costs for damaged lading, poorer and poorer passenger service. Falling revenues and deteriorating public support accelerated the contraction of the system, even when, as we have seen, the railroads were still delivering both passengers and fright at an energy costs well below that of their more profitable and prestigious competitors.

This brings us to the critical point of our presentation. The potential of the railroads'in reducing both energy costs and pollution can be realized only through tremendous efforts capable of overcoming those of all the sets of interests who have gained and hope to continue to gain by keeping the railroads from becoming more competitive. Others don't want railroads to reduce costs by cutting off money they now must pay to these opposing interests.

If the more energy efficient railroads are to be used, it can come only as a result of a major shift in the hierarchy of values of Americans, who have sufficient power to require it, or a modification of the power position of the railroads vis a vis those who up until now have prevented changes that would allow railroads to move toward greater efficiency in terms of energy.

One change supports a major shift as we assess the evidence it appears that, at least in the next decade or so, the economy of the United States will rely heavily upon increased use of coal and of atomic fission to offset the declining availability of gas and petroleum and to provide increases in the use of power. The latter fuels will be used for off the road vehicles and for petrochemicals, particularly plastics, synthetic fiber, pharmaceuticals, fertilizers and pesticides. The industries that produce these products will be able to pay prices sufficiently high to secure petroleum and gas. Coal and nuclear energy will be cheaper as fuels for power and heat.

The return to coal will restore in part the dominance the railroads once had in the transportation of fuel. Some electricity will be generated at mine mouth and distributed by high tension grid. Some coal will be distilled to furnish gas and liquid fuels that can be carried by pipeline. However, much of the coal will be mined in arid regions where the value of water for other uses will preclude its extensive use in the generation of electricity or gas. A great deal of the coal used as heat and in power generation will be shipped by rail.

The pollution produced by coal burning can be reduced most effectively in stationary plants. This dictates the electrification of the railroads, most of which now use the diesel electric locomotive. Their motors can be supplied with current from a catenary system, thus slowly reducing dependence on their diesel engines. New coal burning power plants yield efficiencies far above those of the diesel used on the railway or the highway. This increased efficiency can cut the costs of fuel. Whether this will result in net dollar decreases depends upon the increased overhead costs that come with electrification. But the shift to coal or the atom will probably increase the disparity between the ton mile energy costs of highway freight and those of the railroads.

The fact that it can reduce pollution at the same time will bring the railroads new allies from the ranks of environmentalists.

A further shift in public support is very likely to come from motorists. It is highly probable that liquid fuel for the automobile will continue to receive a low priority among those who allocate energy use. The motorist who, when gasoline was plentiful and cheap willingly produced through fuel tax a large part of the costs of truck freight, is likely to become highly critical of the use of scarce liquid fuel to carry loads that might be carried more cheaply both in terms of money and energy, by the railroads.

Among the first signs of the division appearing between proponents of the private automobile and the truck is the struggle of the latter to retain access to cheap fuel, and to operate at higher speeds than automobiles. Since every motorist on the road is in position to see whether the authorities are in fact yielding to that demand it is not likely that government will acquiesce.

This may result in increased regulation of trucks, but more probably to the abandonment of the common carrier concept. The truck is not only inferior to the railroad in terms of the energy it uses, it is enormously more costly in terms of ton miles per man hour. Forced into open competition for all kinds of lading this handicap would probably doom not only private haulers for others, but also many of the private truck fleets now favored by some firms.

There are already efforts to investigate the past efforts of the manufacturers of automobiles and trucks to sabotage the movement toward support of mass transit and less stringent control of the railroads. If it is widely believed that the alliance of these manufacturers with those of highway builders, tire manufacturers and others who profit from highway traffic have in fact resulted in expensive transportation, pollution, congestion and other ills industrialized areas face, there might be a revival of the image of these businesses that Ida Tarbell and muckrakers gave big business in the early years of the century that produced anti trust acts and the controls from which the railroads suffer.

Technology will not hold back railroad revival. New techniques of railroad operation already exist. They have proven effective in other countries and in small segments of the United States. The public is hardly aware of the enormous change in rolling stock that has come since World War II. A great deal of its potential is denied it by the deterioration of road bed and other fixed structures. Unless sufficient capital is forthcoming a great deal of effort to design new cars and locomotives will be wasted.

Foreign experience and that of the better U. S. railroads demonstrated that we are capable of creating track on which heavy trains can be safely operated at high speed. With such track high speed passenger trains can also quickly become a reality. Experiments with exotics such as magnetic levitation and air cushion trains offer alternative techniques that may prove to be economical energy-wise as well as in terms of money costs. The Department of Transportation is being financed to design new trains and is likely to be much more heavily financed as the rising costs of liquid fuels makes the use of the automobile for long trips more prohibitive.

The barriers are beginning to fall. The fate of railroads in the Northeast sector has resulted in a breakthrough. Bankruptcy has threatened to produce catastrophe, revealing how heavily industrial society still relies on the railroads.

A new corporation to finance these railroads has been authorized and funded. Heavy government subsidy will provide new capital. Labors' claims on jobs have been recognized and funded, but in turn the unions have modified many of the rules which prevented efficient use of railroads in the past. The Interstate Commerce Commission is being forced to loosen its control. This fact is more significant for the evidence it gives of shifting political support than for any substantive concessions yet won. State public utilities commissions have recently been denied the right to force the continuance of service unless the state will itself provide the subsidy necessary to have the railroads carry out some socially desirable but costly function. Again it is the evidence this gives of shifting power that is here most significant.

The battle to shift to the use of railroads in the effort to conserve energy is far from being won. The effort will be opposed by some of the most powerful elements of our society. They will not willingly submit to changes that reduce their profits or other life chances. Already it is clear that the anti-pollution movement was weakened when it collided with the effort to secure more gasoline. The use of sulfurous coal is for the moment preferred to reduction of petroleum for use on the roads. A rather strange element of American values is shown in the fact that while it was impossible to slow road traffic to save lives, it has been done to save gasoline!

Whatever is done to shift freight and passenger traffic to the railroad will come piecemeal. The more prosperous roads can finance necessary changes from their own funds. But many railroads are on the brink of bankruptcy, and it will probably take a crisis like the one which threatened the whole Northeast region to force the changes necessary to permit them to pick up the load.

There are barriers in terms of government budgets where the priority that must be given to the railroads if they are in fact to do the job will displace that of many other values long given a high place in politics. What we are saying is that a shift in the sources and costs of energy forces changes in all kinds of choices for action. But inevitability of decline in available energy imposes new unavoidable limits on what we can do and forces us to make new choices as to what we will do.

In a paper of this length it is impossible to more than hint at what the consequences of the probable changes are likely to be. But perhaps we can mention some probable directions in which the United States is likely to go.

The most significant will probably result from the return to fixed lines of travel both for freight and passengers. Heavy trains, particularly high speed trains cannot be stopped and started without very great expenditure of energy. So, to the degree that transportation by rail increases there will be increased penalty for widespread location of industry and residences into areas far removed from the tracks. If and as the costs of fuel for the internal combusion engine rises, those nearest rail stopping points will have advantage over those who must transport themselves and their goods away from those points. This revives the viability of the kind of population distribution the railroads set before off-the-rail-vehicles became ubiquitous. Mass transit vehicles too are likely to use electricity for propulsion, and make fixed routes a necessity. The kind of urban sprawl that the automobile produced in Los Angeles for example will become monstrously expensive.

Battery powered small cars will reduce pollution and energy cost, but they

will be short haul vehicles. The use of the liquid or gas fueled cars or trucks
for trips between distant points will be largely confined to roll-on-roll-off railroad trains.

There will be the same kind of destruction of values in the enormous motel,
hotel, fuel station, food service industry that has burgeoned with the building
of the interstate system as originally was felt when the same industry was
forced to leave the older rail network and the communities through which it
passed. To mention this is also to contemplate the tremendous resistance to
rates that really reflect energy costs on the railroads is likely to provoke.

For the investment banker, and particularly for great complexes like General
Motors and Ford that have had so much to do with the emergence of the present system the probable changes will shake the foundation.

Farsighted managers will, however, see the profit opportunities in a revived
railroad industry, as they did when they provided the means to a wholesale
shift by the railroads from steam to diesel. They provided the capital, the
research, and most of the repair and maintenance facilities necessary for a
revolution on the railroads that is still altering the nature of the economic
and social relationships among communities, labor organization, and individuals. For a time they were able to shape the character of many complexes of American and world society. So long as the fiction of the "Free
Gift of Nature" could be sustained, it seemed reasonable to act on the assumption that "Demand Creates Supply" and avoid questions raised by those
who saw diminishing resources as real unavoidable limits.

Perhaps it is too much to expect that managers schooled in the idea that by
regulating the supply and velocity of money all economic problems can be
solved, or that values can be created without reference to their ecological
consequences, will change. But it is certain that unless they do they will
be replaced by managers who do take into account more than a money-cost
analysis of their decisions.

FOOTNOTES

1. ENERGY AND SOCIETY. Fred Cottrell, McGraw Hill, New York, 1956.

2. To compare energy consumption with indices of well-being see Cross Polity Survey by A. S. Banks and Robert S. Textor. The M. I. T. Press, Cambridge, Massachusetts, 1963.

   Cross Polity Time Series Data by A. S. Banks. M. I. T. Press, Cambridge, Massachusetts, 1971.

3. For size and rates of growth of population see the Statistical Yearbook of the United States. Statistical Offices of the United Nations, Department of Economic and Social Affairs, New York, United Nations Publications.

   M. K. Hubbert, 1971 Energy Resources for Power Production, I. F. E. E. Transportation. Nuclear Science NS 18: 18-29 indicates mounting rates of energy consumption and future sources.

4. Discussion of the way American institutions were changed in the effort to deal with changes brought on by the railroads, EXPLORATIONS IN SOCIAL CHANGE, George K. Zollschan and Walter Hirsch, eds., Houghton Mifflin, 1963, chapter by Cottrell.

5. Eric Hirst, "Energy Intensiveness of Passenger and Freight Transport Modes: 1950-1970," National Science Foundation, Oak Ridge National Laboratory, ORNL-NSF-EP-44, Oak Ridge, Tennessee, April 1973 (page 17).

   Battelle Columbus Laboratories, "A Final Report on a Study of the Environmental Impact of Projected Increases in Intercity Freight Traffic to the Association of American Railroads," Columbus, Ohio, August 1971 (pages 2, 13). Hereinafter cited as "1971 Battelle Report."

   Battelle Columbus Laboratories, "Topical Report on Energy Requirements for the Movement of Intercity Freight to the Association of American Railroads," Columbus, Ohio, December 15, 1972 (pages 2, 3).

6. In a chapter in THE FUTURE OF THE RAILROADS published by the Transportation Center, Northwestern University, Evanston, Illinois, 1961. Cottrell discusses Sociological Barriers to Technological Change on the Railroads.

SESSION V

INCREASING THE SUPPLY: PETROLEUM AND
NATURAL GAS

Chairman: Glen Werth
           Associate Director, Lawrence
           Livermore Laboratory, Livermore,
           California

OUTLOOK FOR PETROLEUM-BASED ENERGY SOURCES

L. G. Stewart*

Rising prices will reverse the declining trend in
domestic crude production. However the U.S. will
require significant crude and product imports
well into the 1980's.

Energy is a vital question for our day. It is the prime
subject of public debate and is of increasing concern to every
American. I would like to congratulate the many organizations
co-sponsoring this meeting on the excellent timing of this discussion of the energy "delta".

Since coal beneficiation, shale oil and gas, as well as
other sources of energy, will be addressed following my presentation,
I will generally confine my remarks to conventional oil.

Energy sources are now a matter of major concern to fore-
casters, the government and the free world, to all who are concerned
with future social and political development and with the problems
and prospects of economic growth. Man's command over energy has
made it possible for him to rise above mere subsistence standards
of living. If the standards attained by the U.S. are to be preserved
and improved, we must be in a position to guarantee an adequate and
clean energy for the future.

---

* Manager, Planning and Economics Energy Forecasts, Shell Oil
Company, Houston, Texas

There are certain natural resources for which substitutes
are not easily available, whose depletion rightly gives cause for
concern today. For energy as a whole, this is not so. There is a
good chance that in the next century nuclear energy, the use of
hydrogen and possibly the harnessing of solar power will be as
prolific as those energy sources known during the last few decades.
To meet the present need and smoothly transpose to this presumed
future energy base, investment decisions and massive research
commitments are called for now.

To be more specific about conventional oil, I believe
some discussion should be made of oil reserves. In considering
the role of conventional oil in supplying a quota of our future
energy and petrochemical needs the only oil with which we need
concern ourselves is that which can be produced and used within
the time under review. This is the meaning applied herein to
the term "recoverable reserve".

Part of the public's lack of understanding of the problem
of defining oil reserves is lack of knowledge of the fundamental
nature of oil's occurrence, its distribution and its production.
It is unnecessary to consider here the origin and geochemistry of
oil formation; suffice it to say that it is formed in the sedimentary rocks of the earth's crust by transformation of entrapped,
fossil, organic material. It helps to gain a perspective of the
problems to realize that the rocks of the earth's crust have open
pore spaces that both contain fluids and allow fluids to move
within them. With only a few exceptions, it is the sedimentary

rocks that contain the necessary porosity and permeability to allow significant quantities of oil to have accumulated, and more importantly, to be produced at economic rates.

Most of the fluids found in sedimentary rocks are saline solutions, varying from brackish to saturated with salt. Where marine sediments were originally enriched with organic matter, hydrocarbons may develop over geological time spans if appropriate conditions for metamorphosis occur. The hydrocarbons and aqueous saline phases redistribute themselves to achieve capillary and gravitational equilibria.

Without an adequate seal to fluid flow in the source sediments, the hydrocarbons would migrate through permeable rocks until they reached a trap, such as a shale-sand interface (a pinch-out) or an anticlinal fold (a structure) wherein the oil would come to equilibrium at the top of the structure. Alternately, encountering no trap, the hydrocarbons could migrate into the surface environment (oil and gas seeps). It is obvious, therefore, that an oil or gas accumulation is found when there existed a conjunction of hydrocarbon-rich sediments, proper metamorphic environment, and a reservoir trap.

Because most sediments were laid down in water, the minerals are wetted with water and accumulating hydrocarbons usually never displace all the water. Oil and gas saturations rarely exceed 65 to 75 percent of the reservoir porosity. A well drilled into such a highly saturated rock will produce a stream of 100 percent oil or gas. If the saturation of hydrocarbon is less than 30-35 percent, only water will be produced. The ratios of hydrocarbon to water at intervening saturations depends on the nature of the fluids, the

wettability of the rock, and the lithology and pore structure of
the reservoir.

Very soon after the discovery of an oil field, samples
or cores of the reservoir are analyzed to determine the porosity
and oil and water contents. Taking the cores from the reservoir
results in invasion of the pore spaces by drilling fluids and
changes in pressure; the accuracy of these determinations leaves
much to be desired. The industry has developed other techniques
such as measuring electrical resistivity, neutron capture, and
making measurements under restored state conditions which can
provide fluid saturations and lithological parameters with far
greater accuracy, albeit at far greater cost. With this information and additional deductions about the configuration of the oil
accumulation, the amount of oil in place in the reservoir can be
estimated. This can be corroborated under certain conditions by
material balance techniques which rely on reservoir pressure
measurements as a function of early fluid withdrawal.

Once the oil in place is estimated, modern statistical
correlations and mathematical simulations of the physics of fluid
flow in porous media will let the reservoir engineer calculate an
estimated recovery efficiency for the reservoir when producing
under its own natural potential energy. Supplementary energy
can be provided by injecting water, gas or steam to prevent
natural declines in pressure, physically displace the oil or
to lower its viscosity. Here, again, the reservoir engineer
resorts to historical data and calculations on reservoir fluid
flow to estimate potential additional production.

Conservation measures and scientific reservoir management which have been practiced in past years have led to an average
recovery of only 32 percent of the discovered oil in place. In
some specific reservoirs the recovery efficiency has been as high
as 60 percent, as in the great East Texas field, and in others the
recovery will amount to only 10 or 15 percent. This variation is a
function of the viscosity of the oil, the lithology of the rock
(carbonate or sandstone, porosity, tortuosity, wettability and
permeability), the gas content of the oil, the pressure on the
reservoir and the presence of adjoining aquifers.

The industry is not stopping in its quest to produce
more oil. Research and development activities, estimated at
$100MM over the last decade, and increasing rapidly with the
years, have been devoted to tertiary methods - primarily involving
the use of chemicals to alter miscibility, wettability, and the
ratio of viscous to capillary forces - most investigators believe
they will meet with success and boost recovery to an average of
40 percent. A 50 percent average recovery appears to be difficult
to achieve because of the nature of oil reservoirs and their
contained fluids.

Finally, it should be noted that reservoir life, the
life of an oil field, depends on its size, the producing mechanism,
and the number of drainage points (wells) that can be economically
drilled into the reservoir. The life of an oil field can vary from
just a few years to a span approaching a century. Reservoir engineers

have developed techniques for estimating maximum economic recovery rates of production and Reserves/Production ratios have tended to average about ten. In most recent years the falling reserve situation in the country has occasioned this ratio to begin to fall significantly.

With this background, I will confine my remaining remarks to the current problem facing the United States. This country is presently relatively a small net importer of energy materials. It is forecast that this position will change as indigenous conventional petroleum and gas resources become depleted. The domestic alternatives (oil from shale, gas and oil from coal, and nuclear power) will all take a long time to develop.

In 1970 liquid hydrocarbons contributed 44% of our total energy supply or 14 million barrels per day. By 1990, although the petroleum share of our supply will increase on a percentage basis only to 47%, the absolute volume demanded will be 33 million barrels per day. That is a 19 million barrel per day increase during this 20 year period. Virtually, all of this growth will have to be supplied by imports of overseas oil.

These projections were based on Shell's 1973 forecast on basically a surprise-free projection. Our latest look at the future in which we give more weight to the emerging energy conservation ethic, does not change these figures significantly. Oil will continue to be our major source of energy.

The cut of the barrel in 1990 will be 37% used for gasolines and naphthas and 40% for distillate fuels and residual oils. The major markets will continue to be transportation, followed by the industrial market, electric utilities, and the commercial/residential sectors.

Going back as far as 1967, the U.S. has used more oil than it has produced. This gap has been widening and the difference was made up by imports. Until 1967, we produced about as much oil as we found in the U.S. However, since that time, we have produced more oil than we have found. Domestic crude oil production increased from 1946 to a peak at about 9.1 million barrels per day during 1970. We expect a continuing decline until Arctic crude and production from discoveries in South Alaska and the lower 48 states reverse this trend in 1978. Development of Prudhoe Bay along with other dis-coveries in the Arctic, South Alaska, and lower 48 should help to increase U.S. production during the 1980's.

Drilling activity in the U.S. over the past 17 years has been on the downward trend, going from about 55,000 wells drilled in 1956 to approximately 28,000 wells drilled in 1971. In 1971, our success ratio of exploratory wells was only 16%. U.S. reserves are expected to continue the declining trend which started in the 1960's, broken only by the 1968 Arctic Alaska discoveries, until reaching 32 billion barrels in 1979. Based on our forecast of the finding rate, that is the volume discovered per unit foot of drilling and the footage to be drilled annually, we project that in 1990, U.S.

crude production will most probably be around 11 million barrels
per day with a reserve to production ratio of about 9. This forecast assumes development of Prudhoe Bay, South Alaska, Naval
Petroleum Reserves, additional offshore activities and improvements in supplemental recovery.

In 1990, the U.S. is forecast to require a total petroleum
supply of approximately 33 million barrels per day. Of this 7.9
million barrels per day or 24% will most probably come from the
lower 48 states and South Alaska. Another 4.1 million barrels per
day or 12% will come from the U.S. Arctic. Synthetic crude oil from
coal and oil shale, will account for 1.5 million barrels per day or
5%. However, 19.3 or 58% will be imported crude and products. Of
our total crude imports in 1990, most will come from the Eastern
Hemisphere. As we are only too well aware, political disagreements
with major Middle Eastern or African producers will have major
implications on the U.S. energy supply system.

Continuing need for overseas product imports during this
period is primarily attributable to a sudden increase in demand for
distillate and residual fuel oils and to insufficient domestic
refining capacity. We attribute the current shortfall in U.S.
refining capacity to: (1) environmental constraints, and (2)
uncertainty over the foreign supply of crude oil. Environmental
constraints have delayed the construction of numerous refineries
in the U.S. over the past five years. In addition, the uncertainty
in the administration of the oil import quota system made it difficult

to obtain an assured supply of crude oil over the economic life
of the refinery. When President Nixon replaced this system with
the import license fee system in May, 1973, the oil industry
responded by announcing plans for several refinery projects.
However, until the quantity of oil available on international
markets becomes better known, many of these projects are likely
to be delayed.

I would like to emphasize that these petroleum supply
figures were the result of our analysis of coal's potential
contribution to our sources of energy as well as our assessment
on inter-fuel economics of nuclear, natural gas, hydro power,
and geothermal power. More exotic forms of energy are also being
investigated, such as harnessing of solar energy or fusion. But
none of these are forecast by Shell to become commercially significant before the end of the forecast period. More rapid
development of solar energy for residential and utilities use
is possible, if the economics develop favorably and the technology
is improved.

In considering measures to ease the energy supply situation the importance of long lead times cannot be overemphasized.
In the oil industry planners must think in terms of several years,
not months. Geophysical work to find commercial fields can take
from one to three years. The offshore requires one to two years
to drill exploratory wells, 6 to 18 months to construct the platforms and two to three years to complete development drilling.

Refinery construction takes three years to obtain the site, design and get permits, and then two to four years to construct. Marine terminals require three years, while tanker construction is about two to three years.

Environmental concerns play an increasingly important role in assessing our future energy supplies. The primary and secondary air quality limits have a direct impact on coal consumption as do the recently enacted health and safety acts. These have caused electric utilities and industrial customers to switch from coal to distillate and residual low sulfur fuel oils. In addition, as any owner of a new car knows, the auto emission controls have resulted in an increase in fuel consumption and the proposed 1976 standards will further aggravate this situation.

All these environmental pressures caused increases in petroleum consumption (oil in place of coal). Other environmental actions also create decreases in supply. The delay in the construction of the trans-Alaskan pipeline to transport crude oil from Alaska's North Slope is the obvious example. There have also been delays in development of the outer Continental Shelf and in lease sales. We have still not reached an accommodation with our demand for energy and our desire for environmental purity.

The price of crude oil will have a direct impact on our balance of payments, but as it increases significantly, it may have the beneficial side effect of reducing our demand. Until recently, the real prices of both imported and domestic crude were declining.

According to predictions prepared in 1972, a downward trend was
expected to be arrested and reversed with moderate price increases
resulting from participation agreements negotiated by the Organization of Petroleum Exporting Countries. However, in the past year,
OPEC have unilaterally increased crude oil prices to levels far
greater than anticipated.

Similarly, crude prices in the U.S. are rising to world
levels. Increases in domestic crude prices has been allowed under
the two-tier system established by the Cost of Living Council for
so-called new and old crude oil. The goal of increased energy self-
sufficiency calls for improved incentives. It is interesting, I
believe, that from 1961 to 1971, the price of domestic crude was
higher than the price of imported crude.

These higher crude prices may now make supplemental energy
supplies more economically attractive; that is, gas from coal and
increased shale and tar sands production.

Where is the world's oil? The estimated proven world
reserves of crude oil amount to 562.3 billion barrels and of this
quantity 352.8 billion barrels or 63% are in Asia and the Middle
East. As a comparison, North America, including Canada, Mexico,
and the United States have 47.2 billion barrels or 8% of the total
world proven reserves. It does not appear likely, at least until
1990, we will see any significant change in this balance.

Since it appears that we shall remain dependent on imports
to a large extent for the immediate future, the question arises

whether the oil exporting countries will supply the quantities of
oil that we need. Will these countries increase their R/P ratio
to the extent to which we would like? To what extent will their
absorptive capacity, that is, an assessment of the monies that
they can reasonably spend or invest in their own economy, affect
their desire to produce the oil that we would need? There seems
to be a reasonably good chance that oil could be physically available to meet the requirements of the free world, at least through
1980, but at what price? In 1980, the free world could produce 72
million barrels per day with the free world outside North America
having an unconstrained demand of 58 million barrels per day.
About 50 million barrels per day of the total 72 million barrels
of productive capacity potentially available comes from countries
with possible absorptive capacity problems. Thus, there is reason
to doubt that unconstrained demand will be met. Furthermore, a
look we have taken at world productive capacity indicates that even
without artificial economic and political constraints during the
1980's the oil available for import to the U.S. may fall below
forecast needs due to the peaking of world production.

All these sobering considerations took us back to the
drawing board in order to see what changes we could make in order
to decrease our dependence on foreign energy sources. We are
fortunate in the U.S. in having a large supply of coal and a
potential oil shale reserve. Nuclear energy is well developed
and is progressing rapidly. We have some constraints on these

sources due to manpower and capital limitations, the general economic climate, and environmental laws. These tend to place an upper limit on any supply increase, at least until the late 1980's. We still are convinced that oil will continue to play an important role in our energy consumption picture. However, we believe that energy conservation is the only answer that can buy us time, until the difference between adequate supply and demand can be brought into a reasonable balance.

We have used energy wastefully and the main reason for this is that the price of energy has been so low. Though an increased energy price will have a corrective role on demand, a national dedication to the conservation ethic is nevertheless required.

The largest potential for energy saving is in the transportation field. To effect a greater gasoline saving that we've already considered in Shell's forecast, substantial further reduction in auto size and weight will be necessary. If, for example, second and third cars were replaced on a large scale by much smaller vehicles getting perhaps 35 to 45 miles per gallon, we estimate that by 1990 an additional direct savings of 2.4 million barrels per day might result. A large scale switch to very small cars further reduces energy demand in that less steel, plastic, paint and other materials would be used. Carpooling is another possibility for savings when properly utilized, as well as a higher load factor on the airlines.

Industry consumes more energy than any other sector of the American economy. We believe savings are possible by the recovery and use of heat and power formerly lost in plant operations.

Better designed homes and commercial establishments for both heat and lighting are another possibility. Certainly, appliance efficiencies can be improved. So in 1980, we think we can save annually more than 3 million barrels per day. This increases in 1990 to about 8-1/2 million barrels.

Such savings will call for a commitment to efficiency in conservation entirely new to this country, coupled with an acceptance of the trade-offs involved. We must realize that these full savings may only come through more government intervention than at present seems probable or perhaps desirable. With conservation it is possible our imports could amount to only 7 million barrels per day by 1985 and we might achieve virtually energy self-sufficiency with imports of only 1 to 2 million barrels per day.

In conclusion, a period of short supply of hydrocarbons now exists and it is impossible to change the facts. But certain lessons are being learned. First, I believe the U.S. now realizes that a number of isolated decisions is no substitute for explicitly formulated energy policy. Adequate energy is dependent upon adequate investment. That investment will only be made if there is some degree of confidence within the enterprises responsible for the investment on the shape and stability of energy policy. Secondly, many of the delays to investment have been occasioned by the supposed conflicts between environmentalists and those responsible for providing energy. To the extent that these do exist or are believed to exist, they must be resolved -- with much sensitivity but also with dispatch.

For the present, the era of cheap energy which we have experienced for the last 20 years is, I believe, behind us. We shall have to become used to less prolific and higher cost energy and learn new ways of using these basic resources more efficiently.

In the period up almost to the end of this century, the trend will be for fossil fuel to go increasingly to those markets for which they are distinctively suited. The role of providing basic energy resources will steadily be taken over by the electricity industry, with electricity derived from nuclear fuel playing a more and more important part.

FUTURE AVAILABILITY OF NATURAL GAS

F. D. Hart[*]
L. W. Fish[**]

With natural gas providing about one-third of the total energy needs of this nation and 43% of its stationary energy requirements, it is obvious that the importance of adequate gas supplies cannot be underestimated. Preliminary estimates indicate that the natural gas industry provided about 22.6 TCF of gas to consumers in 1973. This is about 3.5 TCF or 13% less than the forecasted economic demand. This figure compares to a shortage of 2 TCF or 8.5% in 1972. For 1974, we anticipate the shortage to grow to approximately 4.5 TCF or nearly 17% of forecasted demand. And, none of these figures include any impact of the current Arabian oil embargo.

The shortage of natural gas is serious, but certainly not hopeless. We believe adequate potential gas resources are available and that we can, in time, meet the legitimate demands of our consumers, <u>provided</u> reasonable business incentives are allowed.

The nation's energy crisis is certainly one of the most discussed topics of the present time. Unfortunately, the vital role the gas industry can play in solving the problem is not adequately understood. There are some of us who contend that the current energy crisis started with natural gas. Back in 1968 many major gas pipelines started to have trouble contracting for new gas supplies for growth. Soon thereafter, major industrial customers, notably power plants, who could not buy more gas from the gas utility for their growing needs, turned to other

---

[*] President, American Gas Association, Arlington, Virginia

[**] Sr. Vice President, Planning, American Gas Association, Arlington, Virginia

fuels, primarily oil. This added oil demand soon became more than the
historical patterns of supply could handle, and the oil shortage mushroomed.

To illustrate the importance of gas, last year's shortage, as indicated
earlier, was 3.5 TCF. This is equivalent to 1-3/4 million barrels of
oil per day. The government's quoted figure for the oil shortage in the
fourth quarter of 1973 was slightly less than 1-1/2 million barrels of
oil per day. The oil shortage for the first quarter of 1974 has been
estimated by the Conference Board to be between 2.2 and 2.4 million
barrels of oil per day, which can be compared to our estimated gas
shortage in 1974 or 4.5 TCF or, coincidentally, also 2.2 million barrels
of oil per day equivalent.

It would be presumptuous for me to claim that we would have <u>no</u> energy
problem if adequate gas were available, because gas is not completely
interchangeable with oil in all markets. Nevertheless, I do believe the
problem would be much less severe and quite possibly energy prices
would not have escalated so rapidly.

My conclusion from observing the trends in energy supply over the past
few years is that the natural gas shortage has been a major contributor
to our present energy problem, and new gas supplies will be essential
to developing a national self-sufficiency in energy.

But, what can we do? Can new supplies of gas be developed? How
fast, and at what price?

These and many other questions have been studied in a virtual flood of
energy studies over the past few years. And, as you might expect,
there is a variety of answers.

Just to give you an idea of the uncertainty involved, let me quote the
range of gas supply forecasts that have been published.

For 1980, for example, the figures vary from 21 TCF to 32 TCF - that
is a difference of 11 TCF or 50% of the lower figure. For 1985, the

forecasts range from 20 TCF to 42 TCF -- that is a difference of 22 TCF, or over 100% of the lower estimate. It would almost appear as if you could pick any number you want and you could find someone to support it for you.

Actually, the wide differences are not due to any calculation errors but rather they reflect different basic assumptions about the response of gas supply to changes in technology and economics. For example, the availability of gas will be affected by wellhead price, by the amount of land which can be leased for exploration, by the development of new technology for supplemental gas, by environmental considerations, by the availability of capital, by our ability to import liquefied natural gas, and many other factors. Each energy study used different assumptions for these factors and, therefore, ended up with different estimates of gas supply.

Today, I do not intend to stick my neck out and pick a most likely set of assumptions. Rather, I want to turn this forecasting business around a little bit and look at the _effect_ of some of the variables on gas supply so we can see what needs to be done if we are to make the maximum amount of gas available in as short a time as possible. I will be referring occasionally to work done by others, but for the most part, I am going to rely on our own A.G.A. studies.

At A.G.A., we have developed our own model of the gas industry. We call our model TERA - which stands for Total Energy Resource Analysis - and it gives us a computerized, simulation capability to study a wide range of variables quickly. The model also lets us look at both gas supply and demand.

Although international political considerations have lately been imposed on the energy picture, the problem is basically an economic one. Adequate new gas supplies were not developed during the '60's because it was not economically attractive to do so. U.S. wellhead gas prices were kept so low under federal regulation that the major producers chose to invest their capital in other ventures.

But, if economics was the key to the <u>problem,</u> then economics must be
a key to the <u>solution.</u> If we hope to develop new sources of gas, we
must examine those factors that affect business incentives. Clearly,
one of the most important of these is the wellhead price of gas.

A.G.A. supports the deregulation of wellhead price for new gas. We
believe higher prices will provide an incentive for increased gas exploration. The present area rate price in South Louisiana is 26¢/MCF for
gas sold to the interstate market. At that price, looking for additional
supplies is simply not as attractive as other investment opportunities.
To test the effect of gas pricing on supply, we used our model TERA.
We first looked at gas production if new gas were priced at 50¢/MCF in
1974 and 1975 and escalated at 5¢/year, reaching a price of $1.00/MCF
in 1985. In the second case, our gas production increased by 2.4 TCF
in 1985, which was better than a 13% increase in supply from conventional sources.

I should add here that I am well aware of the risks in quoting selected
figures from our computations without going through the list of all the
assumptions used in the study. But to try that would take more time
and get more involved than is appropriate.

However, I think it is valid to say that higher prices will bring forth
significant new gas supplies. At the same time, these higher prices
will not impair the competitive value of gas to the consumer. For the
same two cases mentioned earlier, the model indicates that the higher
prices would reduce gas demand by only about 3%.

Let us turn now to another step that can be taken by the federal government which will result in new supplies of natural gas. Some of the most
attractive areas for explorating for gas lie on the offshore areas of the
Continental Shelf. These lands are controlled by the government and
should be leased at an accelerated rate.

President Nixon had called for making three million acres available
each year by 1975, and the Department of Interior announced a schedule

which complied with this request. To show the value of this program,
we tested the impact of the accelerated leasing schedule using our TERA
model. We compared the administration proposal to an earlier leasing
schedule which would have had 1100 acres leased in 1975 increasing to
only 2260 acres by 1985. Incidentally, this is the leasing schedule that
was used in the very thorough energy study prepared by the National
Petroleum Council.

Our model shows an additional 1.3 TCF of gas available in 1980 and 2.7
TCF in 1985 with the higher leasing rate. These are significant improvements which can come from something as simple as a government
decision to lease more land.

Incidentally, on January 23, the President directed Secretary Morton
to increase the land available for leasing to 10 million acres by 1975,
and this is a very encouraging prospect.

Obviously, you cannot extrapolate these numbers endlessly. Eventually
more land could be leased than could reasonably be worked. But we are
not at that limit yet, and it is essential that the offshore areas be made
available. The A.G.A. is especially concerned about the Atlantic Offshore areas. We have known for years of the enticing geological formations on the Atlantic Continental Shelf. So far, not one exploratory well
has been drilled off the U.S. Atlantic coastline. Promising discoveries
have already been made off the coast of Canada. The government's
environmental impact statement on this potentially prolific area is due
to be released in April 1974. We hope some leases can be offered as
soon as legally possible.

So far, I have talked about incentives for increasing the supply of conventional natural gas. Unfortunately, it looks as if the demand for this
cleanest of fossil fuels will continue to grow faster than our ability to
develop "normal" supplies. Therefore, the gas industry is also developing new supplemental supplies of gas. These include moving gas from
the North Slope of Alaska, increasing our imports from Canada,
importing LNG from overseas, and producing synthetic gas from oil
and coal.

The Alaskan oil pipeline project has been delayed for three long years.
As many of you probably know, the natural gas on the North Slope is
associated with oil, which means that the oil must be produced first,
before we can start producing the gas. Apparently the oil pipeline is
going to go ahead and we believe movement of gas to the lower 48 states
will follow at a reasonable interval.

There are two proposals for bringing the gas to the lower 48 states.
One involves construction of a gas pipeline through Canada to the mid-
western states of the U.S. Another proposal consists of a gas pipeline
across Alaska paralleling the oil pipeline with liquefaction facilities on
the southern coast and transportation by tanker to the U.S. west coast.
Both projects have their own unique advantages and I am not prepared
to say which one will be used first - perhaps eventually both. However,
A.G.A. estimates that some gas from Alaska might flow to the lower 48
states by 1980 and the volume should reach 1.5 TCF by 1985.

Predicting gas imports from Canada has turned out to be more difficult
lately than we previously thought it would be. Canada, quite logically,
is concerned about her own self-sufficiency in energy and has adopted a
policy of only exporting energy in surplus of her needs. However, there
seems to be some confusion on the definition of surplus. In 1972, Canada
exported 1.0 TCF to the U.S. At A.G.A. we have taken a conservative
growth of 0.1 TCF growth per year, which results in a figure of 1.8 TCF
in 1980. This is certainly a _physical_ possibility; whether it is _politically_
reasonable remains to be seen.

Importing liquefied natural gas, or LNG, still appears to be a viable
source of gas in spite of the current Arab oil problem. One reason is
that natural gas is still a waste product in most oil exporting nations.
Gas is produced with the oil and since no market exists, the gas is
burned. No nation likes to see this amount of waste. Conservation plus
the abililty to earn some money on a disappearing asset are powerful
incentives to enter into LNG contracts. From the United States' standpoint, developing a variety of LNG sources plus restricting the amount
of U.S. investment in the project reduces our risk. Similarly, we do

not believe that LNG imports will reach the level where they would be effective instruments of international political blackmail. Therefore, we believe LNG projects will move ahead and LNG imports will amount to 1. 7 TCF of gas by 1980 and 2. 7 TCF by 1985.

About a year ago, supplemental gas supplies from gasification of liquid hydrocarbons looked a lot better than they do now. The tragedy of reducing our estimates of gas available from this source is that this appeared to be the quickest way to get more gas. The technology is well known and, last year, feedstock was available at the right price. Now, the anticipated shortage of feedstock coupled with runaway prices changes the outlook. Some plants have already been built and others are under construction; however, we doubt that synthetic gas from liquid feedstocks will grow to more than 1 TCF per year, if that high.

Coal gasification is a different story. This process continues to have long-range attractiveness as a source of synthetic gas. Two plants are planned for the Four Corners area of New Mexico. These plants will use the German Lurgi process which is suitable for the coal of that area. A joint A. G. A. /Office of Coal Research R&D program is aimed at developing a process which can be used for any coal.

Our nation's coal resources are tremendous and can provide a supply of gas for as long as any of us care to forecast. There is much research to be done, however, and it will be the early 80's before appreciable quantities of gas from coal gasification will reach the marketplace. However, pipeline gas from coal could be approaching 1. 5 TCF/year by 1985 and could exceed 3 TCF/year by 1990.

The final potential source of new gas supply that I want to mention is the tight gas formations in the Rocky Mountain area of the United States. Some geologists anticipate as much gas can be recovered from these formations to make gas recovery economical. We expect that gas from this source can be made available by 1980.

If you really want to take a quick glance at truly long-range gas supply, I should mention the hydrogen economy. The theory is to break the water molecule into hydrogen and oxygen. The hydrogen is pumped through the gas utility underground transmission and distribution system to the consumer where it is burned, and the product of combustion is water. The fuel cycle is water to hydrogen to water, and the oceans become our supply source. Our gas industry research is aimed at developing new ways to break apart the water molecule and studying problems that might be associated with distribution. Preliminary calculations show the system to be attractive, but a long way off. Hydrogen will not be a gas supply source within the time period I have been using in this paper, but I mention it to assure everyone in the audience that we have no intention of going out of business.

But let us look at what happens in the next 10 to 12 years and ask whether or not the combination of increased incentives plus supplemental sources will be enough to satisfy our forecasted demand. The answer is a tentative yes. Our model shows that supply can be increased to satisfy demand by 1980. For the gas industry, at least, Project Independence seems to be a real possibility.

Moreover, it can be done under the free enterprise system. We must have less government control, not more. As a matter of interest, we also asked our model what would happen under stricter government control such as we understand Senator Stevenson is proposing. Wellhead gas prices were rigidly controlled but we did permit the luxury of high offshore leasing schedules. The results were rather unpleasant. In 1985 gas production was virtually the same as 1973 while gas demand grew 62%. Increased government control does not appear to be a viable solution to our problem. After reviewing our work so far, there are three conclusions I would like to leave with you:

- First, the gas industry has the ability and the resources to meet the growing demands for gas. But, we need government action to provide better business incentives. The wellhead price of new gas supplies must be deregulated. More offshore lands must be available for leasing -- especially along the Atlantic coast. We

must be free to develop supplemental sources. Unshackling the gas
industry could be one of the most important steps that could be taken
toward reaching a national goal of energy self-sufficiency.

- My second conclusion is that, even though we forecast increased
  supplies of gas, we cannot waste what we have. Every business
  should have an energy audit just as regularly as it gets a financial
  audit. And, every industrial operation should develop its own
  energy contingency plan. Energy costs should be identified for
  every end use and criteria established for knowing under what
  conditions there would be a switch to an alternate fuel.

- My third conclusion is that future supplies of natural gas are neither
  guaranteed nor predestined. Much work needs to be done by many
  people. Every interested individual should examine the gas industry program to develop new gas supplies, and I hope you will want
  to support us. Do not stand on the sidelines watching and waiting.
  We believe our program is logical and truly in the best interest of
  the consumer.

Natural gas is here to stay. Not because I say so, but because it is a

premium fuel with superior characteristics and because the resources

are available for development. Your support is a vital element in

turning these resources into an adequate supply of flowing gas.

THE OUTLOOK FOR SHALE OIL[*]

Arthur E. Lewis[**]

The extent to which oil from domestic oil shale will provide energy in the future depends as much on national and regional political decisions, as it does on technical or economic issues. Technology available to industry in this decade can be used to ensure a modest supply of oil from oil shale at present prices and with a total reserve of about 54 billion barrels of oil, if some government land is made available and environmental requirements are defined. In addition, we have the capability of developing technology within this decade to produce oil in large enough quantity and at low enough cost to exert a major influence on the national and international oil market.

In situ processes and/or massive open-pit technology would open up a much larger resource (200-800 billion barrels) within a small area (600 mi^2) of Colorado. Development of this technology is beyond the capability of industry alone and will require a definition of the role of government and industry in the planning, management, production, and ownership of the resource.

INTRODUCTION

The Green River formation in the states of Colorado, Utah, and Wyoming contains as much as 2000 billion barrels of oil in oil shale. It will probably never be practical because of economic and technical limitations to recover all of this resource. The fraction of this oil that can be obtained and the rate at which it can be obtained is limited by economic, technical, social, and political considerations. Although no significant production of oil from this oil shale has yet been obtained,

---

* Work performed under the auspices of the U.S. Atomic Energy Commission.

** Lawrence Livermore Laboratory, University of California Livermore, California

the presently identified need for energy and the projected prices of oil
seem to promise some production beginning within 5 years.

To accelerate the development of this resource to the extent that it
could supply a significant amount (10% of our needs or 2 million barrels per day) of oil in 10 years or less is technically and economically
possible, but it would require political decisions, as well as an integrated, well-organized program to plan and manage all aspects of the
program - environmental, social, technical, and economic.

With some understanding of the nature and distribution of the resource
and the various methods that might be used for its recovery, it will become apparent that the methods selected will determine how much of
the resource can be recovered, at what rate and at what cost - economic,
environmental, and social.

DESCRIPTION OF RESOURCE

The area underlain by oil shale is shown in Fig. 1. Although the total
area is about 16,500 mi^2, the distribution of oil in place is by no means
uniform. The oil shale that will be economically recoverable in the
foreseeable future is determined by the grade, thickness, and depth.
The resource available at various grades and thicknesses is shown in
Table 1.

For a grade greater than 30 gal/ton, more than 25 ft thick, and less
than 1000 ft deep, there are 160 billion barrels in place. The bulk of
this resource is in Colorado. If the grade is set at 20 gal/ton and more
than 400 ft thick, the oil in place is 720 billion barrels. All is in the
Piceance Creek Basin of Colorado, in an area of approximately 592 mi^2
(See Fig. 2). It is apparent that in the future the oil shale industry will
be confined to a relatively small area, compared to the total area of oil
shale occurrence.

The exact location and extent of this area is dependent on the method of
recovery used. For example, room and pillar mining of thin, high-
grade layers will probably be practical in selected areas on the margins
of the Piceance Creek Basin, especially the southern part of the basin,

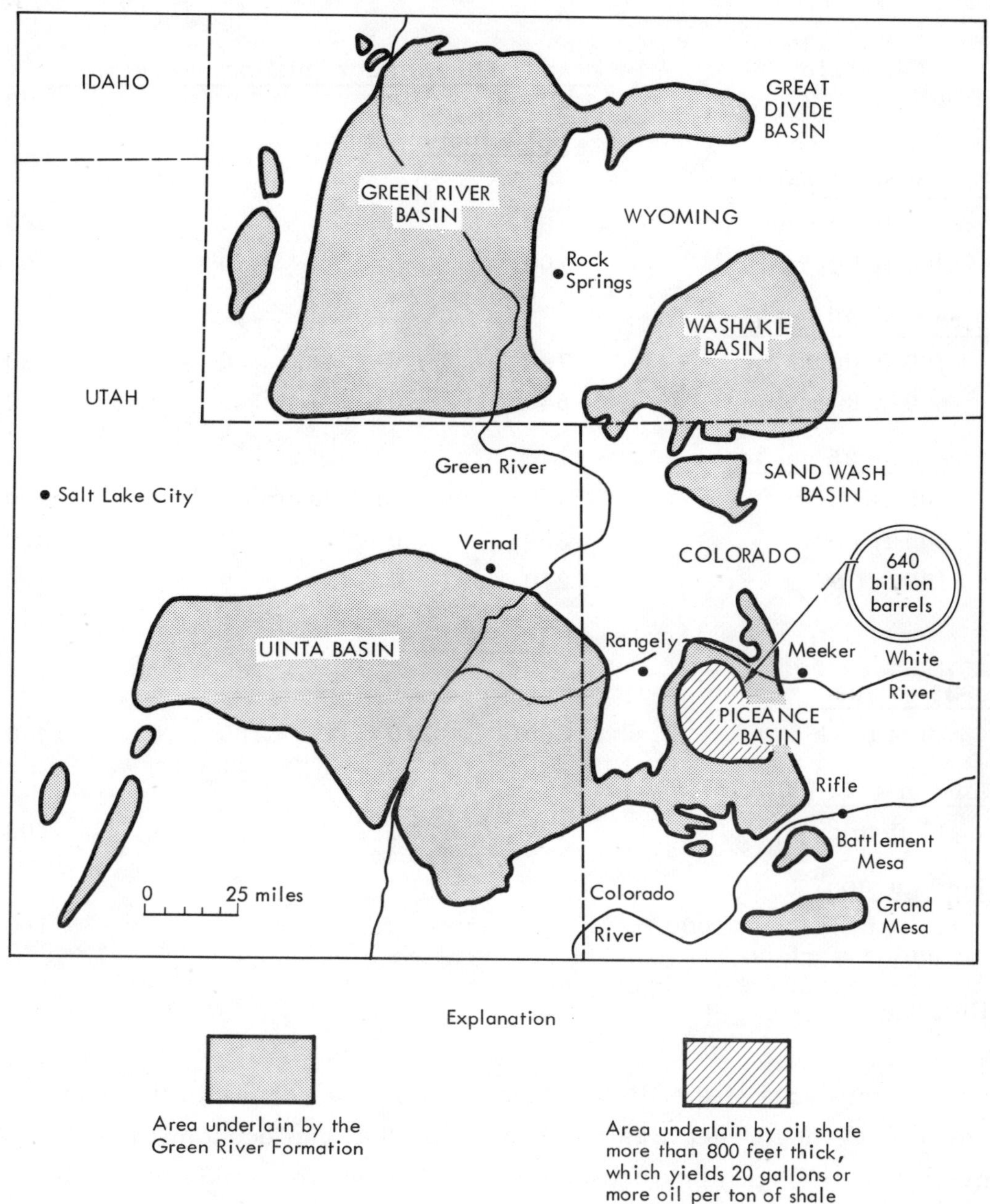

Fig. 1. Location of Oil Shale in Colorado, Utah, and Wyoming.

243

Table 1

ESTIMATES OF OIL IN OIL SHALE OF GREEN RIVER FORMATION.

Oil shale grade and thickness	Oil in place (billion barrels)			
	Colorado	Utah	Wyoming	Total
This Report				
More than 15 gal/ton				
>400 ft thick	790	70	0	860
>800 ft thick	700	0	0	700
More than 20 gal/ton				
>400 ft thick	720	0	0	720
>800 ft thick	640	0	0	640
More than 25 gal/ton				
>100 ft thick	470	(not estimated)		≥490
>500 ft thick	350	0	0	350
>1000 ft thick	270	0	0	270
USBM				
10-25 gal/ton[a]				
>10 ft thick	800	230	400	1430
More than 25 gal/ton[a]				
>10 ft thick	480	90	30	600
30-35 gal/ton[a]				
>25 ft thick and <1000 ft below surface				160

[a]Ref. 1.

and in some parts of the Uinta Basin in Utah, as well as in the north-central Piceance Creek Basin. Processes that can economically recover oil only from thick beds (>400 ft) of oil shale containing more than 20 gal/ton will be restricted to the relatively small north-central Piceance Creek Basin in Colorado. Evaporite beds are contained within the oil shale formation in the Piceance Creek Basin. Groundwater has dissolved some of this material resulting in the presence of saline water. The saline water, as well as permeable aquifers occur in the

244

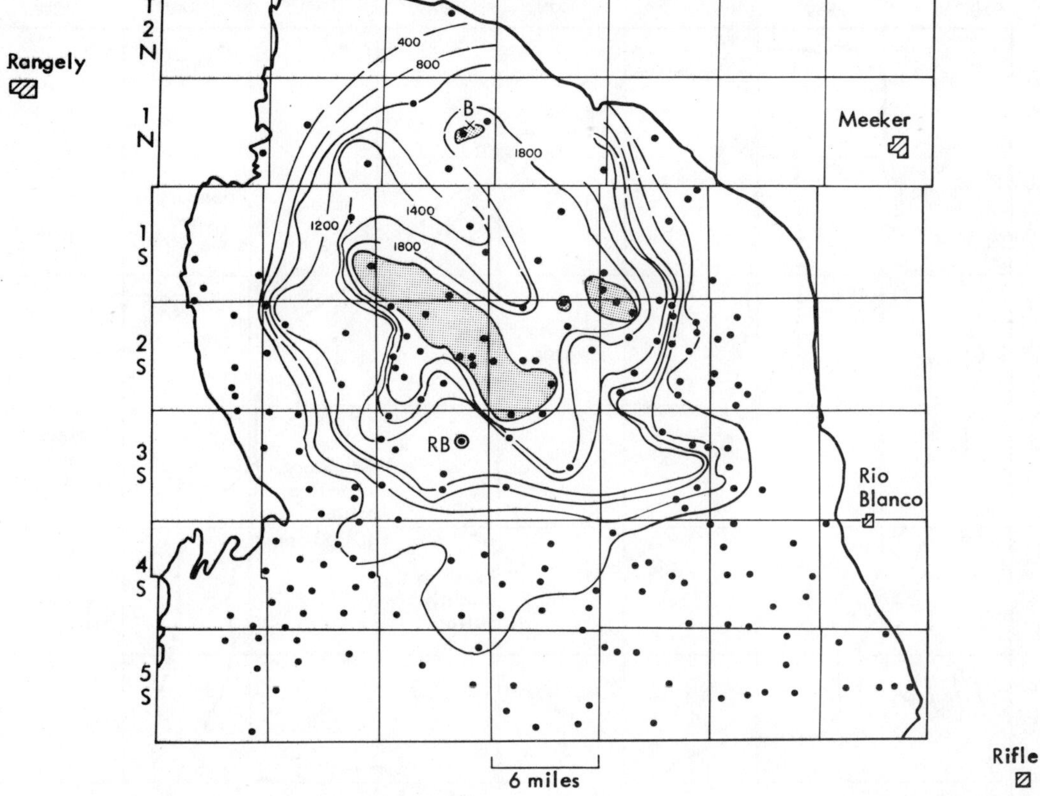

Fig. 2.   Isopach Map of Oil Shale Containing 20 gal/ton (Colorado).

same area as the thickest and richest oil shale (Fig. 3).   Although the
water may be pumped out during production, the salt content is too high
to permit disposal into the surface drainage system.

PROCESSES FOR RECOVERY OF OIL FROM OIL SHALE

Oil shale is an impure marlstone containing solid organic material
bound in the rock structure.   When this rock is heated above 400°C, the
material decomposes and a major fraction is released as a petroleum-
like liquid.   It contains a little less hydrogen and more nitrogen and
oxygen than an average crude, but it can be refined to yield the same
range of products as natural crude.

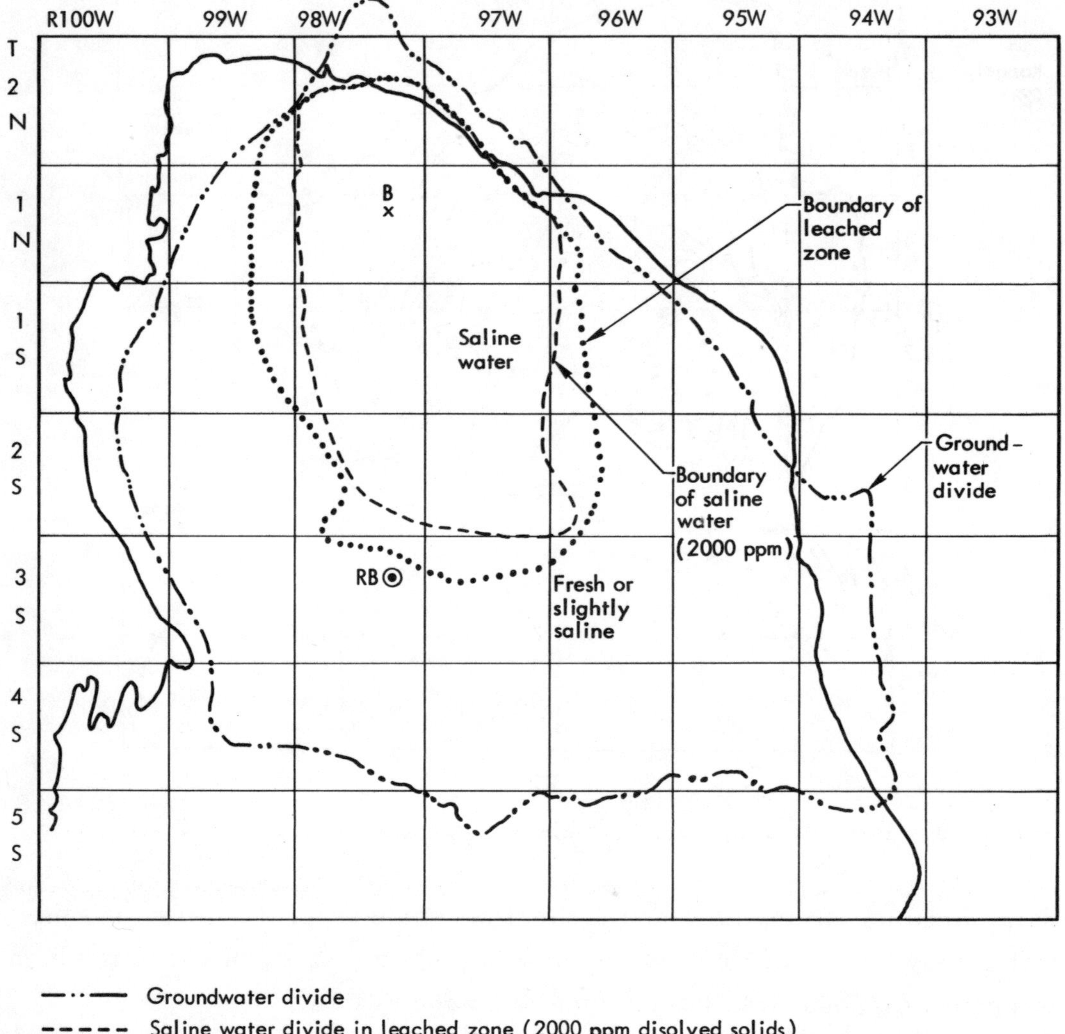

Fig. 3. Groundwater Map of Piceance Creek Basin.

Oil shale as it occurs in the ground is low in permeability and, like most rock, is a relatively good insulator. It is difficult, therefore, to heat the rock to the required temperature unless the rock can be broken or fractured enough to allow a heat-transfer fluid to be circulated through it. Various processes have been developed to accomplish the breaking and heating of the oil shale.

SURFACE RETORTING

In this process the oil shale is mined by one of several techniques, crushed, and hauled to a retort, where it is heated and the oil recovered. The spent shale is then dumped on the surface or partially returned to the mine. There are various kinds of retorts that have been designed or operated on pilot-plant scales. Among these are the Tosco II Process,[*] the Union Oil Company Process, the gas combustion Process, the Petrosix Process, and the Paraho Process.

In the Tosco II Process, heat is transferred to finely crushed ($<\frac{1}{2}$ inch) oil shale from hot ceramic balls. The process has been demonstrated at about 1000 tons/day and presumably engineering design for scale-up to commercial size is close to completion.

In the Union Oil Company process, a rock pump forces the rock upward through the retort. Air and gas move downward through the bed; the rates are adjusted to place the combustion zone near the top surface of the bed. As the hot gases from the combustion zone move downward through the bed they heat the rock, and oil is collected from the bottom of the bed as well as from the exhaust gases. Rock is crushed to about a 3-inch diameter with the fines removed. Although the process has been operated at about 1000 tons/day, the extent of the engineering design for a commercial scale has not been announced.

In the gas combustion process, the shale moves downward against upwardmoving retorting gases. Air and some additional recycle gas are injected into the gas stream about a third of the way up from the bottom of the retort. Combustion of the gases and solid carbonaceous matter provides heat for transfer to the rock above the combustion zone. This process was developed by the U.S. Bureau of Mines and has been operated at rates as high as 150 tons/day.

The Petrosix process is a modification of the gas combustion process in which the recycled gas is burned outside the retort and the hot gases

---

[*]Reference to a company or product name does not imply approval or recommendation of the product by the University of California of the U.S. Atomic Energy Commission to the exclusion of others that may be suitable.

injected into the retort to heat the oil shale. This process is being
tested in Brazil in a plant designed to process 2500 tons/day.

Another process involving modifications of both internal and external
aspects of gas combustion has been developed by Development Engineering, Inc., and is being tested by a group of companies on a pilot
scale in the Paraho program.

At this time any or all of these methods may be successful, with no
clearly proved advantage apparent for any one. Engineering design
and scale-up to commercial size (approximately 10,000 tons/day)
remains to be demonstrated.

Oil shale is supplied to the surface retorts either by underground
mining or surface mining. The principal underground method considered is room and pillar mining, which appears economically
limited to the highest grades of oil shale, approximately 30 gal/ton
or higher. This means that relatively thin layers of high-grade oil
shale will be mined and only a small fraction of the total resource
recovered. A study[2] conducted by the National Petroleum Council
estimated that approximately 54 billion barrels of oil shale could be
economically recovered by this method from 100 billion barrels in
place (30 gal/ton).

Open-pit mining is an alternative method of supplying oil shale to the
retorts. Near the margins of the Piceance Creek Basin, some of the
high-grade material is shallow enough to be mined by open-pit methods.
Almost all the resource richer than 20 gal/ton could be recovered if
massive open-pit mines as deep as 3000 ft were to be excavated across
the whole basin. This may be economically practical at an early date.
More than 700 billion barrels might ultimately be recovered by this
method.

IN SITU RETORTING

In this process natural fractures or man-made fractures in oil shale or
rubblized oil shale are used to allow heating and retorting in the ground.
If successful, the mining costs as well as the cost of surface retorting

facilities may be reduced and the amount of spent shale to be disposed
of greatly reduced.

Attempts to recover oil from natural or man-made fractures between
drill holes in oil shale have met with little success. It seems unlikely
that a significant fraction of the oil in place can ever be recovered by
this method, although research is continuing.

A more promising method of _in situ_ processing is the retorting of
rubble produced underground by either mining techniques or nuclear
explosives (Fig. 4). In both of these methods broken rock containing
a large amount of exposed surface area and high bed porosity is produced. The retorting process is, therefore, very similar to surface
retorting processes, except that it takes place on a bigger scale underground. The larger rock sizes are acceptable, considering the large
size and longer time available to retort underground. Openings for air
provided into the top of a rubble mass prepared underground. The bed
is ignited, and the hot gases move down through the bed and at the
bottom of the bed are exhausted through openings to the surface. As
the combustion zone advances downward, the hot gases moving ahead
of the combustion zone heat the oil shale to retorting temperature. The
oil at the bottom of the rubble mass is collected through drill holes and
from the exhaust gases. This method was proposed for the retorting of
rubble produced by nuclear explosives and initial pilot-type experiments
conducted by the U.S. Atomic Energy Commission and the Bureau of
Mines at Laramie, Wyoming. Since then, the Garrett Research subsidiary of Occidential Petroleum Corporation has successfully demonstrated the retorting method on a pilot scale in the field on rubble produced by mining.

In the mining variation of this _in situ_ process, space (perhaps 20%) is
created underground by one of various conventional mining techniques.
Rock is then collapsed into this space using high explosives. Suitable
openings for the retorting process are prepared either by drilling or by
mining techniques. If nuclear explosives are used, the space is created
by upward displacement of the surface over a broad area above the explosion and collapse of rock into the cavity produces rubble. Openings

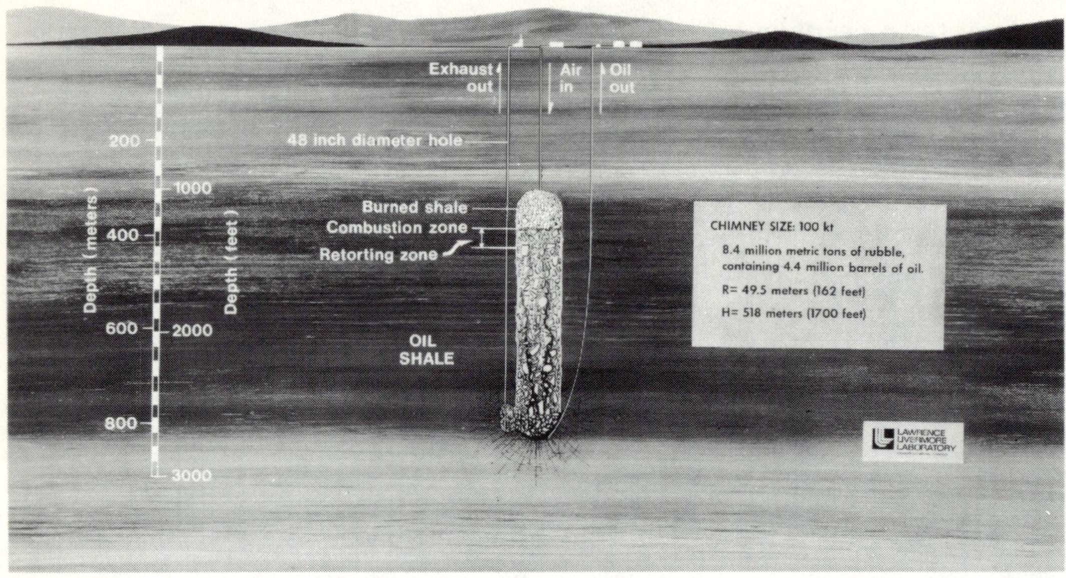

Fig. 4. In-situ Oil Recovery from Oil Shale.

are provided by drilling holes as required. Both in situ methods appear workable but the technology is not as far advanced or as certain as the surface retorting processes. Rubble columns of the required area and height have yet to be constructed and successfully retorted in both the mined and nuclear variations, although the mining technology is probably more advanced than the nuclear method in most respects.

These are the methods considered for development of oil shale. Selection of the best methods requires considerations of the advantages and disadvantages of each, the state of the technology, economic costs, resource utilization, social and environmental consequences and limitations, and other considerations.

COST OF OIL AND SIZE OF RESOURCE BASE

Not only the price of oil produced, but the amount ultimately recoverable, varies with the process used. Estimates shown in Table 2 include a 15% rate of return on invested capital.

Table 2

COST AND RESOURCES RECOVERABLE FOR VARIOUS OIL SHALE
PROCESSES

Method of Oil Recovery	Cost per Barrel ($)	Grade (gal/ton)	Resource Base Billion Barrels In Place	Recoverable
Surface Retort				
Underground	5.00	35	34	20
Room & Pillar	6.00	30	100	54
Open Pit	5.50	25	400	380
	6.50	20	800	760
In situ Retort				
Mined	5.00	20	800	300
Nuclear	3.50	20	650 +	200 +

The richer the oil shale, the lower the cost and the smaller the resource. Underground room and pillar mining is expensive and can be used economically in only the thin layers of high-grade oil shale with a relatively small resource base. Massive open-pit methods and in situ methods afford a lower cost per ton but are generally applicable only to thick sections which are of lower grade.

The estimated costs include about $1.50/barrel for upgrading. Oil obtained from surface retorts must be upgraded to reduce the viscosity so that it may be transported by pipeline and to remove nitrogen, which poisons catalysts used in the refining process. The process assumed is hydrogenation, which reduces the viscosity and removes both nitrogen and sulfur. The in situ processes are estimated to recover only 60% of the oil in the rubble zone, as compared to perhaps 95% for the surface retorting. At this lower recovery fraction there is some evidence that the lighter fraction is preferentially recovered, and therefore the viscosity may be low enough to transport by pipeline. Removal of nitrogen would still be required at the refinery, however.

It may be concluded that a very large amount of oil can ultimately be recovered from oil shale at a price that is attractive by present

251

standards. How soon and at what rate production may be obtained and at what social and environmental cost is the next question to be addressed.

CONSEQUENCES AND LIMITATIONS OF RAPID GROWTH OF OIL-SHALE INDUSTRY

The oil shale occurs in a semi-arid area with a low population density. An area of about 600 mi^2 (20 × 30 mi) contains almost all the oil shale of economic interest. Limited water resources may limit the ultimate size of the industry and population that may be supported. The rate at which production may be accelerated will be limited by the rate at which regional facilities can be expanded, by the availability of trained miners and other workers, and by capital equipment and facilities for mining, retorting, and upgrading. The technology available or the rate at which it can be developed are other factors that can limit the rate of growth. It is important to note, however, that both the limitation and the environmental consequences of a large industry are variable, depending on the method of production selected.

Some of the consequences and limitations of production of oil from oil shale at a rate of 1 million barrels/day are shown in Table 3. Underground room and pillar mining at a rate of 0.5 billion tons per year would necessitate doubling the underground mining capacity of the United States; 0.441 billion tons were mined underground in the United States in 1971 (including metals, nonmetals, and coal). The open-pit option would require increasing United States open-pit capacity by more than one third (excluding sand and gravel). Both equipment and miners are in short supply at present, and rapid expansion must compete with other requirements, such as the need to expand coal production.

Waste disposal required for spent shale is highest for the surface retorting processes, is reduced greatly by the mined _in situ_ process, and eliminated entirely in the nuclear _in situ_ method. The population influx required is estimated to be between 130,000 and 140,000 for the surface processes. The productivity is higher in the open-pit method but lower-grade oil shale as well as overburden must be mined, retorted, and disposed of, with little change in the number of workers

Table 3

CONSEQUENCES AND LIMITATIONS OF PRODUCTION OF 1 MILLION
BARRELS/DAY

Method of Oil Recovery	Disposal Required (billion tons/ year)	Population Influx	Water Use (acre-ft)	Worker Environment and Safety	Other
Surface Retort					
Underground				7100 miners	
Room & Pillar	0.5	130,000	176,000	18 fatalities/yr	
Open Pit	1.2	140,000	173,000	11 fatalities/yr	Saline water disposal
In situ Retort					
Mined	0.23	85,000	96,000	3500 miners 4000 drillers 12 fatalities/yr	Saline water disposal
Nuclear	None	55,000	65,000	5000 drillers 6 fatalities/yr	Saline water disposal ground motion radio-activity

required.   The population required is again reduced significantly for both
in situ methods, with the smallest population required for the nuclear
option. Because a large amount of water is required for disposal of
spent shale, water requirements are significantly lower for the in situ
processes.

Any process which recovers oil from the deep, rich oil shale in the
Piceance Creek Basin, which constitutes the bulk of our recoverable reserves, must face the problem of removing and disposing of the saline
water. The problem can be avoided to some extent on the edges of this
area (i.e., both leases Colorado-a and Colorado-b). At some moderate
production scale, water can be pumped away from an operating property
and reinjected into the formation at some distance away. It cannot be
dumped into surface streams because of its high salinity. The least
expensive and probably the best solution for large-scale production is
to pipe the saline water to either the Great Salt Lake or to the Pacific
Ocean. This would cost no more than a few cents a barrel at a

253

production rate of 1 million barrels a day or more. Effects of dewatering small or large parts of this area on surface water flow, springs, and wells, must be understood, evaluated, and paid for by the oil-shale industry.

Although the nuclear _in situ_ method eliminates the need to dispose of spent shale, and reduces the population influx, it produces some consequences of a different nature. Ground motion is produced by the explosions and may cause damage to nearby structures. We estimate that the largest explosive required for oil shale would produce ground motion in the surrounding towns to Meeker, Rangely, and Rifle which would only be two thirds of that experienced in the recent Rio Blanco gas-stimulation experiment[3] which was in the same area. At present the sparsely settled area seems suitable for development of the nuclear option, but a major population influx into the area could remove this option if the location of such a population is not planned properly.

Radioactivity that would be deposited underground must be studied to assure that it will not be a hazard to anyone. A very slight amount of radioactivity in the form of tritium is expected to be present in the oil produced.[3] Exposure to a using population is estimated to be less than 0.1 mr/yr or less than 0.1% of background.

Table 3 also shows some estimates of the annual fatalities expected for the various options for all operations. Underground mining is the most hazardous occupation. Smaller hazards result from drilling and surface processing.

In situ retorting appears preferable to surface retorting in most respects, if our estimates are correct. Unfortunately, however, the technology for the _in situ_ processes is not as well developed as the surface processes and therefore is not as certain. The earliest production can therefore be obtained using the surface processes. At the same time the magnitude of the investment and the time required to train miners and build facilities will limit the rate at which production can be expanded. The ultimate production, limited by the ability of the area to supply water and other facilities, has been estimated to be approximately 1 million barrels per day for surface retorting.[2]

If adequate support is given to developing the _in situ_ technology, it is likely that the time required to reach a production of 1 to 2 million barrels per day may be no longer, because of the reduced need for building capital equipment and the smaller population influx required. Also, the ultimate production limit could be 3 to 5 times higher and expansion could be accomplished at a much faster rate than the first 1 million barrels per day.

In addition to the economic, technical, and environmental problems discussed, national policy decisions on leasing and development of oil shale lands are critical, as are government policy decisions affecting the present and future price of oil. Essentially all of the best oil shale in the north-central Piceance Creek is owned by the government. Two tracts in this area have been leased as a result of the present government lease program. Aside from these leases, development is limited to the more marginal land in private hands in the southern part of the Piceance Creek Basin and to private and state lands in eastern Utah. Two additional tracts in Utah will be put up for bidding in the next few months. The two tracts in Wyoming will follow but are probably of little interest to most bidders.

DEPARTMENT OF THE INTERIOR OIL-SHALE LEASING PROGRAM

The pace and size of the present oil-shale leasing program is not consistent with an assumption that we need to move rapidly toward significant production in the next 10 or 15 years. However, it must be recognized that the leasing program was largely designed before the present public awareness of the energy shortage, at a time when environmental concerns and fears of a "giveaway" of public lands were paramount.

The lease program requires that after sale of the lease, 1 year of environmental base-line surveys be conducted. A plan for development may then be submitted and may be approved by the department after public hearings are conducted. If the plan is not acceptable, two 1-year periods are allowed for revisions, with the final plan due by the third anniversary of the lease.

After acceptance of a plan, construction can begin on a mine and facility to produce oil. Plans for a 50,000-barrel/day operation by 1981 and 100,000 barrels/day by 1982 were reported in the Denver Post for the successful bidders on the Colorado-a lease. There have been other reports that engineering on the Tosco II process is well along and a 50,000 barrel/day plant might be ready as soon as 1978 on private land. Members of the colony group were among those successful in obtaining the Colorado-b lease. Should they decide to proceed there, rather than on private land, development could be delayed beyond 1978.

A further assumption of the lease program was that lease sales will be sufficient to induce industry to begin an oil-shale industry. The in situ processes described in this paper as the most promising were not discussed or seriously considered in the Environmental Impact Statement written for the leasing program. It was assumed, probably correctly, that surface retorting processes would be used by industry initially.

It is further stated that no further leases will be made available until all "the environmental effects of the prototype program indicating feasibility of a mature industry are fully evaluated."[4]

The oil-shale leasing will probably help to establish an oil-shale industry but its development will be slow and deliberate, and little if anything will be done to determine the potential of the in situ processes.

PROSPECTS FOR PRODUCTION OF OIL FROM OIL-SHALE

Oil shale contains a vast amount of economically recoverable oil. The rate of development and the rate of production are matters of speculation. As a society, we have the technical capability to develop production from this resource at a rate of 2 to 5 million barrels/day in the next 10 to 15 years. Production at the rate of 1 million barrels/day could be reached in less than 10 years. Nothing like this will happen

without a national decision to mount a well-organized effort, including
the parallel development of surface retorting and <u>in situ</u> technology.
Major participation by all levels of government, as well as participation,
by industry, will be required. A master plan for development of the
resource and the area must be designed to solve a broad range of problems
including water use, saline water disposal, land use for industry and
urban development, and minimization of environmental effects in addition
to technical and economic problems of oil recovery. Priorities for men
and materials may also be required.

Initially, production would come from surface retorting while technology
for the in-situ processes is developed. The <u>in situ</u> option would then
be available for further expansion of production using the optimum
methods. Although this kind of development is possible, it requires a
national program of a kind not yet in sight.

If there is no significant government involvement other than the lease
program, the development will be much slower.

I would expect that a production rate of 50,000 or 100,000 barrels/day
might be achieved in 5 years by one or more cooperative industry groups
spreading the financial risk. Successful operation of these plants would
then lead to the design and construction of second-generation plants that
would lead to perhaps five times as much production in the next 5 years.
The unavailability of prime land could limit this further development and
will probably greatly inhibit the development of the mined <u>in situ</u> process
unless one of the present Colorado lease holders develops an interest in
it. The nuclear <u>in situ</u> process cannot proceed without government
involvement, and continuing unplanned development in the Piceance Creek
Basin may make it impractical because of ground-motion damage to
poorly located or designed structures.

Although recent prices of crude are high enough to project profitable
recovery of oil from oil shale, companies have limited capital and
must invest it where it will bring the greatest return with the least
risk.

Large-scale investment in oil shale would require management and
technical talent to be diverted into a new unfamiliar technology as
compared to competing alternatives that are conventional and also
profitable at present higher prices. Technical and managerial risks
are deemed to be higher for oil shale, and future prices are uncertain,
since they depend on government regulation and decisions by foreign
governments.

CONCLUSION

A small part of the Piceance Creek Basin (600 mi^2) in Colorado contains
most of the economically recoverable oil contained in oil shale and is
almost entirely owned by the United States Government. The amount
estimated to be recoverable ranges between 20 and 760 billion barrels,
depending upon recovery methods and price. Hundreds of years will
be required to exploit this resource. Beginning of production and slow
expansion to a few hundred thousand barrels a day is predicted in the
next decade without further government involvement. A significant
increase in production to a level of 2 to 5 million barrels/day is possible
in 10 to 15 years, provided that a well organized program by government
and industry is initiated to solve a broad range of technical, environmental,
social, and political problems.

ACKNOWLEDGMENT

Much of the information used in this paper was contributed by members
of the staff of the Lawrence Livermore Laboratory, especially
A. Rothman.

REFERENCES

1. <u>Mineral Facts and Problems</u>, 1970 Ed., U.S. Bureau of Mines, Bull. 650 (1970).

2. National Petroleum Council, Committee on U.S. Energy Outlook, <u>An Initial Appraisal</u> by the Oil Shale Task Group, 1971-1985 (1972).

3. A. E. Lewis, <u>Nuclear In Situ Recovery of Oil From Oil Shale</u>, Lawrence Livermore Laboratory Rept UCRL-51453, (1973).

4. <u>Environmental Statement for the Proposed Prototype Oil Shale Leasing Program VI-III</u>, U.S. Dept. of Interior (1973).

SESSION VI

INCREASING THE SUPPLY: COAL DERIVED FUELS

Chairman: Arnold H. Packer
           Senior Economist, Committee for
           Economic Development, Washington, D.C.

COAL GASIFICATION: A REVIEW OF
STATUS AND TECHNOLOGY

Jack G. Conner*

Techniques for coal gasification as a means for increasing the supply of clean energy derived from coal are reviewed. Applications for coal gasification are discussed as well as those economic, technical, and environmental considerations that can be expected to influence the decisions to employ gasification. High-, intermediate-, and low-Btu gasification processes are reviewed as to status and potential, and problem areas are identified. The production of low-Btu gas for industrial use is considered along with the large-scale operations for high-Btu-gas production. Basic technological needs associated with gasification are identified, and conclusions are drawn as to likely trends in U. S. development work on coal gasification.

INTRODUCTION

Coal gasification is that process whereby coal is converted to a gaseous product suitable for use as a clean fuel or as a synthesis gas in residential, commercial, industrial, and utility applications. The basic purpose for gasification is to convert a problem material, namely coal, into a convenient, clean, and environmentally acceptable material, fuel or synthesis gas, at an economically acceptable cost. Or, to put it another way, the purpose of gasification is to give coal a new look for the future — a look which appears aesthetically pleasing to the responsible environmentalist, practical to the convenience-oriented home or store owner, attractive to the profit-oriented industrialist, and economically feasible to the capital-intensive utility operator. Coal gasification is a conversion route which will not be easy to develop, nor is it one without profound economic ramifications; but it is one which appears reasonable in view of today's circumstances. I am reminded of the good news/bad news sequence with which all of you are familiar. The good news is that coal gasification has been around a long time and that we know quite a bit about it. The bad news is that it does not come free — the coal product is going to be expensive. But such is true of any coal conversion process; gasification is not unique.

____

* Associate Director, Battelle Energy Program, Battelle Memorial Institute, 505 King Avenue, Columbus, Ohio

Perhaps a brief mention of the other alternative coal conversion or clean-up processes should be made before we become so engrossed with gasification that we forget that there are, indeed, other alternatives. For example, coal liquefaction, as discussed earlier by Neil Cochran, is an alternate route which may have an edge over gasification for applications requiring a liquid fuel. Or, for the utility that currently operates a direct-coal-fired furnace, an alternative route to employing a fuel produced by gasification might be continued direct-coal combustion followed by stack-gas cleaning, or, perhaps, might involve employing a coal from which the sulfur has been removed prior to combustion via a chemical refining technique such as solvent refining or chemical leaching. Actually, selection from the potential available alternative processes is at present a difficult decision, involving technology evaluations, systems engineering, conomics, and other considerations concerning processes still in the developmental stage. My point in bringing this to your attention is to place in perspective the fact that utilization of coal gasification is but one of several alternatives to the enhanced use of coal, and that many factors and considerations will be involved in the ultimate decision as to which process, i.e., gasification, liquefaction, solvent refining chemical leaching, stack-gas cleanup, etc., should be employed. With this perspective as a basis, the remainder of the paper will be devoted to coal gasification.

For purposes of this paper, I have chosen to divide the topic of gasification into three nominal product-based categories: high-Btu (950–1050 Btu/scf) gas, intermediate-Btu (280–450 Btu/scf) gas, and low-Btu (100–220 Btu/scf) gas. In each category, I will attempt to provide you with an understanding of the intended uses(s) for the gaseous product, a general overview of the development status of the major processes within the category, and the important engineering or developmental difficulties facing the developers of the processes as they move into pilot or full-scale applications.

GASIFICATION PROCESSES UNDER STUDY

High-Btu Gas

The high-Btu product of coal gasification, commonly referred to as "synthetic natural gas (SNG)" or "pipeline-quality gas," is intended as a direct substitute for natural gas. Energy (Btu) content is similar to that of natural gas (NG) and, consequently, its applications will be similar. Its composition is basically methane (CH_4). As currently envisioned, pipeline-quality gas from coal could augment the dwindling supply of NG for high-priority uses — it is not proposed nor foreseen ever as a replacement. Compared with the other products of gasification processes, the high-Btu product is the most difficult and costly to make.

The production of SNG from coal involves several process steps:

      1. Gasification to produce a synthesis gas*

---

* Synthesis gas is predominately a mixture of CO and H_2.

2. Removal of particulates, carbon dioxide, sulfur, and nitrogen compounds

3. Catalytic shifting of the gas composition to give the desired ratio of hydrogen to carbon monoxide

4. Catalytic methanation to produce the pipeline-quality gas.

No coal gasifiers are presently dedicated to the production of high-Btu gas on a commercial basis, although all of the process steps defined above, except methanation, have been demonstrated on a meaningful engineering scale. Coal gasification processes for which an experience base exists include the Lurgi fixed-bed high-pressure process, the Winkler fluidized-bed atmospheric-pressure process, and the Koppers-Totzek atmospheric-pressure entrained- or suspended-bed process. Industrial operating experience appears greatest with the Lurgi system, although all have been used commercially. Experiments taking a portion of the gaseous product from an existing Lurgi gasifier system and methanating it to produce a gas rated at 950+ Btu/scf are presently underway in Scotland, along with experiments using various U. S. caking coals in a modified Lurgi gasifier.[1,2]

Several large-scale projects for producing pipeling-quality gas have been announced for siting in the U. S. These projects would utilize a number of Lurgi gasifiers to achieve gas outputs approximating 250 million cu ft/day. Coal consumption is estimated to be on the order of 25,000 tons/day. Construction has not yet begun for these planned facilities, which are scheduled for operation beyond 1975.

"Advanced" processes for the gasification of coal for feed stock for subsequent operations yielding high-Btu gas are being investigated jointly by the U.S. Government (Office of Coal Research) and the American Gas Association. Processes being developed and the organizations assigned responsibility for development include:[3]

1. Hygas, by the Institute of Gas Technology

2. CO_2 Acceptor, by Consolidation Coal Company

3. Synthane, by the U.S. Bureau of Mines

4. Bi-Gas, by Bituminous Coal Research, Inc.

5. The Ash-Agglomerating Process, by Battelle Memorial Institute.

Other advanced processes for coal gasification are being or have been investigated with either private or government support.

Time does not permit a review of each of the advanced processes, most of which involve the same basic gasification chemical reactions. Rather, it seems instructive to treat the advanced processes as a group and to suggest that they all are seeking to improve the overall efficiency (and, therefore, the economics) of the gasification process through two means:

1. Obtaining greater production of methane in the initial gasification step via contact of the coal volatiles with a hydrogen-rich gas, normally under high pressures. This hydrogasification attempt to produce methane early in the

process will, if successful, improve overall process efficiency by involving a
heat-liberation reaction (i.e., methane formation) in the initial endothermic
gasification step, and obtaining a reduction of the heat liberated in the
subsequent highly exothermic methanation step as a consequence of the
smaller quantity of gas that will require methanation at that stage.

2. Developing process design concepts which will effect a reduction in the
amount of oxygen required to supply heat to the gasification reaction.
The ash-agglomerating and CO_2 acceptor processes seek to accomplish this
by recirculation of hot solids into the gasification reactor, thereby using a
portion of their sensible heat. The reheating of these solids is accomplished
in a separate combustor which uses air instead of oxygen.

Several of the advanced processes employ operating pressures on the order of 1000 psi or more to
achieve greater methane formation in the gasifier and improved efficiencies in the methanation step.
This use of elevated pressure in itself seems relatively innocuous until the superimposed difficulties of
elevated-temperature operation, materials, and coal feeding/ash extraction under pressure are also considered. The pressure feeding system currently in commercial use is the lock hopper. Indications are
that this system is high in power and gas consumption and that the resulting economics leave something
to be desired. The development of improved pressure feeding and extraction processes is an area where
new and innovative concepts may be able to make a significant economic impact on coal gasification.

Whether or not the advanced processes succeed in the race for improved economic performance and
subsequent commercialization remains to be seen. Experience suggests that not all development work
will be successful and that some processes or concepts will ultimately fall by the wayside. Experience
also suggests that new or improved processes for production of high-Btu gas will continue to emerge
and that the industry must remain alert and flexible to the incorporation of these processes in the overall program when their analysis indicates substantive technical and economic promise.

Methanation of CO and H_2 mixtures is a critically important step in the production of high-Btu gas and
one which requires further investigation and development. Problems that have been encountered in the
methanation step include poisoning of the Raney nickel catalyst, overheating of the catalyst by the
highly exothermic reaction of hydrogen with carbon monoxide, and carbon deposition on the methanation catalyst. All processes for production of pipeline gas would benefit from development of new,
long-lived methanation catalysts and novel methods for heat removal from the catalyst particles. Efforts
along these lines are underway by several organizations, supported by both government and industry.[4]

Materials problems can also be foreseen in the plants for manufacturing high-Btu gas. Only limited
information exists with regard to ceramics operating in a reducing atmosphere containing substantial
quantities of CO and H_2, along with steam and H_2S, and particulate matter composed of char and ash.

Gas velocities will vary up to 60 ft/sec and temperatures will range from 1200 to 3300 F, depending upon the specific process. Regarding the steel components of the gasification plant, problems can be anticipated with hydrogen attack and condensate corrosion. With steam, H_2, CO, CO_2, CH_4, H_2S, NH_3, chlorides, and complex organic compounds making up the atmosphere, the condensates can be expected to be corrosive toward most structural steels. In addition, materials problems can be expected in the valves and controlled-rate feed devices where severe abrasion from particulate coal will be encountered. Special linings, facings, or claddings may have to be employed to attain the necessary wear resistance and crushing strength in such devices.

Intermediate-Btu Gas

The intermediate-Btu product is one which finds application either as a fuel gas or as a synthesis gas for the production of various chemicals, such as ammonia and methanol. In either case, the gas is produced in a gasifier which is blown with oxygen and steam. Some of the same gasifiers employed for the production of high-Btu gas are applicable to production of intermediate-Btu gas, but the expensive and troublesome methanation step is eliminated, thereby making unnecessary the associated steps of gas conversion and carbon dioxide removal. In addition, attempts to increase methane formation in the gasifier are neither necessary nor desirable for certain intermediate-Btu applications; thus lower operating pressures and simplified gasifier designs are permissible. The water-gas-shift reaction is also eliminated if fuel gas is the desired product, but removal of sulfur and particulates will probably be required.

Experience with producing an intermediate-Btu gas from coal is substantial. The so-called "town-gas," for many years a household word in both Europe and the U.S., is an example of an intermediate-Btu gas which is composed principally of hydrogen and carbon monoxide. Unfortunately, not all of this experience is applicable to coal since most of the early operations were based on coke as the fuel source.

The major techniques for making intermediate-Btu gas, which were investigated in Germany during the 1940's, include the Winkler, Koppers-Totzek, Lurgi, Thyssen-Galocsy, and the Leuna processes.[5] In addition to production of the town gas noted above, the Germans turned to the production of chemicals — liquid fuels, fertilizers, dyes, pharmaceuticals, plastics, and munitions — using the synthesis gases from coal (especially the German brown coal) and coke as the basic raw materials. Three of the processes from this early work for production of intermediate-Btu gas are available today on a commercial basis: the Lurgi, the Koppers-Totzek, and the Winkler. It should be noted, however, that experience with the Lurgi process has involved, basically, use of noncaking coals, and that operating experience with the Koppers-Totzek and Winkler has involved atmospheric pressures. Such conditions may not be acceptable for certain intermediate-Btu applications in the future, and gasification-system modifications are being considered by the manufacturers concerned, as well as being incorporated in the advanced processes referred to earlier under the high-Btu category. The improvements in thermal efficiencies and economics expected in the advanced gasification process currently under development for production of pipeline-quality gas will pertain equally to the intermediate-Btu product.

Gas transmission economics suggest that intermediate-Btu gas could be transported to utilities or industries operating within a limited area. Long-distance transmission is unlikely, however.

An important problem area which deserves development attention in the intermediate-Btu processes, and also in the low-Btu processes to be discussed next, is hot-gas clean-up. Existing gas purification processes make it necessary to perform the chemical clean-up operation for sulfur (i.e., H_2S) removal at temperatures below about 200 F. The temperature of the hot gas exiting the gasifier must be reduced, thereby wasting significant sensible heat and reducing process thermal efficiencies. If this clean-up process could be accomplished at or near the gas exit temperatures, e.g., 1200–2000 F, thereby permitting use of the sensible heat contained in the gas, a substantial improvement in overall process efficiency could be expected. For high-capacity intermediate-Btu gas production processes operating at elevated pressures, coal-feeding and ash-extraction operations can be expected to present problems similar to those mentioned in discussing production of high-Btu gas.

Low-Btu Gas

The gasification of coal to produce a low-Btu gas is being considered for applications ranging from raising industrial process heat to electricity generation via combined-cycle plants. The basic process technology to make a low-Btu gas is founded upon experience dating back to the early 1900's and involves an air (rather than oxygen) blowing operation during the gasification process. There is no requirement for an oxygen plant, nor is there inclusion of shift or methanation steps. The "producer" gas thus produced is a mixture primarily of hydrogen, carbon monoxide, nitrogen, and carbon dioxide.

The use of low-Btu gas in U. S. industry flourished during the 1920's when about 11,000 gasifiers were in operation, half of which were supplying producer gas to the steel industry for open hearths, soaking pits, and various heat-treating operations. Competition from other fuels subsequently diminished the industrial popularity of the low-Btu gasifier to a point where only a very few are operating today in the U. S. However, with dwindling or questionable natural gas and oil supplies, good arguments can be presented for a low-Btu gasification unit for industrial use, and renewed interest in the fuel produced is evident.

As noted in the discussion of intermediate-Btu gas, the sulfur-removal step requires lowering the gas temperature and thereby losing much of the sensible heat; the same situation applies in the manufacture of the low-Btu product. However, the envisaged application of low-Btu gas for electrical energy generation in combined-cycle plants also necessitates removal of extremely fine particulates from the feed gas in order to reduce erosion of gas-turbine components. Thus, both rigorous chemical and particle clean-up for the low-Btu product becomes necessary, and, so far, neither process has been accomplished successfully at elevated temperatures under operating conditions.

Hot gas clean-up is one of the most technically challenging problems facing the developers of low-Btu processes intended for combined-cycle applications. Of possible advantage to gas clean-up for the low-Btu process as compared with the intermediate-Btu process is the utilization of a lower overall process temperature because of the air (as opposed to oxygen) blow. However, this apparent advantage is offset by the greater volume of gas to be purified as a result of the air blow.

An interesting pilot-plant operation using low-Btu gas and a combined turbine/boiler cycle for electricity generation is underway in Leuna, Germany, using a Lurgi gasifier, and several combined-cycle projects are being planned in the U. S.[6,7] The basic impetus behind the combined-cycle concept for electricity generation is the expected high overall efficiencies of the plants. Net plant efficiencies of 38 and 42 percent have been estimated for combined-cycle plants with gas-turbine temperature capabilities of 2400 F and 3000 F, respectively. Improved turbine materials, coupled with hot gas clean-up, are basic to the success of this concept.[2]

The conversion of industrial combustion systems from natural gas or oil to producer gas will not be without problems. Boiler deratings may become necessary because of the lower heat release per unit volume of gas in the furnace. However, various approaches to modified burner and furnace design to minimize the potential derating problem are being evaluated. For direct-fired applications, as for example in continuous ceramic kilns, the retrofit requirements for operation with a clean low-Btu gas are minimal.

Another consideration that accompanies the use of low-Btu gas is the impracticality of storage or long-distance transport. Economic considerations suggest simultaneous manufacture and use of the low-Btu gas at the plant site. In cases where this is impossible or impractical, other fuel alternatives will probably have to be sought.

ANCILLARY CONSIDERATIONS

The characteristics of the particular coal to be gasified, as well as its availability, are important considerations to all gasification processes. Certain of the processes depend upon the use of a noncaking or only slightly caking coal; others are capable of using any bituminous coal, caking or noncaking, or lignite; another depends solely on lignite for its feed. Because of this dependence on the characteristics of the coal fed, coal beneficiation and/or pretreatment steps may be necessary prior to gasification. Pretreatment may consist of a simple physical separation operation, or may involve a more complex chemical treatment. In addition, dedicated mines will be a necessary part of the operation of large-scale gasification facilities, thereby assuring availability of feed material and providing at least some degree of coal property uniformity.

Some of the planned high-Btu gasification facilities are expected to feed as much as 1000 tons/hr of coal. Such a throughput is calculated to produce on the order of 250 million cu ft/day of pipeline gas.

One estimate[8] suggests that 36 such plants may be constructed by 1990, which, in turn, implies a coal requirement of some 300 million ton /yr. This is about one-half of the present annual tonnage of coal mined. From these figures, it should be obvious that the success of the larger gasifier operations will depend heavily upon the development of associated mining operations.

Siting for gasification facilities will have to be selected with care and due respect for environmental concerns. Water consumption will be an important factor in the larger plants, which are calculated to consume upwards of 6000 gal/min. Thermal discharges from such facilities can be expected to be large, consistent with their estimated 65 percent thermal efficiency. Outputs of by-product materials will also reach gargantuan proportions in the large facilities. Ash, char, tar, and sulfur utilization or disposal must be factored into the total operation in an environmentally and economically acceptable way. It is estimated that about 1000 persons will be employed in each plant and associated mine complex.

Capital- and operating-cost data for gasification plants present the economist with numerous uncertainties. Although some gasification plants are currently in operation in certain parts of the world, only a few small facilities are operating in the U. S. It is not possible to extrapolate realistic costs from these small operations to the larger ones being planned, but it is interesting to note some recent experience and extrapolations. The Glen-Gery Corporation has returned to service one of its old McDowell-Wellman low-Btu gasifiers for its brick-firing operation near Reading, Pennsylvania, has places three similar gasifiers on a standby basis, and has ordered a new unit for delivery in August, 1974. The new unit was originally planned as a standby for a new oil-fired kiln, but, because of supply difficulties with the basic fuel, it is now scheduled to become the primary fuel source. The gasifier currently operating was pressed into service using anthracite coal (Buckwheat) at $21/ton as its feed. At the present time, on the basis of the Btu values, the total cost for gas produced by this fully depreciated gasification system is said to be equivalent to natural gas at $0.80/1000 cu ft; natural gas is presently priced between $0.85 and $0.90/1000 cu ft at the plant. Capital cost of the new unit, which is rated at 50 to 60 million Btu/day, is given as $300,000.[9]

Capital cost comparisons for the large-scale gasification plants have been estimated recently using the Lurgi system as a basis for producing 235 million Btu/day of pipeline-quality, intermediate-Btu, or low-Btu gas.[2] The figures must be regarded as approximations subject to change as the experience base broadens and are provided solely for purposes of making "ball park" comparisons of capital costs. For a pipeline-quality-gas plant, costs are estimated at $290 million; for the intermediate-Btu plant, $183 million; and for the low-Btu plant, $164 million. Note the significantly decreasing capital costs in going from high-Btu to low-Btu plants. The reduction in costs can generally be attributed to the increased engineering simplicities and higher efficiencies of the processes for producing gases of decreasing Btu contents. Note also that the capital costs associated with the mining operation are not included and that these could well run into the $100 million level.

As a means of providing guidance to Battelle's own coal research and development efforts, a computerized cost model has been constructed which enables one to compute the total cost of energy delivered from an energy delivery system if the costs and efficiencies of the various modules comprising the delivery system are specified. A module is defined as a distinct step in the mining, transportation, or processing of coal to produce energy. The cost model has been applied to various hypothetical coal-based systems, and attempts have been made to compare the various coal conversion and clean-up processes, i.e., production of high- and low-Btu gas, liquefaction, chemical treatment, and stack-gas cleaning. On the basis of present knowledge, no firm cost differentials were found for any of the processes sufficient to eliminate them from the research and development program. Programs dealing with each of the aforementioned processes are currently under way at Battelle on the basis of the proposition that within each process there exist leverage points where technological developments could have significant impact on process economics. Our attention is focused on these points, with full recognition that substantiation of our laboratory developments will come only through meaningful engineering-scale tests.

Assistance and contributions of H. R. Batchelder in preparation of this paper are gratefully acknowledged.

CONCLUSIONS

The following conclusions may be drawn from this overview of coal gasification as a means of enhancing coal as an energy source:

1. Coal gasification will be utilized in the U. S. as a technique for providing a clean fuel from coal in the form of high-, intermediate-, and low-Btu gas.

2. For many applications, coal gasification will have to compete with alternative techniques for sulfur removal, including liquefaction, chemical treatment, stack-gas cleaning, etc. The economics of these competing processes are at about the same early stage of definition as those for gasification, thus preventing definitive comparisons at this time.

3. Technological developments for enhanced economics are being pursued in various advanced coal gasification systems; however, the economics associated with these developments have yet to be established on any meaningful engineering scale.

4. Small gasification systems producing either low- or intermediate-Btu gas for on-site industrial use could provide a clean and reliable source of energy from coal. An immediate and expanding use is predicted for such applications.

5. The development of low-Btu gasification processes for use in combined-cycle electrical generating facilities will become an increasingly important aspect of the total U. S. coal utilization program.

6. Large gasification plants for pipeline gas manufacture appear as long-term possibilities for augmenting the natural gas supply, but demonstration plants are needed to answer critical engineering and economic questions. In addition, legislative, regulatory, financial, and environmental questions will require immediate attention and resolution if such plants are to become a reality.

REFERENCES

1. Henry J. Linden, "The Role of SNG in the U.S. Energy Balance," special report for the Gas Supply Committee of the American Gas Association, May 15, 1973.

2. *Evaluation of Coal-Gasification Technology,* Part II Low- and Intermediate-Btu Fuel Gases, prepared by *ad hoc* Panel on Evaluation of Coal-Gasification Technology, Committee on Air Quality Management, National Academy of Engineering, Washington, D.C., 1973.

3. *Evaluation of Coal-Gasification Technology,* Part I Pipeline-Quality Gas, Idem, 1972.

4. *Clean Energy From Coal—A National Priority,* 1972 Annual Report, Office of Coal Research, U.S. Department of Interior, Supt. of Documents, Washington, D.C., 1973.

5. J. F. Foster and R. J. Lund, *Economics of Fuel Gas from Coal,* McGraw-Hill, 1950.

6. "Development of a New Coal Gasification Method in Essen," (In Essen wird neues Verfahren der Kohlevergasung entwickelt) news item published in Germany (F.R.), June 1973.

7. John Papmarcos, "Gas from Coal," *Power Engineering,* February, 1973, pp 32–39.

8. Final Report prepared by Synthetic Gas-Coal Task Force for the Supply-Technical Advisory Committee, National Gas Survey, Federal Power Commission, Washington, D.C., April, 1973.

9. "A Revival for Producer Gas," *Business Week* (Industrial Ed.) November 3, 1973.

ENVIRONMENTAL IMPACT OF LARGE SCALE
DEVELOPMENT OF COAL AND OIL SHALE

Joseph P. Brennan[*]

By 1980, coal production should be more than 900 million tons per year.
By 1985, the American coal industry will produce 1.5 billion tons
annually.

Much of this production will come from the traditional mining areas in
the East, and the Midwest. A growing amount -- somewhere above
200 million tons by the early 1980's -- will be mined in the now virgin
fields of the West, especially in the northern great plains area, and in
the southwestern part of the United States.

Both surface mining and underground methods will be used, and in
approximately the same ratio as today with some fluctuations on a
year-to-year basis. In the surface mining portion of the industry,
reclamation will become an increasingly important consideration under
the stress of both governmental regulations and enlightened self-interest
on the part of the operators. Underground, new mining methods will
gradually take their place alongside advanced conventional technology,
although increasing automation will be a central feature of the mining
operations. So too will be systems analysis of the entire production
and consumption cycle, aiming at the optimization of the complex
infra-structure of a modern coal delivery system.

The trend toward long-term supply contracts tying together a producer
and a consumer will accelerate. The underlying imperatives of capital
requirements and security of supply clearly point in this direction. So

---

[*] Vice President, Economics & Planning, National Coal Association,
Washington, D. C.

too do the sheer dimensions of the logistics of mining the quantities of coal that will be required by 1985. In effect, producer, transporter, and consumer will be bound together in a complicated energy-delivery system which begins with a coal reserve and ends at a conversion plant where coal is made into electricity, one of a variety of liquid products, or coal gas of one sort or another.

Consumption patterns will begin to change toward the end of the present decade, but the pre-eminence of the utility market will remain for several decades at least. By 1985, we will be supplying relatively substantial volumes of coal for gasification and liquefaction. But such tonnages will still be a minority part of the total. The bulk of the coal consumed will be as coal, under boilers, to make steam for the generation of electric power or for industrial purposes. Steel will still account for a substantial tonnage, although obviously a much smaller share, and exports may grow to a significant degree depending upon both conditions here at home, and the vagaries of international economics, resource availability, and politics.

Manpower, now central to coal development, will become even more so. Large numbers of men will leave the industry in the next decade through normal attrition. Our present work force has more than 50 percent of its number over age 45 -- with possible retirement under UMW Welfare Fund rules at age 55. Thus, as these men leave the industry, they will have to be replaced. Replacement, however, will mean much more than exchanging a younger man for an older one. For, with the paucity of men in the mid-age ranks -- 30-45 age group -- the industry is faced with the task of infusing experience on an accelerated rate along with the normal personnal functions of att raction, retention, and motivation.

The role of the exogenous variables -- particularly those dealing with government -- will assume a central role in coal development. The variables must move from negative or neutral to strongly expansionary if coal is to achieve the level of development now indicated. Of particular importance i s the degree of predictability necessary to reduce the

uncertainties inherent in long-term commitments to a level commensurate with more or less normal entrepreneurial risk. Coal must be
given these assurances so that the development of the industry can take
place free of the constant specter of legal or environmental execution.
Long lead times necessary for mine development, as well as the locked-
in nature of the investment, add to the absolute requirement for an expansionary governmental climate.

The outline we have sketched is that of a brand new industry superimposed on the old but still impressive structure of the present one.
The imperative to the new industry is the now clearly urgent need for
at least the option for self-sufficiency. The raw material for its
growth is abundantly present in our vast coal-reserve base. Its own
internal infrastructure is coperative in the mines themselves, in the
manufacturers of equipment, and in the vast though somewhat inadequate
transport system.

But as coal moves into a new position of primacy, careful attention
must be given to the ingredients of growth. In part, this will involve
the removal of the negative influences to expansion, in part to the
creation of the clearly expansionary climate to which we have alluded
above.

One point is obvious. Energy growth will continue, although the rate
of growth may slacken somewhat. To meet this growth, we have only
two choices for the immediate future: importation or the use of conventional indigeneous fuel resources with existing or nearly developed
technology. These alternatives impose upon our nation and its citizens
the need to balance costs and benefits, and to make choices in terms of
what is better, and not necessarily what is the absolute best. The
initiation of the Arab oil embargo marked an end to our age of innocence
in energy. From now on, for a decade at least, we are going to have
to rebuild the structure of energy adequacy, a structure sadly neglected
for one hundred years. We as a nation are going to have to regard our
energy resources in true economic terms -- finite resources balanced

against infinite wants. And, we are going to have to invest our money and our resources to build an energy base adequate for the national goals. Finally, attention must be given to energy conservation.

The focus of attention in conservation will have to be on both the supply and the demand side. Recognition is already being given to the dampening of demand ranging all the way from the current boom in small cars, to turned-down thermostats, and to a conspicuous absence of energy-expanding advertising. Bending the growth curve downward makes sense so long as it does not distrupt other socially desirable goals. On the other hand, I personnaly feel strongly that conservation must also look at the supply side of the equation. For too long we have been profligate in our use of scarce fuels in markets for which they had no unique application. Natural gas under utility boilers is one classic example of this. Last year, for example, about 26 percent of the energy used by utilities was natural gas. Oil, for boiler purposes has grown phenomenally in recent years, accounting for 19 percent of the total in 1972. In that year more than 493 million barrels of oil were used for electric generation.

At the same time, for a variety of reasons, coal has stagnated. Much of the growth of energy for power purposes has gone, not to coal, an abundant resource, but to oil and natural gas, scarce resources. Even the most cursory examination of this inverted supply/demand relationship highlights its perversity.

This lack of care for our natural patrimony was tolerable during an era of abundance. But now, in our emerging period of inadequacy, a re-ordering of our national energy consumption pattern is very much an urgent priority. Coal - the ideal boiler fuel - the resource base for a growing synthetic fuels industry - the keystone of energy self-sufficiency - must now come into its own.

As we have said, coal's new role is primary. Logically then the growth of coal must be one of urgent concern to all Americans, and encouraged

by every possible means. Further, as we have previously pointed out,
growth means both the removal of negative activity, and the creation
of a positive framework conductive to expansion.

The first -- removal of negative constraints -- is a first order of
priority. Principal among these restraints are environmental regulations.

Coal is a mineral. At every step from production to ultimate consumption, contaminents of one type or another are created. Even the physical process of coal extraction creates environmental problems, not
only in surface mining, but also in underground operations and in the
processing incident to coal preparation.

In recent years, the sulfur content of coal has been much in the news,
and in the minds of government officials, environmentalists, and
industry officials. Much of the impetus to foreign oil use as opposed
to domestic coal has come from the simple fact that oil is, or can be
made, sulfur-free with far less effort than coal. Large sections of
coal's traditional markets have been lost because of the sulfur content
of the coal supplying them.

Today, in the midst of every crisis, there seems to be a re-examination of the rationale surrounding SO_2 controls. These controls were
imposed in our era of abundance, under the assumption of alternative
energy availability. This assumption is now clearly inoperative and
thus the regulations flowing from it are also open to question.

Most of the U.S. coal production destined for utilities is still centered
in high-sulfur coal areas. For the immediate future this will continue
to be the case, especially for utility demand occurring east of the
Mississippi River, a region long served by eastern high-sulfur coal.

I am not suggesting that America abandon its struggle for a clean
environment. What is needed, however, is a re-examination of the

goals we have set and the means of achieving these goals that we have adopted. I am suggesting that since we can no longer assume unlimited energy availability, our cost/benefit analysis must take energy adequacy into account when setting our pollution abatement regulations.

Two things are evident for the immediate future:

1. Much of our present utility capacity now on coal will have to use high-sulfur coals simply because nothing else is available. Also, the conversion program from oil and natural gas to coal will continue with relatively high-sulfur coal playing a major role

2. While abatement devices, or pre-combustion cleanup, seem close at hand, most of the existing stations and many of the new ones will continue to burn coal in its conventional form, using abatement techniques to minimize pollution rather than utilizing the expensive and very complicated pre- or post-combustion abatement technology.

My thesis on this point is simple. We as a nation must use coal for power plant purposes. Much of the coal most readily available is high-sulfur which can be used in its conventional form at existing stations without impacting on human health. Moreover, much of the new growth, in the near term at least, can be with coal used in the conventional manner, without health impact.

This is not to say that the current impetus for pollution abatement must cease. To the contrary, as more and more development takes place, technology to reduce the environmental cost of increasing power demand must come increasingly to the fore. Scrubber systems must be installed and used. Solvent refined coal, low Btu gas, and other low sulfur fuels must be used, especially in conjunction with improved power systems, and even newer and less polluting power technology must be developed.

The last point is significant. Today, we waste sixty percent or better of all of the coal Btu inputs to a power station. Light water reactors are even more wasteful. This loss has widespread ramifications from the standpoint of resource utilization, environmental quality, and economic progress. In the era of energy innocence, waste was unfortunately acceptable. Today, it is clearly not so.

In sum, the environmental question has become enmeshed with the problem of resource availability, lagging technology, and conflicting national goals. As energy demand grows, and as the importance of self-sufficiency becomes more manifest, reconsideration of the dichotomy will become even more crucial.

Of immediate interest to many Americans is the question of surface mining. Approximately 50 percent of our national coal production comes from surface mining methods. The actual percentage by region will vary up to almost 100 percent in the developing areas of the West.

There are many reasons for the growth of the surface mining industry. For one, it is easier and more economic. For another, it is far less labor intensive, and therefore less affected by the emerging labor shortages characteristic in underground mining. Surface mining is inherently safer than is underground, but strong efforts by the underground operators should narrow this differential. Finally, surface mining, especially for the small eastern operator, is far more amendable to rapid expansion than are underground mines.

There is no question that the production of coal from surface operations will expand. Indeed, there is no available alternative to such expansion, in the short run at least, if the industry is to meet its projected targets. This is especially true in the West where very large surface reserves are available and have been committed to the broad spectrum of coal markets including the initial gasification plants.

The question, therefore, is not do we have increased stripping operations, but rather how can the environmental costs be minimized?

Cold rational analysis suggests that there is an environmental cost associated with surface operations. But so too are there environmental costs with deep mining, with road building, with industrial activity of any kind, and even with the development of an expanded tourist industry.

In the case of surface mining, such disruption of the natural environment can be minimized with proper reclamation. In fact, in many cases long-term land value can actually be enhanced by proper reclamation. Happily, for at least three reasons, the present legacy of surface mining is not prologue to its future.

1. Neither public opinion nor law will permit mining without reclamation. Almost every producing or potentially producing coal state now has a surface mining law. The Congress is in the final stages of enacting a federal law which will guarantee proper reclamation.

2. The consumer, prodded by public opinion and a changing institutional structure in coal, is now willing to include reclamation as a part of the cost of production. The era of "cheap" coal is over and the industry can internalize environmental costs without the inherent risk of going broke.

3. The industry itself has moved to implement sound reclamation practices. Most coal companies, especially those who look to coal for their future, now include reclamation as an integral part of the mining cycle. Many are carrying on extensive research programs where needed to improve their reclamation potential, especially in areas traditionally thought to be hard to reclaim. We do not suggest that reclamation problems are easy, but rather that the environmental impact is now both temporary and well within tolerable limits.

In recent years much attention has been given to the problems associated with mine refuse. The tragedy at Buffalo Creek, West Virginia, highlighted the situation, although it represented only one aspect of the overall dilemma. Much of the difficulty of gob piles, today's impoundments, etc., is a carryover from the past. Dealing with it will be both difficult and expensive.

However, as the demand for "clean" coal increases, so will the refuse, and thus the problem inherent in it. Parenthetically, the refuse from scrubber systems aimed at removing SO_2 from coal represents a new and unique disposal problem. In any event, much work remains to be done in the area of refuse disposal. Suggestions range all the way from underground storage to more soundly engendered surface methods. Whatever method is used, the operator faces an increased financial burden, and a vastly increased social responsibility.

Finally, the growth of coal inevitably involves the use of water. Water
is needed for power generation, for gasification, and in actual mining
operations. We can anticipate even greater water usage in the years
ahead. A great deal of attention has been expended in recent years on
water, both to minimize the formation of "acid mine drainage" and to
eliminate or control the other pollutants. Sharp acceleration of this
work is needed in the years ahead, as well as more efficient conversion systems which use water as a part of the conversion process.

Finally, we come to the question of manpower. Although, in the
strictest sense, this topic is beyond the parameters of my paper, it
is nonetheless central to the need for expansion of coal production.
The modern coal miner -- whether surface or underground -- is a
well-trained, well-paid, and highly skilled specialist. In the underground sector, the special nature of the job adds both demands upon
the employee and a certain fundamental kinship unknown in other
industries.

During the 1950's and 1960's, the period of coal decline, manpower
declined sharply in the coal industry. In fact, a whole generation was
lost. Today, therefore, we are faced with a bi-polar demographic
distribution, the grandfather, grandson syndrome. Now, with coal
expansion on a scale never before experienced, coal must cope with
large numbers of men leaving the industry through retirement, large
numbers being recruited, with little of the stability associated with
the 30-45 year age group. Obviously, manpower is key, and the industry response must be both innovative and progressive. To date, some
progress has been made, although the surface has hardly been touched.

Finally, I would like to comment briefly on coal's capital demand.
Currently, the coal industry has a capital value of about four billion
dollars. By 1985, we estimate that the capital value of the industry
must be about fifteen billion dollars. Obviously, in order to attain
this level, both profit margins of the industry must be increased and
the uncertainties which continually surround the production of coal

must be reduced to tolerable proportions. Coal will have to go into
financial markets for much of the capital and compete with other investment opportunities. Success will be determined by the relative attractiveness of coal as an investment opportunity, a circumstance which
must be considered by those charged with the development and implementation of a National Energy Policy.

Coal is now a new industry, new in potential -- potential for growth --
potential for the national well-being. To realize that potential is one
of the single greatest challenges of America for the rest of this century.

SESSION VII

INCREASING SUPPLY: NUCLEAR FISSION AND
FUSION SOURCES

Chairman: Rolf Sinclair
Program Director, Physics Section,
National Science Foundation,
Washington, D.C.

DEVELOPMENT OF THE NUCLEAR-ELECTRIC ENERGY ECONOMY

P. N. Ross[*]

> If we are to deal effectively and realistically with
> the energy crisis, we must sharply reduce our over-
> whelming dependence on oil and natural gas by shifting
> to an electric energy economy based on coal and uranium.
> To do this, we must begin now to substitute electricity
> for direct combustion of oil and gas in all those end
> uses where this is technically and economically feasi-
> ble. The question is not where we are going, but
> rather the timing and the route by which we make the
> transition from the oil-gas energy economy of today to
> the nuclear-electric energy economy of tomorrow.

The flood of rhetoric and debate on the energy crisis is still filled with many widely different perceptions as to what the problem is, let alone what the solution should be. The environmentalist sees the situation in terms of excessive use and waste of energy (resources); the gas industry sees it in terms of well-head prices; the auto-maker in terms of exhaust emission standards; the oil industry in terms of refining capacity and import restraints; the electric utility in terms of licensing delays; the university in terms of R&D; the Treasury in terms of balance of payments; the coal industry in terms of strip mining regulations.

This diversity in perceptions is actually less remarkable than the fact that almost everything you read and hear is focused on the near-term fuel supply-demand imbalance. As a result, most proposals are directed towards cutting back on present consumption and increasing our productive capacity. That is, we must simply ration gasoline; lower the thermostat; build smaller cars; undertake more exploration; drill more wells; build more refineries and tankers; open more mines; lift the Arab boycott; and the crisis will go away and we can return to "business as usual."

Unfortunately and tragically, that is just not true. The central fact which all of these assessments and proposals seem to overlook is that there is a large body of data that points overwhelmingly to the conclusion that exhaustion of U.S. and World oil and gas resources is within sight. In other words, at current growth rates, we are simply running out of oil and gas. Since these are the fuels on which we depend for nearly 80 percent of our energy needs, their potential depletion, as well as their current shortage, defines the true nature of the energy crisis.

---

[*] Manager, Westinghouse Electric Corporation, Power Systems Planning, East Pittsburgh, Pennsylvania

If we are to deal effectively and realistically with the energy crisis, we must sharply
reduce our excessive dependence on oil and gas by shifting to energy sources that are
more plentiful -- uranium and coal.   For most purposes, these two fuels have to be
converted to electricity -- a requirement that applies to practically every other
energy source.   Geothermal steam in California won't heat my house in Pittsburgh.
The wind or tide in Maine won't illuminate a farmhouse on the plains.   The sun in
Arizona isn't going to run a production line in Gary, Indiana.   All the energy sources
that might replace oil and gas must first be converted to electricity.

In other words, the way to solve our energy shortage will be to use more electricity
in the years ahead -- not less.   Understand that this is not a call for _wasting_ electricity, but for a shift _to_ electricity from other energy forms.   Our society must put
more of its energy into the form of electricity, and reserve oil and gas for such
vital tasks as only these fuels can handle.

Creating the electric economy means that we should begin now to substitute electricity for the direct combustion of oil and gas wherever this is technically and economically feasible.   The consumers of energy will have to be persuaded of the feasibility
and advantages of switching to electricity; the products must be made available to
permit them to make the switch; and our electric system capacity must grow to
meet these new demands.

If that sounds like a big order, it is.   But I believe that there is really _no_ alternative.
The question is not where we are going, but rather the timing and route by which we
make the transition from the fossil fuel energy economy of today to the electric
energy economy of tomorrow.   If no one takes the lead -- if we try to sit back and
let it happen -- the result will be massive shortages, a jerry-built energy economy,
and an ever-tightening tangle of regulation, fuel allocations, etc.

I'm convinced that if there ever was a time to avoid the promotion of electricity,
that time is past.   We need to begin now to create an electric economy in the national
interest.   I believe that a positive program of coordinated action by both government
and industry is needed -- neither can do it alone.   We need especially to avoid public
policies that work at cross-purposes or produce severe distortions in our energy
economy such as arose out of fixing the well-head price of gas 20 years ago.

Let me take a few minutes now to review the energy problem and develop more
fully the reasons why I believe the shift to an electric energy economy is the only
viable solution.

Examining the total consumption of energy in the U. S. in 1972, it is evident that
ours is a fossil fuel energy economy with direct combustion of oil and gas the
dominant mode of end use (Figure 1).

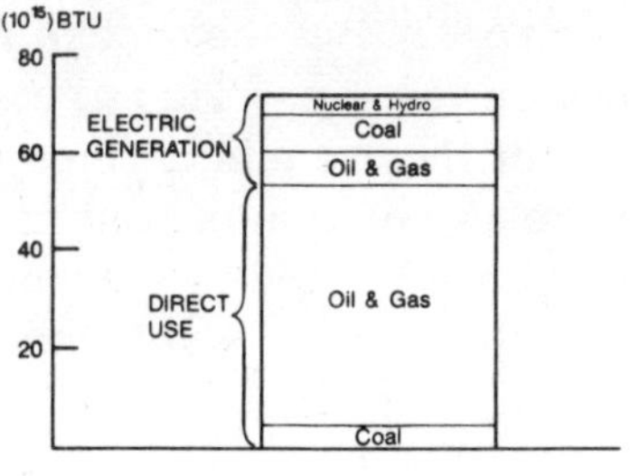

1

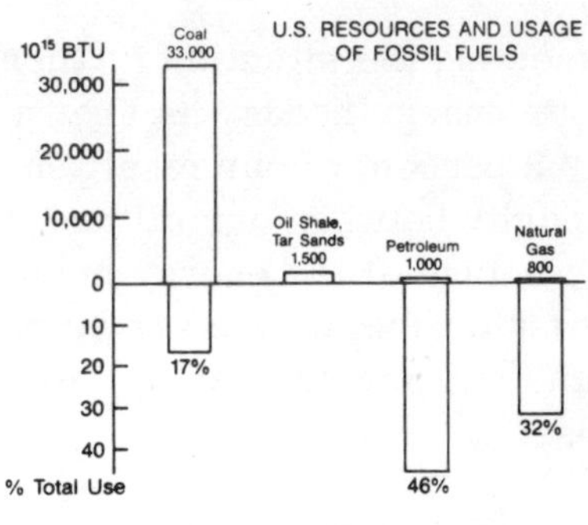

2

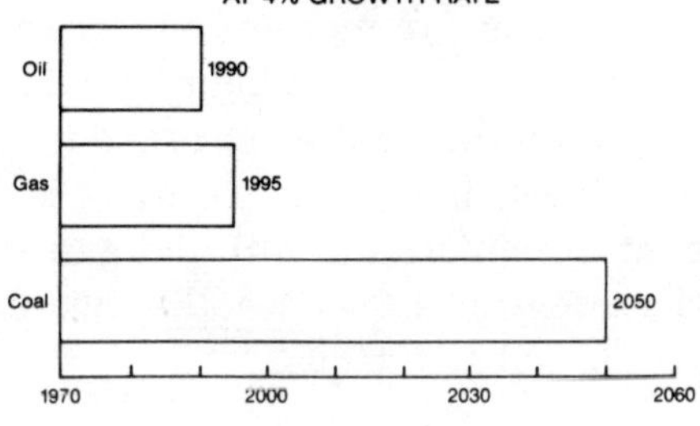

3

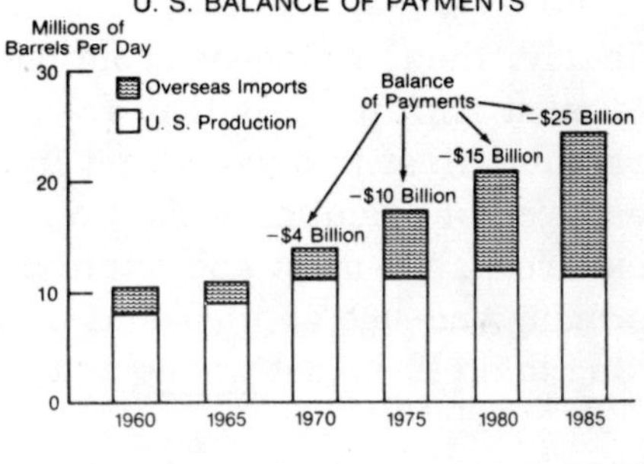

4

MAJOR WORLD ENERGY RESOURCES
(QUINTILLION BTU'S)

Coal	170
Oil	33
Gas	10
Uranium (Fission)	70
Breeder Reactor	420,000
Fusion	10,000,000,000

5

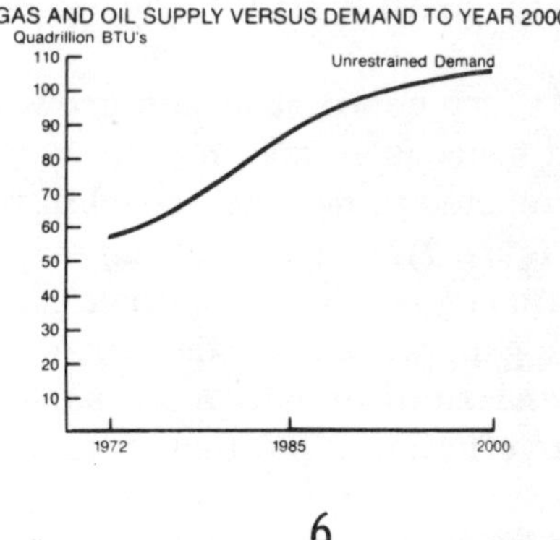

6

Dissecting the situation further, we find that oil and gas furnish nearly 80 percent of our energy needs, as shown on the bottom half of Figure 2. But they make up only 3 percent of our fossil energy resources, as shown on the top. This great disparity between our oil and gas resources and our use of these fuels defines the true nature of the energy crisis. We are in imminent danger of running out of our most vital fuels. We are burning up oil and gas as though we wanted to get rid of them, and saving coal as though it were scarce. This pattern just doesn't make sense.

If our consumption continues to increase at a rate of 4 percent, and we rely on our own domestic resources, we will run out of oil by 1990 ... out of natural gas by 1995 ... and out of coal by the middle of the next century (Figure 3). To make up for our limited domestic supply, we will have to be importing more than half of our oil at a cost of 25 billion dollars a year by 1985. As recent events have made all too clear, that would threaten our economy and our national security, besides placing an intolerable burden on our balance of payments (Figure 4).

Clearly, then, we cannot continue to depend on oil and gas for most of our energy. We must shift to fuels that are more abundant. This means -- relying on coal, and uranium, first in our present light-water reactors; then the vastly increased resource of uranium in the breeder; and finally, fusion (Figure 5). To use these resources, we must substitute electricity for direct combustion of oil and gas in as many end-use applications as feasible. And the question is not whether we are going to do this, but where and when we will make the substitutions.

Let's see what effect on reduced demand for oil and gas such electric substitutions could have in the period between now and the year 2000, and what requirements for additional electric generation capacity they would produce. For this purpose, let me first define a "base" case which assumes no severe fuel "crunch" and every-thing seemingly going along as before (Figure 6). Under these assumptions, the projected growth in the demand for oil and gas would just about double, going from 55 quadrillion BTU today, to 90 in 1985, and 110 in the year 2000.

In sharp contrast to this growing demand, our domestic supply of oil and gas from all sources -- known, and unknown reserves, Alaska, and synthetic fuels -- is projected to decline, reaching only 80 percent of present levels by the year 2000 (Figure 7). The resulting ever-widening gap between the supply and demand for oil and gas defines the true nature and magnitude of our energy crisis. To close the gap, we must either drastically reduce our use of energy and therefore our standard of living, or prepare to import more than one-half our needs, or improve our efficiency and find suitable substitutes for these fuels.

Adoption of significant conservation practices -- such as smaller cars, more effi-cient equipment, better thermal insulation, etc. -- will alone produce savings of as much as 20 percent (Figure 8).

GAS AND OIL SUPPLY VERSUS DEMAND TO YEAR 2000
Quadrillion BTU's
110
100
90
80
70
60
50
40
30
20
10
Unrestrained Demand
Synthetic Fuel
Total Domestic Supply
Alaska
Known Reserves
Unknown Reserves
1972
1985
2000

7

GAS AND OIL SUPPLY VERSUS DEMAND TO YEAR 2000
Quadrillion BTU's
110
100
90
80
70
60
50
40
30
20
10
Unrestrained Demand
Conservation
Synthetic Fuel
Total Domestic Supply
Alaska
Known Reserves
Unknown Reserves
1972
1985
2000

8

GAS AND OIL SUPPLY VERSUS DEMAND TO YEAR 2000
Quadrillion BTU's
110
100
90
80
70
60
50
40
30
20
10
Unrestrained Demand
Conservation
Electric Utility
Coal Substitution
Synthetic Fuel
Total Domestic Supply
Alaska
Known Reserves
Unknown Reserves
1972
1985
2000

9

10^15 BTU
20
15
10
5
0
OIL & GAS USAGE IN THE U.S. 1972
Auto
Trucks
Others
Space Heating
Process Steam
Direct Heat
Feedstocks
Water Heating
Cooking
Others

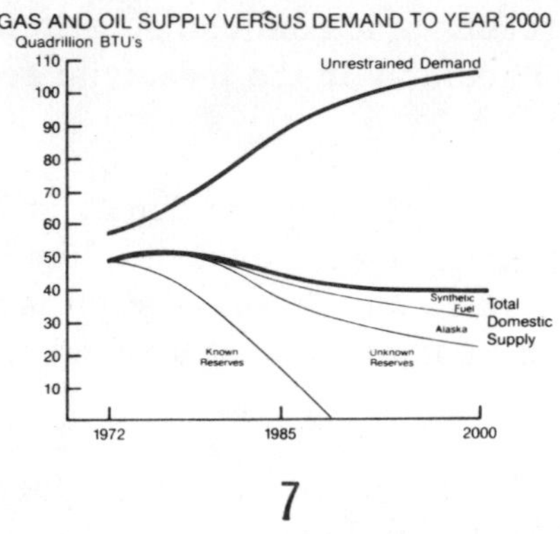

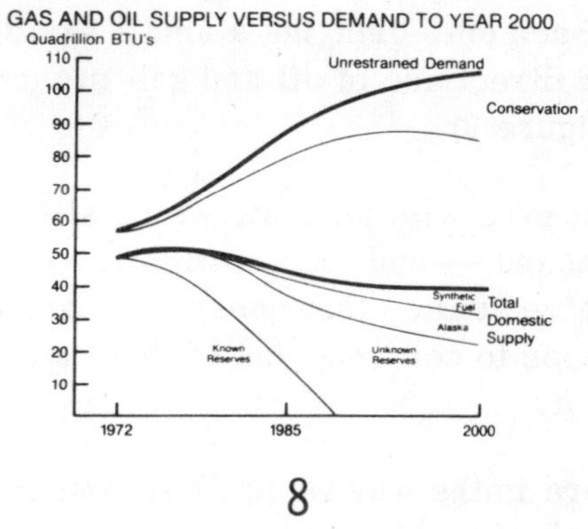

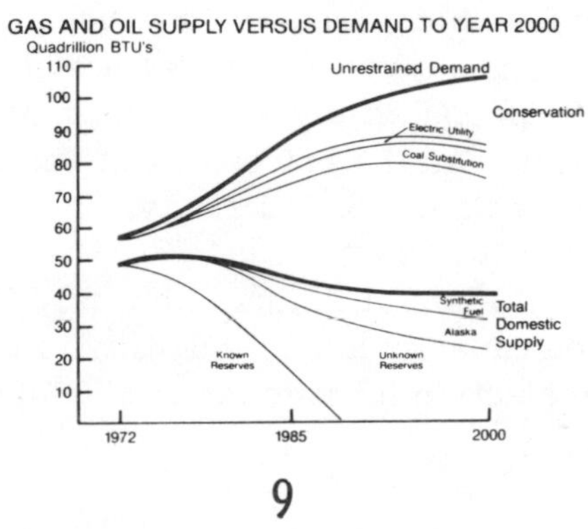

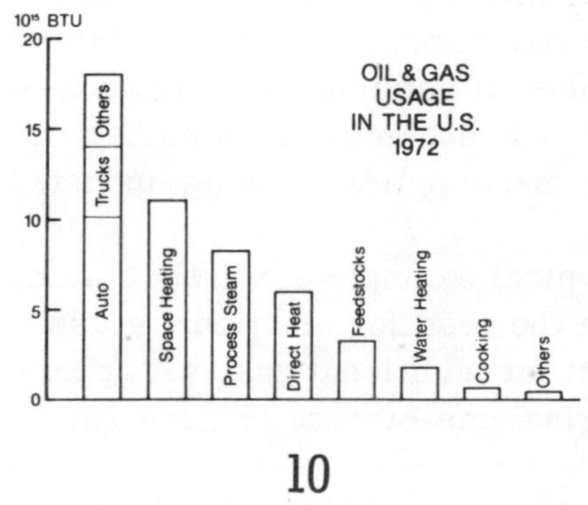

10

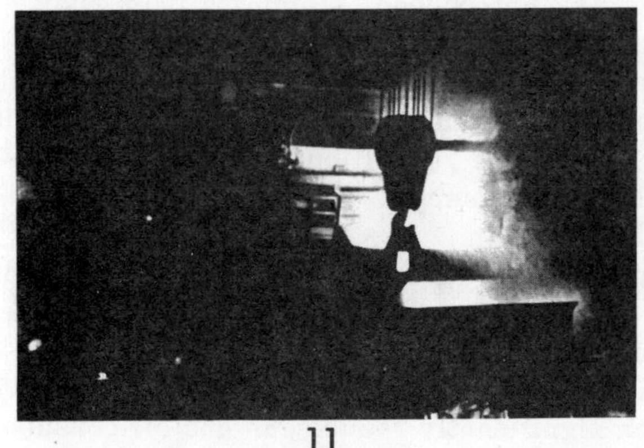

11

12

In addition, the use of gas and oil in the production of electricity will be gradually phased out, bringing a small saving of about 2 percent. And coal can substitute for direct use of oil and gas in many end uses, particularly in the industrial sector (Figure 9).

But even with all that, we're still left with a significant gap between supply and demand -- and the question is -- "Can we fill that gap by electric substitution?" We've studied that question in great detail, and we're convinced the answer is yes. I hope to convince you of that, and to discuss just a little bit of what's required to do it.

Here is the way we used oil and gas in the U.S. in 1972 (Figure 10). If we focus on the first four items -- transportation, space heating, process steam, and direct heat in industry -- we are looking at nearly 80 percent of the total use of oil and gas. If we are to achieve any significant reduction in the demand for oil and gas, we must do it in these areas.

Start with direct heat. Substitution is obviously feasible. Industry now uses electric resistance, induction and dielectric furnaces, and ovens for direct heat in a number of applications (Figure 11). They are clean and efficient, and often provide both economy and a higher quality end product. As the prices of gas and oil rise and supplies dwindle, industry will turn increasingly to electricity.

A typical example of what's coming is in our Lamp Plant at Salina, Kansas. Faced with the need for additional glass melting furnace capacity and an inability to contract for additional gas, we're now installing an electric furnace along side of the original gas furnace (Figure 12).

Electric space heating is already increasing in use, but is primarily of the resistance type. The heat pump, with more than four times the efficiency of an oil or gas furnace, should become the major means of heating and cooling space (Figure 13). It is now economically competitive over most of the nation (Figure 14). Undeniably, the heat pump had its problems when it was first introduced, stemming from poor reliability, misapplication, and inadequate service. But today we have a new plant in Norman, Oklahoma which is manufacturing a completely redesigned and thoroughly reliable unit (Figure 15). Given a strong marketing effort, the heat pump will regain acceptance in most areas of the country. We foresee 70 million homes heated electrically by the year 2000 -- 30 million with the heat pump.

Turning next to process steam, it may surprise you, as much as it did me, to learn that it is the largest single user of energy in our entire economy, accounting for one-sixth of the total (Figure 16). It is used in a wide variety of applications, with gas- and oil-fired boilers providing 80 percent, and coal-fired boilers the rest (Figure 17). Greater use of coal and electric boilers can significantly reduce gas and oil consumption for process steam, but we are developing a new device for this application which we call a Templifier -- short for temperature amplifier (Figure 18). It functions

13

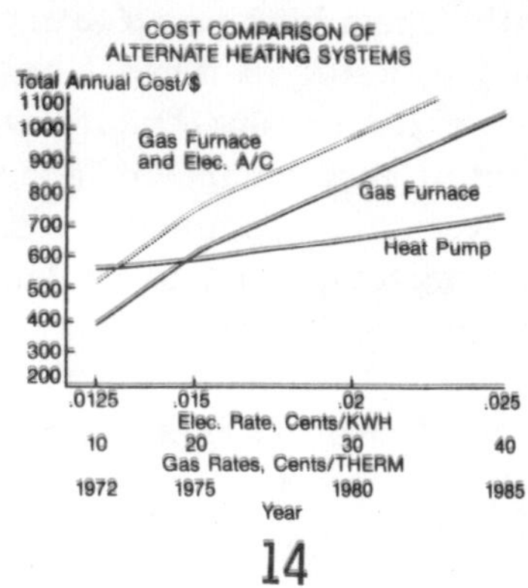

14

15

16

PROCESS STEAM CONSUMPTION
IN THE UNITED STATES—1972
(Trillion BTU)

	Coal	Gas & Oil	Percent of Total
Food and Kindred Products	189	264	5
Paper and Allied Products	482	614	10
Chemical and Allied Products	292	1161	15
Petroleum and Coal Products	–	1166	10
Primary Metal Industry	515	581	10
Transportation Equipment	197	1419	15
Other	407	2940	35
Total	1980	8150	

17

18

essentially like a heat pump except with a higher compression ratio. Utilizing any large body of water as a heat source, the Templifier can produce steam at temperatures from 250 to 400°F which will meet most industrial requirements and at a coefficient of performance of about 300 percent. Any body of water will do -- a lake, a river, the ocean -- or waste heat from electric power plants. We're now building a prototype unit to replace an existing gas-fired boiler in our Transformer Plant in Muncie, Indiana.

Last on our list of major uses of gas and oil is transportation which consumed nearly 60 percent of our oil last year (Figure 19). This is the critical test for electric substitution.

Throughout the world, 20 million dollars a year is being spent on electric battery research. From that will shortly come new batteries with double the performance of the present lead-acid (Figure 20). They'll recharge in less than half the time and last five times as long. By 1985, we foresee batteries with four times the performance of lead-acid, and before the end of the century that figure will be ten.

At our R&D Laboratory, three experimental battery-powered vehicles have been in operation for the past year and a half. Two of these vehicles are designed for light delivery service (Figure 21). They are fully licensed and have accumulated nearly 5000 miles on the road. Included in our test program has been the operation over actual residential mail delivery routes in the very hilly area near the Center (Figure 22). During such test runs they travel an average of 21 miles, make 350 starts and stops, and use only 12 kwhrs (Figure 23). That's only 11 percent of the energy in the three gallons of gasoline used by the regular post office van of the same size. This test program has demonstrated conclusively that battery-powered vehicles are capable of effectively, conveniently, and safely performing missions of this nature.

We expect the electric vehicle will find initial acceptance and application as a light van for urban service in fleet-type operations (Figure 24). By this I mean mail and parcel delivery, appliance servicing, telephone and electric utility installation and maintenance, taxis, and meter reading. I expect there will be several thousand such vehicles in operation throughout the country within a few years. With this as a start, I believe it would be possible to have 5 million electric vehicles of all types by 1985, and 100 million by 2000.

Also, by then, "horizontal elevators," like those now in use at the airports in Tampa and Seattle-Tacoma, will provide free transportation in urban centers, much as vertical elevators do in tall buildings today (Figure 25). In addition, electrified trains will have captured half of the intracity freight from trucks, and 10 percent of the passenger traffic from airplanes. Electrified rapid transit will take about 12 percent of the commuters out of automobiles.

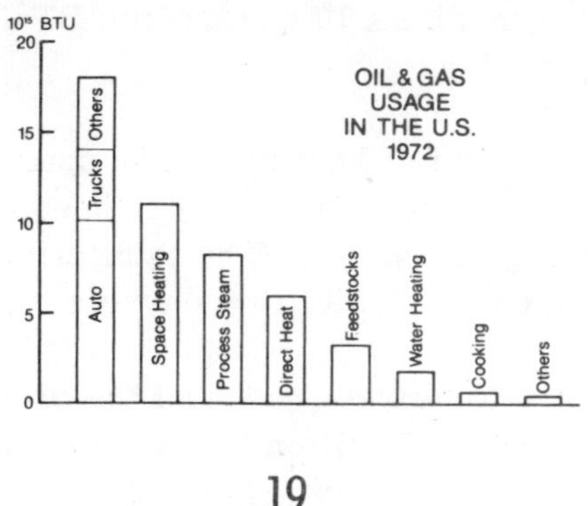

19

	Lead Acid	Battery-A
ENERGY DENSITY Watthours/Lb.	12	20-25
CHARGING TIME	8 hrs.	3 hrs.
LIFE	1 yr.	5 yrs.

20

21

22

ELECTRIC POST OFFICE VAN TEST MISSION

Distance	20 Miles
Starts & Stops	318
Energy Consumed	12 KWH

23

24

Returning now to our oil and gas demand/supply imbalance, let's look at the reduction in oil and gas demand these various substitutions will make possible (Figure 26). In transportation, the savings could amount to as much as 10 quadrillion BTU equivalent to 1.5 billion barrels of oil per year (Figure 27).

In the residential and commercial sectors, application of heat pumps to space heating, coupled with easily achieved electric substitution for other energy needs in these sectors, offer major savings in oil and gas (Figure 28). The total effect would be to reduce demand by the equivalent of another 1.5 billion barrels of oil.

In the industrial sector, greater use of electricity for direct heat and the production of process steam offers potential savings of more than 2 billion barrels of oil -- bringing total savings to more than 5 billion barrels annually (Figure 29).

From all of this it is clear that the electric economy can have a tremendously favorable impact on the oil and gas supply/demand imbalance and the imports problem (Figure 30). It is even possible to contemplate that we could become self-sufficient in our oil and gas supply by the year 2000.

Now let's examine what all this means in terms of requirements for new electric generating capacity. Here is what we call the "base case" -- which is the current conventional projection of generation capacity, assuming that there is no shortage of fossil fuels (Figure 31). This base case puts total capacity at 2000 gigawatts by the year 2000. But to take care of all the substitutions we have discussed, total capacity rises to 3000 gigawatts -- an increase of 50 percent over the conventional projection (Figure 32).

The greatest impact on new electric power plant construction will probably come about the middle of the 1980's -- the period for which we should be making plans about now. During those years, we would need to add generating capacity at about twice the annual rate assumed by the conventional forecast. Such a job is not to be taken lightly. But I believe it is well within the capability of the industry. The financial requirements are heavy but manageable, and the equipment manufacturers have the resources to produce the greater volume.

Developments in equipment will help cut the task down to size also, and one of these may be of special interest to you. I refer to the offshore floating nuclear power plant (Figure 33). This concept offers great promise towards solving the cost, siting, and environmental problems; and I am very encouraged by its growing acceptance by the electric utilities.

Taking into account the fuels used to produce the electricity, here then is a summary picture of the total energy economy for the base and for the nuclear-electric economy in the year 2000 (Figure 34). The total input of 200 quadrillion BTU is about three times the 1972 level. The major shifts in fuel usage are evident. In the nuclear-electric economy, direct combustion of oil and gas account for only 15 percent of

25

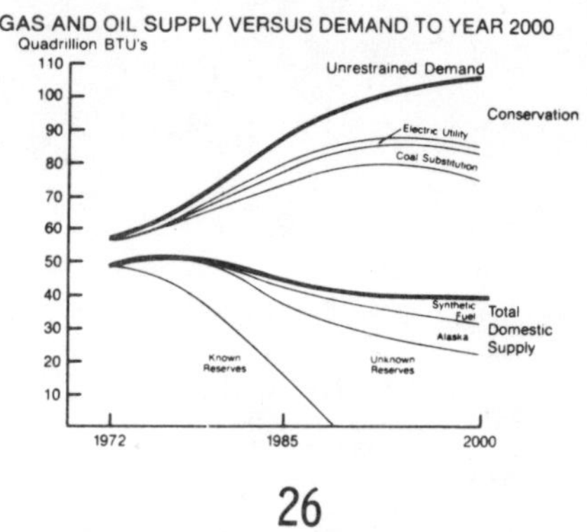

26

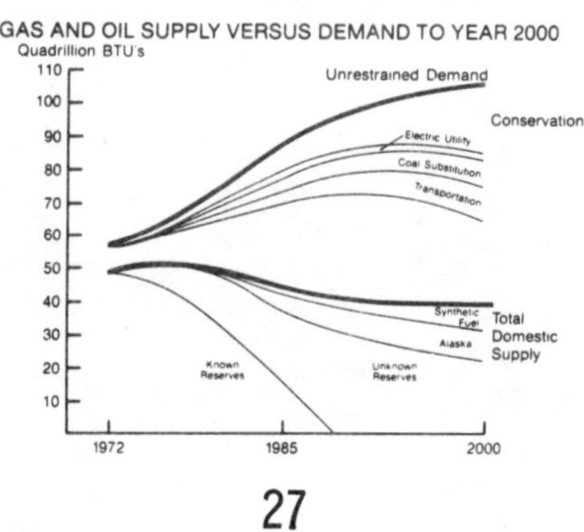

27

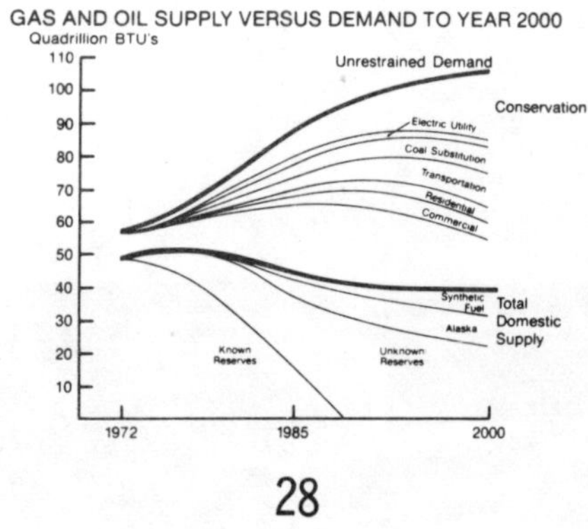

28

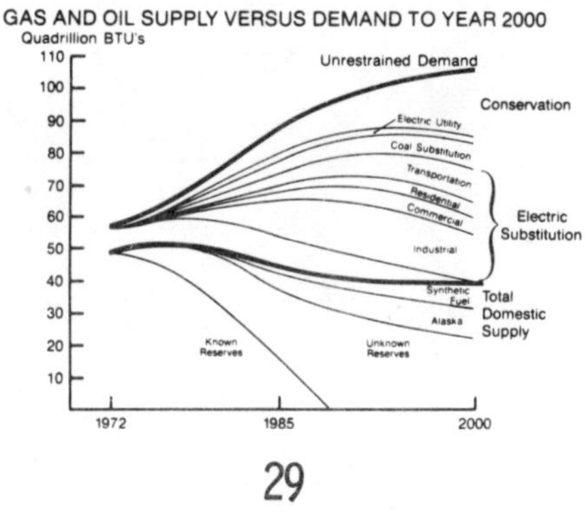

29

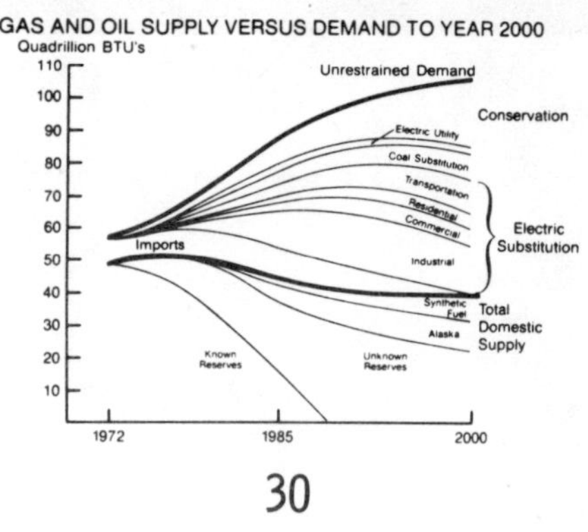

30

GIGAWATTS
3000
2500
2000
1500
1000
500
0
U. S. INSTALLED GENERATION CAPACITY
BASE CASE
1972
1985
2000

31

GIGAWATTS
3000
2500
2000
1500
1000
500
0
U. S. INSTALLED GENERATION CAPACITY
Industrial
Commercial
Residential
Transportation
BASE CASE
1972
1985
2000

32

33

TOTAL ENERGY ECONOMY—YEAR 2000
Quadrillion BTU
200
150
100
50
0
ELECTRIC GENERATION
Nuclear
Other
Oil
Coal
Coal Gasification
Natural Gas
DIRECT USE
Oil
Coal
Syn. Fuel from Coal
BASE
Nuclear
Natural Gas
Other
Oil
Coal
Coal Gasification
Natural Gas
Oil
Coal
Syn. Fuel from Coal
NEE
ELECTRIC GENERATION
DIRECT USE

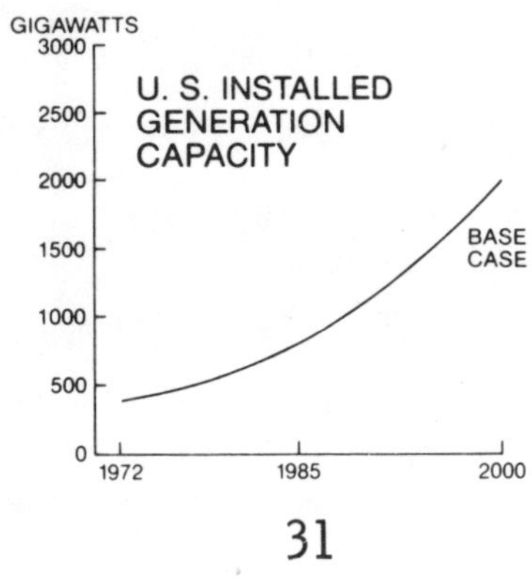

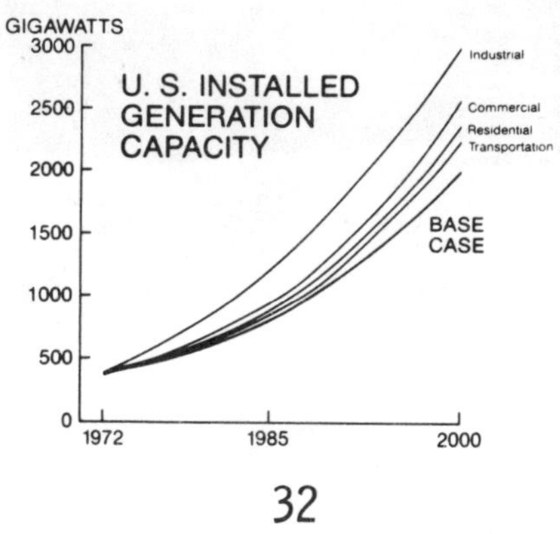

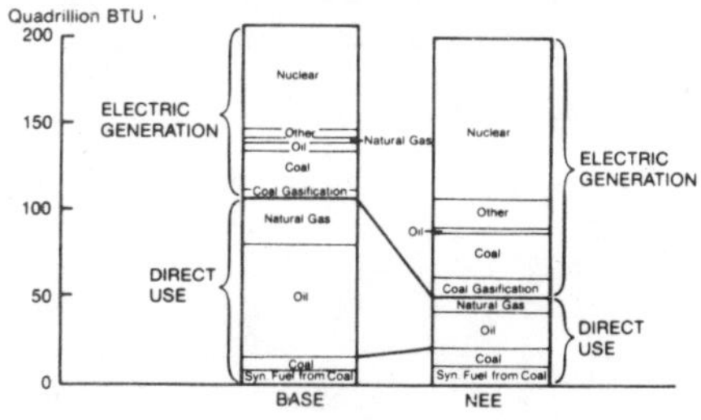

34

the total energy input, sharply down from 45 percent for the base case. Uranium's
contribution grows from 30 percent of the total in the base case to almost one-half
in the nuclear-electric economy. Coal is clearly of key importance to the nuclear-
electric economy, providing 30 percent of the total energy. And most striking of
all is the shift to dependence on electricity for 75 percent of the total energy input
to the economy.

What, then, does all this mean? Well, I think it means that we must take a long,
searching look at re-ordering our energy goals and priorities. We should place
our emphasis

> .... on mass transit, <u>not</u> on MHD;
> .... on coal liquefaction and gasification, <u>not</u> on geothermal systems;
> .... on uranium enrichment, <u>not</u> on fuel cells;
> .... on heat pumps and more uses of electricity, <u>not</u> on fewer;
> .... on the fast breeder and fusion, <u>not</u> on orbiting solar power plants;
> .... on electric cars, <u>not</u> on Wankel engines!

The logic of the case is overwhelming. If our nation is not to settle for a low-
energy, low-vitality world, then our economy must be built on energy sources
other than gas and oil. Only two such energy sources are now available -- coal
and uranium. The electric economy is an absolute corollary of using coal and
uranium. Even later, if we find a way to use fusion, or solar energy, or geothermal energy, or tidal energy, or the wind, electricity is the only way to put
these sources to work. Whatever the source, there is quite literally no alternative to an electric economy -- except a declining economy.

This is the challenge -- we must begin now to develop the electric energy economy
as the answer to our limited supplies and finite reserves of oil and gas. We should
henceforth measure all decisions by the standard of whether they help or hinder the
achievement of this new energy system. Recognition of the dominant roles that
electricity, nuclear fuels, and coal will play in meeting our future energy needs
should be the crux of our national energy policy -- the central thrust of our future
energy planning. The time is short; the matter is of the utmost urgency. It
demands a national commitment akin to that which put men on the moon.

THE PROSPECTS FOR LARGE-SCALE
BREEDER REACTORS*

B. Wolfe and R. Palmer[+]

The 1962 AEC Report to the President on nuclear power
proposed that breeder reactors be developed on a time
scale such that the cumulative use of uranium by the
thermal reactors would be limited to the very high
grade ($10/lb) uranium deposits. This objective can
no longer be met. This paper examines the timing of
the breeder which would avoid the utilization of our
very low grade uranium deposits such as the Tennessee
shales.

The early development of economic breeders and their
large scale application in the late 80's could reduce
cumulative uranium requirements so that use of very
low grade uranium deposits would not be necessary.
Breeder introduction on this time scale could, by the
end of the century, eliminate the need for additional
uranium mining for centuries to come. This objective
appears technically achievable but would require a
change in the pace and orientation of the present
national breeder development program.

* A complete copy of this paper appears in Brochure NEDO-14012, General
  Electric, Advanced Technology Dept., 310 DeGuigne Drive, Sunnyvale,
  California 94086

+ General Electric, Advanced Technology Dept., Sunnyvale, California

THE OUTLOOK FOR FUSION ENERGY SOURCES —
REMAINING TECHNOLOGICAL HURDLES*

R. F. Post**

Controlled fusion research is now in its third
decade. Through this effort most of the critical
scientific issues for the main approach to
fusion — magnetic confinement — have been
resolved. Remaining critical scientific questions
are mainly quantitative in nature, and have to do
with residual instabilities of the confined
fusion plasma. The status of the research is
such that there is a strong basis for believing
that these questions will be satisfactorily
answered in the next 8 to 10 years, to the point
that fusion based on magnetic confinement could
be assured of success. Laser fusion, much
newer on the scene, also has critical quantitative
issues to resolve, issues that will be addressed
in the next few years. Despite the fact that
fusion reactors do not yet exist much of the
technology that will be required for fusion
reactors is perforce being developed to satisfy
the demanding needs of the scientific phase.
Through this entre into fusion technology we
can begin to visualize fusion reactors, their
unique features (vis a vis other power generation
means), and their technological problems. As
power sources they would have many advantages,
both environmentally and operationally (particularly
in lowered hazard potential) over conventional
nuclear power. Their unique problems would seem

---

* Work performed under the auspices of the U.S.
Atomic Energy Commission.

**Lawrence Livermore Laboratory, University of
California, Livermore, California.

to be those associated with vacuum problems,
i.e., with protection of the reacting plasmas
from contamination, with plasma heating and
refueling, and with plasma chamber wall deterioration
under neutron or charged particle bombardment.
These issues appear difficult, but solvable, by
design and materials choices. Their resolution
will probably set the time scale for the practical
relization of fusion power following the completion
of the scientific phase. It is therefore important
to pursue now the technological issues, in parallel
with the present scientific studies, to move the
advent of fusion power to the earliest possible
date.

INTRODUCTION

Fusion research has not yet realized its stated goal. However, the
search for fusion has been underway at a rising level of effort for more
than two decades. As a result, fusion research has reached a degree of
maturity sufficient to allow informed speculation as to its probable
future. In particular, it is widely felt that most of the critical
scientific issues for fusion have by now been resolved and that the
remaining ones should be resolvable in the next 6 to 8 years, given a
sufficiently intense effort. In the discussion to follow I will present
a brief analysis of the present status of fusion research and in the light
of that discussion will attempt to define those scientific-technical issues
that still stand in the way of the achievement of fusion power.

The scientific goal of fusion research is to provide the technical knowledge
required to achieve a net positive energy release from nuclear fusion
reactions. The practical goal of the research is to demonstrate the
viability of fusion as an environmentally acceptable primary energy
source suitable for the generation of electricity at an economically
competitive cost. Important though it is, the achievement of the first
goal, the scientific one, is not enough; the practical requirements of
the environment and of economics must be met or else fusion research will
not have achieved its aim. Fortunately, fusion has many favorable factors

going for it relative to these latter requirements. Fusion fuel would be of near-zero cost and virtually inexhaustible. Furthermore, the adverse environmental effects of fusion would be minimal at worst, and would tend to decrease as the technique evolved. But the scientific-technological hurdles that remain between our position now and the first demonstration of economic fusion power are formidable, carrying with them considerable uncertainties. As a consequence we find that both the U.S. fusion program, and that of the other nations with major fusion programs — the U.S.S.R., Japan, the U.K., and other in the European Economic community — are characterized by a substantial degree of diversity, both as to approaches and as to emphases. Neither the scientific nor the technological features of fusion are so well understood that one approach to fusion can be singled out and all others shelved. Nor should this be the case. Fusion is both too fraught with unanswered questions and at the same time too broad in its potentialities and implications to narrow the scope of fusion research, either now or for the forseeable future. Beyond the elimination of clearly impractical approaches it seems imperative that fusion research should be carried out on a broad front.

THE BASIC APPROACHES TO FUSION

Two general approaches to fusion power are being pursued. These differ radically both as to philosophy and as to the technology required. The older approach is magnetic confinement: the idea of holding a heated fusion fuel plasma by means of intense magnetic fields of special configuration. The new approach is laser pellet fusion: the idea of heating and densifying a tiny solid pellet of fusion fuel by means of intense pulses of energy from focused laser beams of fractional nanosecond duration. The regimes and the physics involved in the two approaches are vastly different; the fundamental issues being addressed are the same.

These fundamental issues are those of _heating_ and _confinement_. For net power to be produced in a fusion reaction, a fuel charge must be brought

up to fusion temperatures (kinetic temperatures of 100 million degrees or
higher) and be held confined without substantial loss of fuel particles or
energy for a long enough time for the energy released by the reactions to exceed
the energy input required to perform the heating. Both the past progress
and most of the yet unsolved problems for fusion can be discussed in
terms of these two issues. Let us first discuss the problem of confinement,
the one that has historically demanded the greatest effort throughout
the search for fusion.

In the two approaches to fusion the problem of confinement is tackled in
entirely different ways. In laser fusion "confinement" means the time-wise
localization of a tiny, densely, compressed pellet charge of fusion fuel
accomplished by scaling down the time duration (and scaling up the
heating power density) for the fusion process to the point that the
entire process — that is, heating, compression, and fusion — takes place
before the reacting fuel particles can escape from the tiny region where
the action is. In other words here the confinement is "inertial." The
focused laser beams (or perhaps, alternatively intense focused electron
beams) then provide the astronomical power densities needed (mega-giga
watts). Instantaneous fusion power rates would be also astronomical,
but of very short time duration, so that the total energy released would
be no more than a fraction of a kilowatt hour per fusion pulse. Repetition
of the process, using new fuel pellets dropped into the reactor chamber,
would then produce power at an average rate high enough to presumably
satisfy practical requirements for economic power generation.

At the other end of the confinement time scale is the "conventional"
approach to fusion. Here, the fusion process is conceived of as being
maintained on a continuous or nearly continuous basis, that is as a
steady combustion of the fusion fuel, carried out a rate consistent with
the steady-state power density limitations imposed by materials and by
heat transfer limitations. Thus while laser pellet fusion requires very
high densities (higher than the density of solid matter) to satisfy the

requirements for inertial confinement, for steady combustion fusion the
fuel densities required are orders of orders of magnitude lower, i.e.,
about 10^{-4} to 10^{-5} of the particle density of the atmosphere. Varying
as the square of the fuel particle density, fusion power densities are
already tens to hundreds of megawatts per cubic meter of reacting fuel
at 10^{-5} atmospheric density (3×10^{14} particles/cm^3). At 10^{-4} atmospheric,
the power densities are thus giga-watts per cubic meter, above the upper
limit for heat transfer, except for intermittantly operating systems.
But not only do power density limitations imposed by engineering and
materials set upper limits on fusion fuel densities for the conventional
approaches to fusion but there is also another, equally important,
limitation. This is the matter of the pressure exerted by the hot fusion
fuel gas. To keep this pressure within bounds, those set by the strength
of materials, the density must be lowered to about the same level (10^{-5}
to 10^{-4} of atmospheric) as the limits imposed by power density
considerations. But under these conditions the problem of confinement
indeed becomes the main issue. This circumstance can be brought into
focus, and the contrast between the technological problems of laser fusion
and those of magnetic confinement approach to fusion can be made clear,
through the so-called "Lawson Criterion" for net fusion power. This
condition simply states that under most circumstances the net positive power
condition is equivalent to the requirement that the product of fusion fuel
particle density and its mean confinement time must exceed 10^{14} cm^{-3} sec.
At pellet fusion densities this would imply times of a small fraction of
a nanosecond, compatible with inertial confinement. But at 10^{-5}
atmospheric particles density (3×10^{14} cm^{-3}) the corresponding required
confinement time is of order 0.3 sec, far too long for inertial effects
to be important. In fact, at their mean velocities of some thousand or
more kilometers per second, the nuclei of the fusion fuel charge would
travel a distance of hundreds of kilometers in this long a time, whereas
localization within at most a few meters would be required for a
practical-sized reactor. As is by now well known the only practical

means by which such localization can be accomplished under fusion conditions
is the use of intense and specially shaped magnetic fields. These fields
act on the ions and electrons of the fusion fuel plasma to keep them from
contacting the walls of the reactor chamber. The problem here is not so
much that of protecting the wall as it is of protecting the plasma; any
contact with foreign matter, either directly or through the ingestion of
even a small percentage of high atomic number impurities would quench
the fusion reaction. From this circumstance stems most of the difficulty
in achieving fusion via magnetic confinement. In the next section we
shall enumerate the specific problems, many of which have by now been
solved, that are the source of this difficulty.

Thus far we have not mentioned the specific fusion reactions that can be
considered for fusion reactors. The primary fuel for fusion will no
doubt be deuterium: heavy hydrogen. Deuterons can react with each other
or with other light element isotopes. Figure 1a and 1b show the two
branches of the D-D reaction. These occur with about equal probability
and lead to reaction products that are themselves fertile, tritium and
helium-3, as shown in Figs. 1c and 1d. The DT reaction is the one with

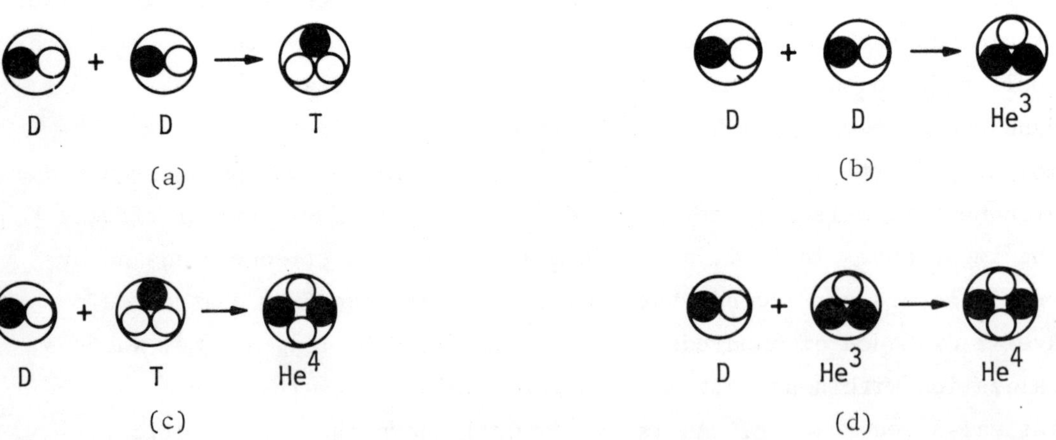

Fig. 1   Deuterium-deuterium, deuterium-tritium, and deuterium-helium-3
fusion reactions.

the lowest ignition temperature and is the one most studied for first-
generation fusion reactors. The D-helium 3 reaction, the second most
probable one, in terms of reaction cross-sections, is of special
interest because its reaction products are charged and its energy
release is high. For these reasons D-^{3}He is a particularly attractive
candidate for a reactor with direct conversion. There are also other
reactions, involving lithium and boron isotopes, that may some day be
useful for their special properties, such as the absence of neutron
reaction products. Perhaps the main point here is that fusion not only
has an almost infinite abundance of cheap primary fuel but that it also
in time may allow the choice of a variety of fuel cycle combinations,
selected for their particular advantages.

THE MAGNETIC CONFINEMENT APPROACH TO FUSION

From the start, the use of magnetic fields to hold a hot fusion plasma
has appeared to be an almost ideal solution to the confinement problem.
That is, magnetic fields should provide a nonmaterial means both for
sustaining the outward pressure (perhaps as high as thousands of pounds
per square inch) of a fusion fuel plasma, and at the same time for
inhibiting its diffusion to the reactor chamber walls, in theory for
times even longer than those required for a fusion reactor. Furthermore,
that same theory (now more than 25 years old) showed that magnetic
confinement should improve with increasing temperature, certainly an
ideal state of affairs for fusion. We have therefore had before our
eyes an apparently almost ideal solution to the fusion problem for more
than 20 years — and still we do not today have a working fusion reactor.
The problem is that the theory that predicts these marvelous results
makes one crucial assumption: that the confined plasma is quiescent,
that is, that it behaves the way you would expect a gas to behave while
sitting quietly in a pressure vessel; quiescent except for the normal
uncorrelated random motion of its molecules. But a plasma is a gas of
charged particles, coupled to each other through long-range electric

forces that can (and do) result in collective currents and electric fields.
When these currents and electric fields become self-regenerative magnetic
confinement can be either rapidly destroyed or at least seriously weakened.

The 20-year search for fusion via magnetic confinement has been first a
search for understanding of the nature of hot plasma and its modes of
instability, followed by the discovery of means and situations that avoid,
or at least weaken, the effect of these instabilities to a tolerable point.
Many of us feel that this task is now almost complete.

Two general approaches to magnetic confinement have stood the test of
time and now form the basis for most of the magnetic confinement fusion
research now in progress. They are, as shown in Fig. 2, either "closed"
toroidal confinement, the idea of closing magnetic fields on themselves
so that there are no ends for leaking, or alternatively the "open ended"
magnetic mirror machine: by capping the ends of the field with stronger
fields particles are reflected back and forth, perhaps thousands of
time, before they leak out through collisional effects. But in their
earliest versions both of these simple, neat, and intuitively obvious
solutions to a complicated problem failed: the reason, dumping of the
plasma in microseconds by violent plasma instability.

Briefly, hot plasma in a magnetic field can exhibit two general classes
of instabilities: gross or hydromagnetic, or wave-turbulence instabilities.
The first kind is the most violent. It can arise whenever the plasma is
given a chance to expand in a direction of weakening magnetic field
strength. All early approaches to fusion fell into this trap, and
failed accordingly. The answer, which was found in the 1960's, lies
in complexifying the shape of the magnetic field so that MHD instabilities
cannot occur. Two general principles were found to work: the <u>magnetic
well</u> idea and <u>magnetic shear</u>.

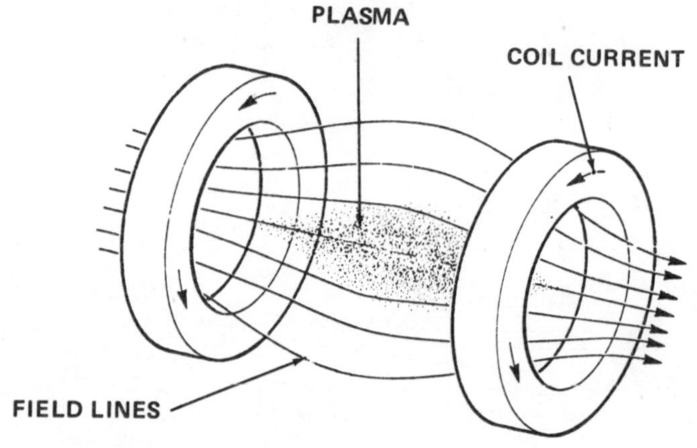

OPEN SYSTEM - SIMPLE MAGNETIC MIRROR

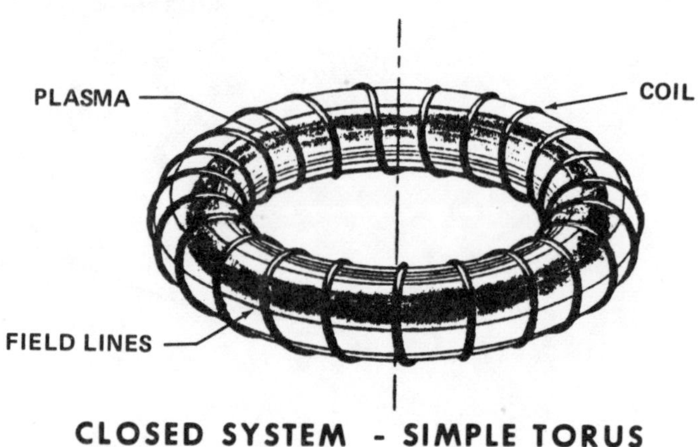

CLOSED SYSTEM - SIMPLE TORUS

Fig. 2 Magnetic containment configurations.

The magnetic well idea works especially effectively for the mirror machine.
Figure 3 shows a magnetic-well mirror field made by a "baseball" coil
(shaped like the seam on a baseball). Plasma sitting inside this coil
finds itself at a low point for motion in any direction. In this way
gross instabilities are made energetically impossible. The plasma can
still grumble, but that is another story.

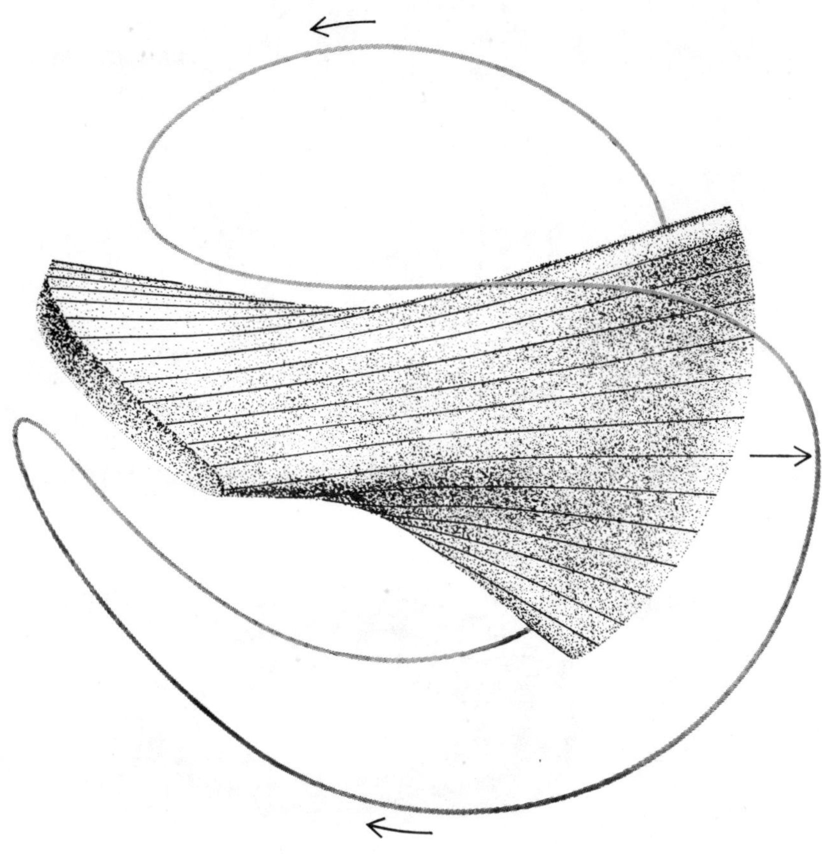

Fig. 3 Magnetic well mirror field as produced by a "baseball" coil.

The second method of controlling MHD (and other) instabilities is <u>magnetic shear</u>; arranging for the field lines of the confining field to have a kind of basket weave character as shown in Fig. 4. This trick can be accomplished either by adding spiral helical windings outside the plasma or by having induced currents flow inside the plasma as well as in outside conductors.

Between magnetic well and magnetic shear effects MHD instabilities can be suppressed. What can remain are plasma turbulences; that is amplified

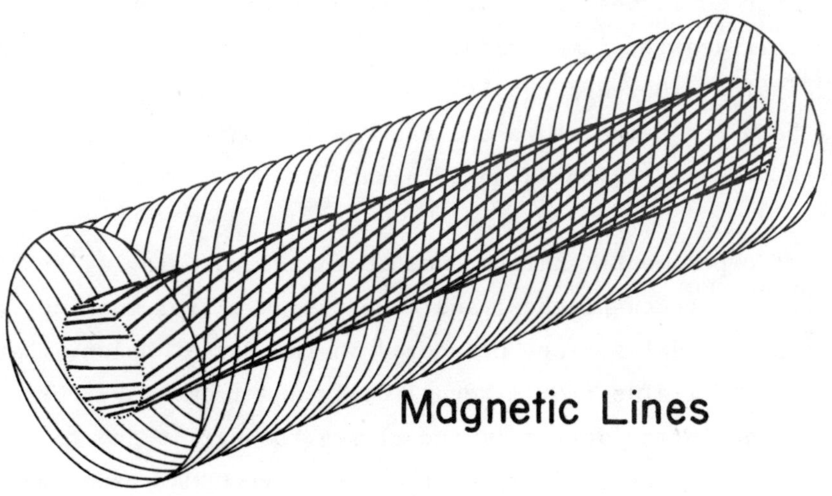

Fig. 4 Magnetic lines of a magnetic field with shear.

waves or fluctuations in the plasmas. Though they cannot "dump" the plasma
as can MHD instabilities, they can vitiate confinement by speeding up
the rate of diffusion out of the confining field. Means have been, and
are being, found to suppress wave turbulence in both toroidal and mirror
systems. These means involve both the shaping of the magnetic field and
the control of the plasma conditions and conditions at the plasma
boundaries.

The upshot of the confinement research today is that experiments in both
toroidal and mirror systems have demonstrated plasma confinement that
comes close, in some cases very close, to the theoretical ideal confinement
times calculated at the densities and temperature of the experimental
plasmas. However, while these results _do_ prove the feasibility of the
idea of magnetic confinement they _do not_ prove the scientific feasibility
of fusion reactors that might be based on these approaches. This is
because the experiments generally are carried out at plasma temperatures
or densities that are still well below those that would be required for

a reactor. For these experiments the density-confinement time product
is thus not yet up to reactor values. As I said earlier, for net positive
fusion power this product must be somewhere in the range of 10^{13} to 10^{14}
(particles per cubic centimeter times seconds) to achieve net power.
Presently this density-time product is still a factor of 100 or more too
small for fusion; 10 years ago it was 10,000 to 100,000 times too small.
But until both the temperature and the confinement factor reach reactor
values, the scientific job is not finished.

The kinds of experiments that have achieved the encouraging results are
called tokamaks, theta-pinches, and compression mirror machines. The
tokamak is a toroidal device, pioneered in the Soviet Union, that uses
a combination of fields from external coils and from heavy currents
induced by transformer action in the plasma to heat it and to shape the
field. Figure 5 shows an example; the Oak Ridge ORMAC experiment. There
are by now many tokamaks in laboratories throughout the world.

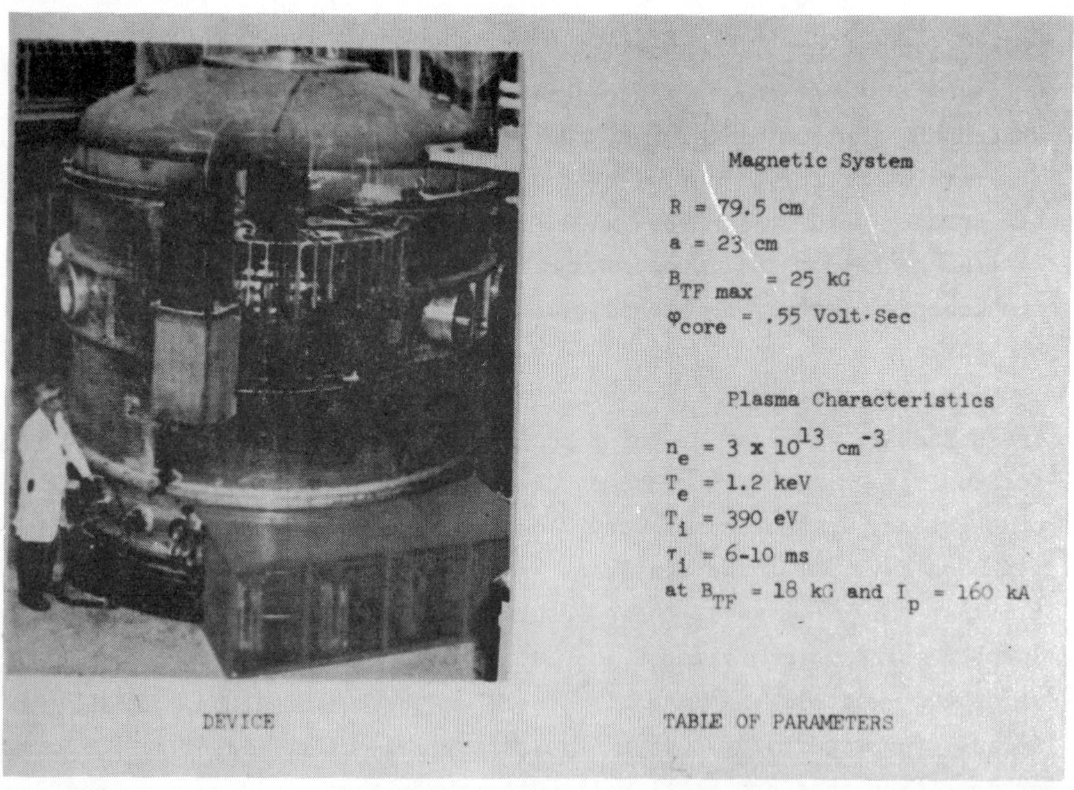

Fig. 5 The Oak Ridge ORMAC tokamak experiment.

The second device to achieve high temperatures and good stability is the
theta-pinch. In this device the plasma is shock heated and compression
heated by a rapidly rising field produced by a pulsed coil. Figure 6
shows a stable compressed column in a linear theta pinch at Los Alamos.
A large toroidal version is now coming on line at the same laboratory.

The mirror device that has produced the best plasma results for that
approach is the 2XII experiment at Livermore, shown schematically in
Fig. 7.

Generally speaking all of the above experiments are still too small in
size or too limited in magnetic field strength to contain a reactor-grade
fusion plasma. Next generation experiments that should come much closer
are either under construction or in planning. An example is the
Princeton PLT (Princeton Large Torus) that is to be completed in about
2 years. The critical question for all of these new experiments will
be: will the presently favorable results continue to be confirmed as
plasma temperature and density are increased?

PLASMA HEATING — LASER FUSION

Before speculating on what technological hurdles are probably yet to be
encountered, I would like to say something about plasma heating. Heating
and confinement are the two issues, and major progress has been made in
confinement. Heating is now receiving greater emphasis and is responding
to this effort.

In one approach, the laser pellet, shown schematically in Fig. 8, the
main issue is heating. Pellet fusion relies on the idea of focusing an
enormous power density, from lasers or possibly from high current
electron beams, onto a tiny frozen pellet of fusion fuel. The laser
beams must then heat and densify the beams so rapidly that fusion reactions
can take place before the pellet flies apart. The main demands here are
quantitative: very high laser powers are required to make the idea work.

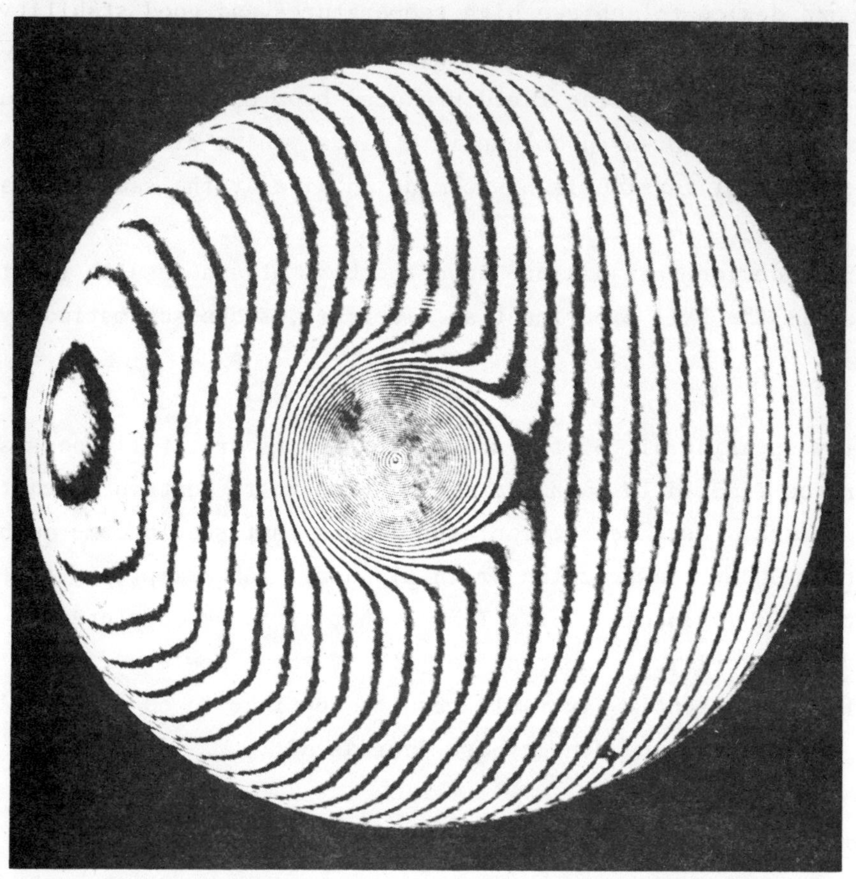

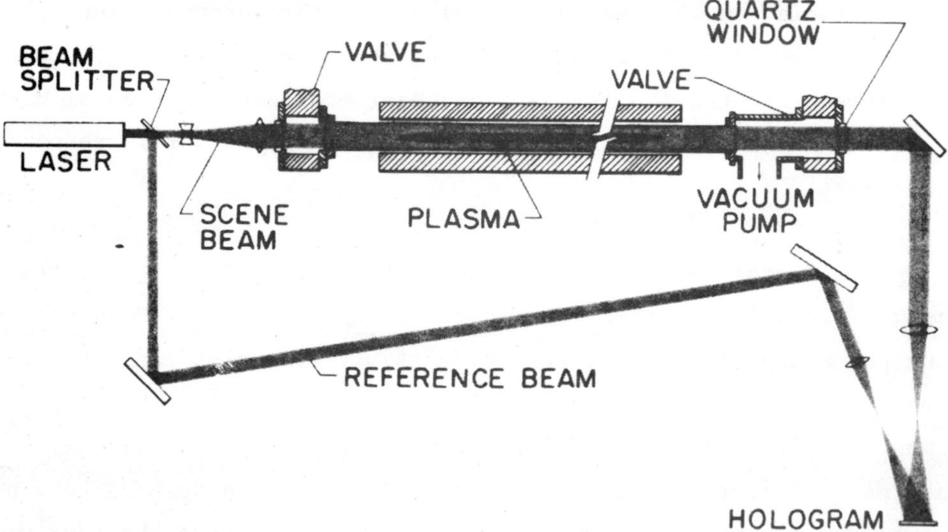

Fig. 6 Laser interferogram of stable compressed plasma column in Los Alamos theta pinch.

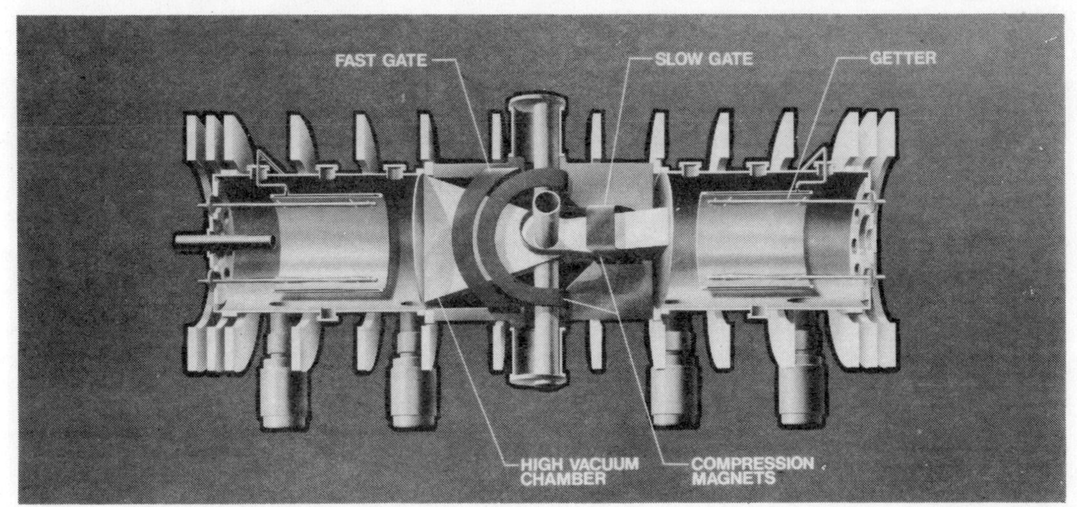

Fig. 7 Schematic of 2XII/mirror experiment at Livermore.

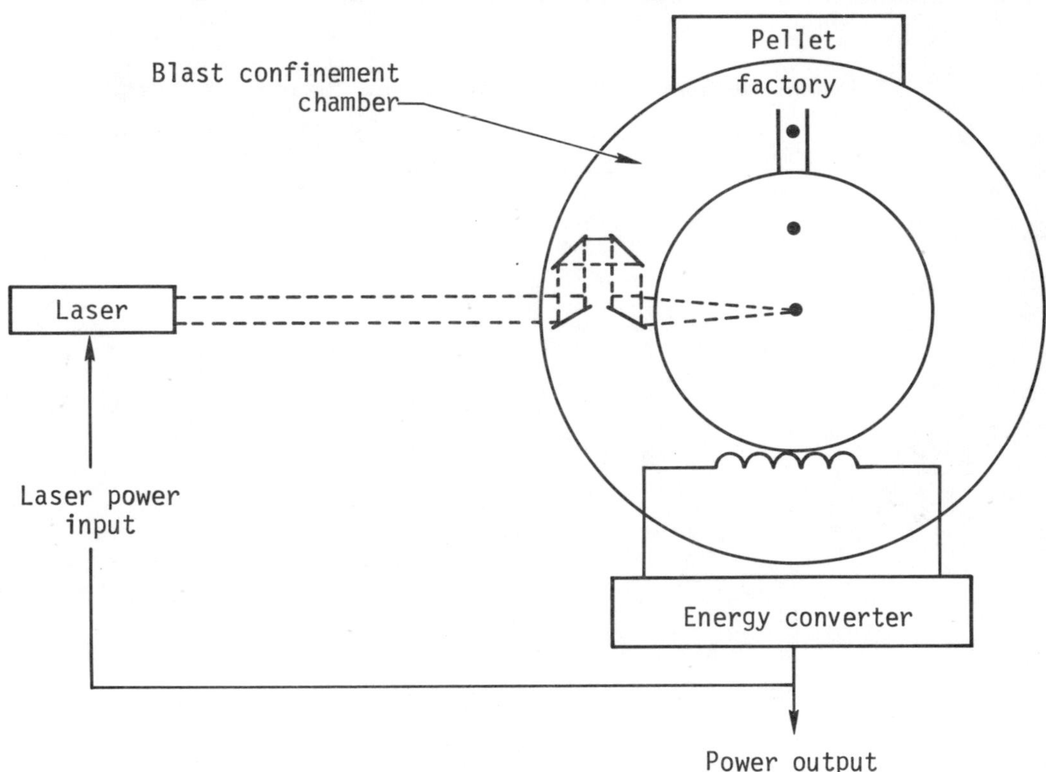

Fig. 8 Conceptual laser fusion reactor.

Figure 9 shows a proposed laser experiment at Livermore and Fig. 10
shows some of the numbers that must be attained in trying to achieve an

Fig. 9 Model of proposed Livermore laser fusion experiment.

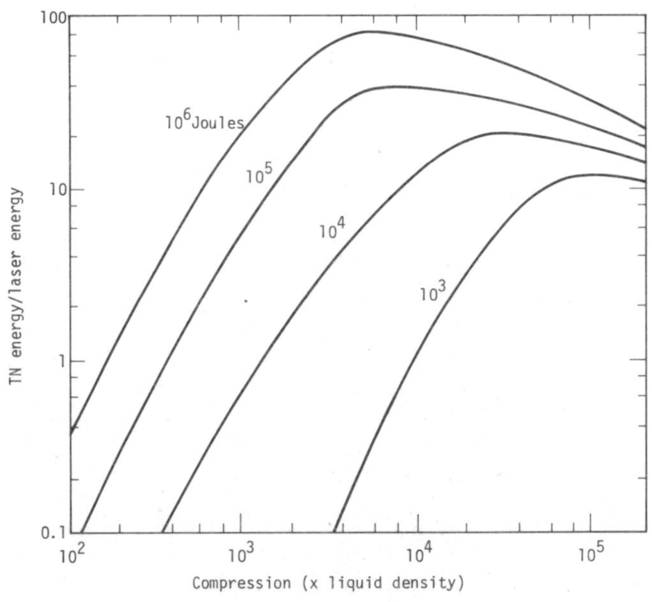

Fig. 10 Example of compression factors required in laser fusion.

316

energetic breakeven condition. The issues are both technological
and scientific: will the heating and densification proceed without
interference from plasma instabilities encountered en route? It will
probably be some years before the answers to these questions will be
known.

In the 2XII and theta-pinch experiments mentioned earlier compression
heating has worked very well. It has also been successfully applied
in the tokamak, in the Princeton ATC (an acronym for Adiabatic Toroidal
Compressor) experiment, shown in Fig. 11. Here a large diameter plasma
is first prepared, as in an ordinary tokamak. Then the plasma is moved
inward and squeezed by controlling the external confining field, thus
both densifying and heating it.

Fig. 11 Model of Princeton ATC tokamak experiment.

One of the most important developments in plasma heating is the
technique of neutral beam heating. This technique was developed for
mirror machines but is now being applied in tokamaks, such as ORMAC and
ATC. The idea is illustrated in Fig. 12, which is a drawing of the

Fig. 12   Baseball II superconducting neutral beam/mirror experiment at
          Livermore.

Baseball II experiment at Livermore. A beam of energetic neutral atoms
is made by neutralizing a focused beam of ions from a high-current ion
source by passing the ions through a low-density gas jet. The neutral
atom beams formed in this way then can freely cross the magnetic field
and plunge into the confined plasma. Here its atoms are broken up again
into ions and electrons and join the trapped plasma, in this way bringing
in both new particles and kinetic energy.

As a result of intensive development some new and powerful neutral beam
sources have been developed and are being applied in fusion confinement
experiments. Figure 13 shows a photo of 50 A (equivalent), 20,000 eV

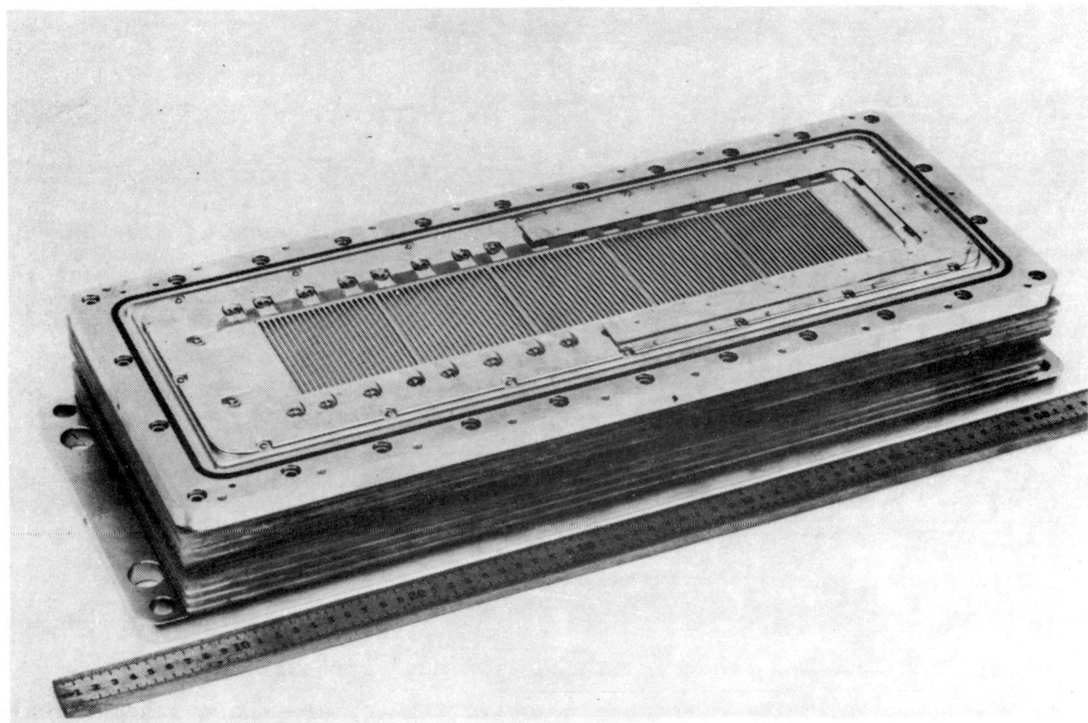

Fig. 13   Fifty-ampere neutral beam source module developed at Lawrence
          Berkeley Laboratory.

source module, that is a 1-MW beam, developed for 2XII. Twelve of these
sources, that is 600 A and 12 MW of beam, will be installed in the new
2XIIB experiment now being assembled. Sources like these and others on
the drawing board should enable us to heat plasmas in mirror machines
and tokamaks up to fusion reactor temperatures if the confinement
conditions remain favorable.

As a result of the developments in plasma heating techniques, either already accomplished or in process, it seems clear that we will have in hand the means for creating and maintaining reactor density plasmas at fusion temperatures. These techniques are being incorporated into present and upcoming fusion experiments. Heating will therefore almost surely not be the main issue in these experiments. The issue will remain that of confinement.

REACTOR CONCEPTS

At this stage of fusion research no one really knows what a fusion reactor will look like. Nevertheless, based on the emerging favorable picture for plasma confinement it is possible to visualize such reactors, assuming continuing success with the scientific phase of the research.

Conceptually, the laser fusion reactor is probably the simplest of all. We have already discussed some of the issues in laser pellet fusion. Probably the next simplest reactor would be some kind of toroidal confinement system, preferably steady-state, containing an ignited DT fusion plasma, that is, one that has somehow been heated to fusion temperature and is then maintained at this temperature by energy from the charged reaction products (the helium nuclei resulting from the DT reaction). The elements of such a reactor are shown schematically in Fig. 14. Note that in this example the tritium is provided by a fuel cycle involving the capture of the DT neutron in lithium, followed by the recovery of the resulting tritium neutron capture product. In this example the cycle involves only thermal conversion, since most of the energy of the reaction shows up in the blanket as heat resulting from the capture of the 14-MeV DT neutrons.

To be frank, we do not even know _conceptually_ how to build a reactor of the kind shown. No steady-state toroidal system has been made to work: Tokamaks are inherently pulsed devices, albeit potentially with very

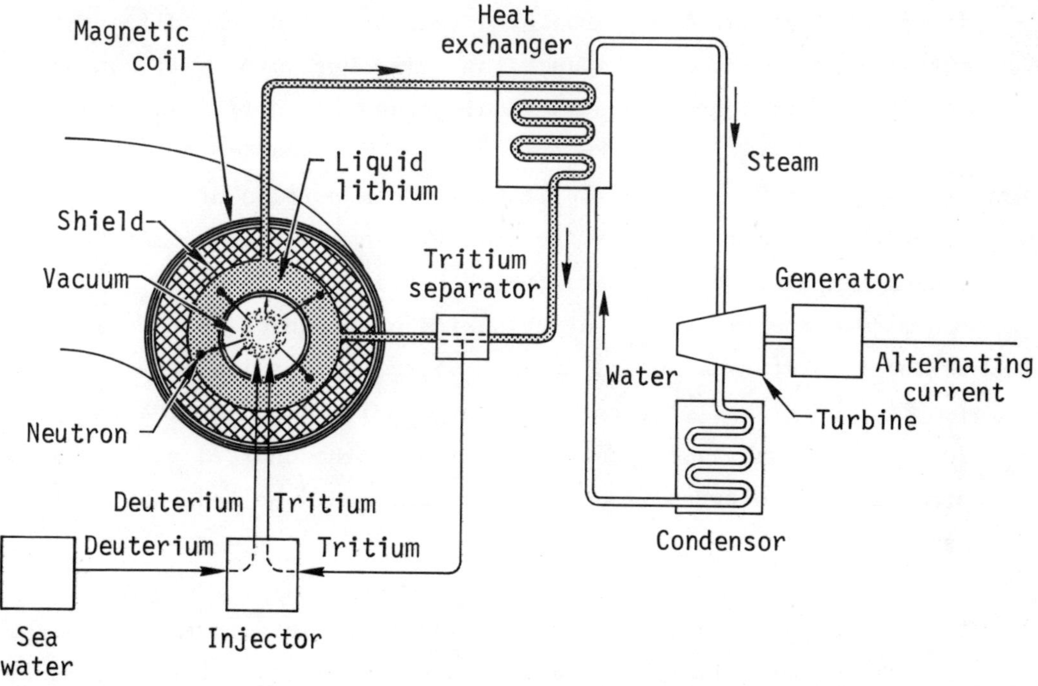

Fig. 14   Schematic of a toroidal fusion reactor using the deuterium-tritium
fuel cycle.

long pulses (several seconds). Furthermore, we do not really know how
to refuel a steadily burning fuel with cold fuel, although the injection
of frozen pellets into the plasma (sometimes called the "snowball in hell"
technique!) has been examined in a preliminary way.

One way around such difficulties is the pulsed theta-pinch reactor. This
approach assumes a repetitively pulsed system involving shock heating of
fuel charge followed by magnetic compression and burning, followed by
decompression and recovery of a portion of the magnetic energy for the
next cycle. Studies of such a reactor concept are being made at Los
Alamos. The most difficult technical questions involved are those
concerning the shock heating system and the pulse energy storage systems.

In studies of possible mirror reactors our problems are seen to be
different ones and our approach to them has also been different. To
obtain a positive power balance in a mirror reactor will not only
require a high degree of plasma quiescence but will also require highly
efficient heating and energy recovery. We visualize a mirror reactor
as being a sort of power amplifier where a portion of the electrical
power generated is fed back to neutral beam sources to maintain the
plasma temperature and to replenish particle losses through the mirrors.
To improve the efficiency of this kind of feedback loop we are studying
in the laboratory a method for the direct conversion of particle energy
leaking out of the mirrors. In our case it would involve the electrostatic
deceleration and subsequent collection of the charged particles to
produce high voltage direct current that would be fed back to the neutral
beam sources.

The general idea involved in this kind of direct conversion is shown in
Fig. 15. It involves an expansion of the particle stream (something
like a turbine expansion nozzle, except that it is magnetic) followed
by separation of the electrons from the ions and the collection of the
electrons at the negative terminal of this "high-voltage battery." The
ions are then decelerated and selectively collected only after they have
given up almost all their kinetic energy.

In small-scale tests we have achieved conversion efficiencies approaching
90%, although in a reactor the conversion efficiencies would doubtless
be somewhat lower. The overall system efficiency would be even lower,
of course, depending on how much of the energy was converted thermally
and how much by direct conversion. With the DT cycle most of the
reaction energy is in the neutron. In this case our direct conversion
system serves mainly to increase the efficiency of the confinement and
heating portion of the cycle. But if we ever learn how to use the $D-{}^3He$
cycle, where the reaction products are charged, the system efficiency
might be substantially higher than present thermal efficiencies.

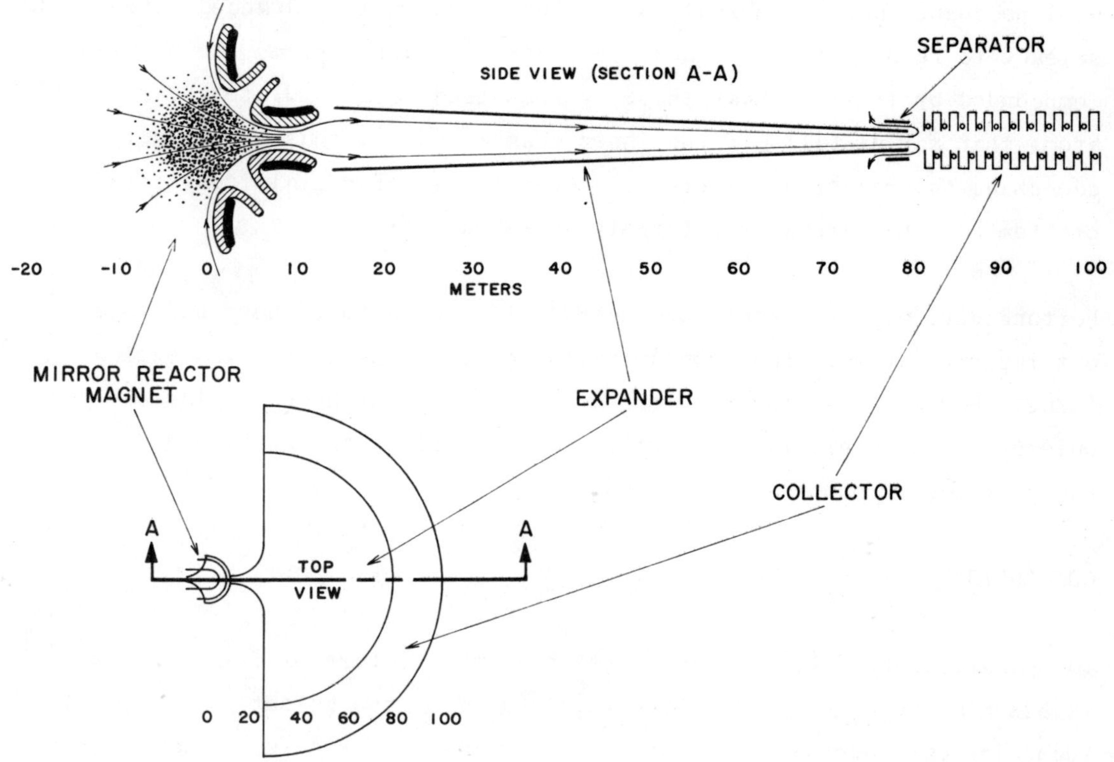

Fig. 15   Schematic of a direct converter system for a mirror fusion reactor.

These reactor studies, including extensive ones on the tokamak concept,
surely cannot represent accurately what fusion reactors will in fact be
like. Nevertheless they have a real value in exposing problem areas
and areas where technological advances are needed. I will briefly list
some of the more important issues that come to mind. First, technology:
The further development of superconducting magnets is essential to many
of the approaches. Neutral beam technology needs to be extended to
higher beam energies and higher efficiency, with long lifetime sources.
In laser fusion the issues are laser power and laser efficiency; experts
in that subject might add other needed developments.

As for problem areas, the main one that comes to mind, after the central
issue of confinement itself, is that of bombardment of the inner chamber
wall of the reactor by ions, by neutral atoms, and by fast neutrons. It
will no doubt take a combination of clever design and advanced materials
science to resolve these issues. For the toroidal systems the problem is
compounded by the fact that these systems tend to accumulate the impurity
atoms that may distill off the inner chamber walls, thus prematurely
quenching the reaction. There are several ideas for minimizing this
problem but they are as yet largely untested.

Fortunately, mirror systems are self-purifying in that they tend to distill
out impurity atoms. Thus contamination of the plasma is not a serious
issue. For mirrors, the problems are those of guaranteeing plasma
quiescence and achieving high efficiency in all of the heating and
energy recovery processes involved.

CONCLUSION

To summarize briefly, it seems to me that the picture of fusion power
is assuming ever greater reality as a result of the scientific and
technological developments that are occurring. We see fusion as evolving
toward a method of energy generation with many possible
advantages and with a variety of possible forms. While it is clear that
we have not yet proved to anyone's satisfaction that fusion power is
just around the corner, there is now a world-wide effort that is intent
on achieving it. This being the case, it is all right if you are "from
Missouri"; we are just as anxious as you to see the first fusion reactor
built. Make your own bets as to when it will happen. Just remember,
the stakes are high: a _permanent_ solution to man's energy needs.

SESSION VIII

INCREASING THE SUPPLY: SOLAR SOURCES

Chairman: Ray Hallet
            Director, Advanced Solar & Nuclear
            Systems, McDonnell Douglas Astronautics
            Co., Huntington Beach, Ca.

SOLAR HEATING AND COOLING OF BUILDINGS

R. D. Bourke*

E. S. Davis*

Solar energy has been used for space heating and water heating for many years. A less common application, although technically feasible, is solar cooling. This paper describes the techniques employed in the heating and cooling of buildings, and in water heating. The potential for solar energy to displace conventional energy sources is discussed. Water heating for new apartments appears to have some features which could make it a place to begin the resurgence of solar energy applications in the United States. A project to investigate apartment solar water heating, currently in the pilot plant construction phase, is described.

INTRODUCTION

The concept of heating water and space heating buildings with solar energy is not new - it has been used - extensively in the case of heating water - in many parts of the world for several decades. Heating water by solar energy is now common practice in Israel and Japan where other sources of energy are scarce. Solar energy formerly commanded a sizeable share of the energy market in California, as well as Florida.

Neither solar space cooling nor solar space heating has experienced the wide use of water heating, but a number of technically sound techniques exist and several full-scale experiments have been conducted or are underway.

With this long history of practice, this paper will not attempt to present revolutionary new developments in solar heating and cooling

---

*Systems Analysis Section, Jet Propulsion Laboratory

technology. Rather, the paper is devoted to a broad survey of established
technologies applicable to solar heating and cooling and discusses the
potential for these techniques to favorably impact the energy problem
in the United States. Emphasis is placed on the non-technological
requirements for resurgence of a viable solar energy industry in the
U. S. It is argued that certain critical criteria must be met before
solar heating and cooling can achieve wide-scale commercial use.
These criteria include obvious factors such as the price of solar-
provided energy in comparison with conventional fuels, and more subtle
items such as the existence of solar equipment manufacturers and the
awareness of a majority of the institutions in the building industry of
the availability and tradeoffs associated with solar energy use.

A project directed toward creating a viable commercial venture in solar
assisted gas water heating is underway at the Jet Propulsion Laboratory and is described at the end of this paper. The project, entitled
SAGE (Solar Assisted Gas Energy), is developing a water heater for
new apartments in Southern California.

BRIEF REVIEW OF THE STATE-OF-THE ART

Techniques for solar heating and cooling of buildings range from hand
operation of venetian blinds and planting of deciduous shade trees adjacent to buildings, to elaborate automatic systems which provide all
environmental conditioning from solar power. Active means for heating
water (as opposed to passive, energy saving methods) are in wide-spread
use in Japan, Israel, Australia, USSR, and on a small scale in the U. S.
Active space heating, as previously mentioned, while technologically
feasible, has not received the attention accorded to water heating, but
has been confined in this country to a few dozen experiments of the
single family dwelling scale. Larger projects, however, are now
underway (e. g., the Massachusetts Audubon Society Headquarters in
Lincoln, Mass.). An excellent, detailed survey of solar heating and
cooling technology, with an extensive bibliography, is found in Ref. 1.

<u>Water Heating</u>

Water heating is the simplest and most straightforward application for
solar energy use as it requires the least number of functions. These
functions, in chronological order, are:

1)   Absorption of solar radiation and conversion to sensible
      heat.

2)   Transfer of heat to a working fluid.

3)   Transfer of heated fluid to storage.

4)   Provision of a heating source when storage is exhausted.

5)   Delivery of stored heat to the user.

Stored hot water obtained in the foregoing manner is used either
directly (i.e., when the working fluid is the potable supply) or indirectly,
with the heat being transferred to the potable supply. A schematic of
two designs, one utilizing a single fluid, the other two fluids, is shown
in Fig. 1. While the former design is simpler, and therefore less
costly, the second may be required where corrosion inhibitors or anti-
freeze agents are needed to protect the solar collector. Even less
complex arrangements than those illustrated in Fig. 1 are possible.
For example, the circulation pump can be eliminated by using the
thermosyphon principal, (this requires locating the storage tank well
above the collector). Also, in Japan, it is common to use a plastic
pillow-like bag as a combination collector and storage tank. The fluid
is discharged at the end of the day for a single period of use.

The schematic diagram of the water heating designs illustrates two
key elements in conventional solar heating and cooling concepts: (1) the
requirements for thermal storage, and (2) the requirements for conventional back-up. Storage permits use of heated water when the sun
is not shining. Storage is also required when the use rate is higher
than that at which the collector can heat water. When stored water is
exhausted, either by high use rates or extended cloudy periods, the
conventional heater provides hot water. The tradeoff between storage size
and use-frequency of the conventional backup is entirely one of economics,
where the annualized marginal cost of additional storage is compared
with the cost of operating the conventional heater. While the latter is
a direct function of the fuel price and therefore subject to change, the
economic optimum generally lies at providing about two days of
storage (Ref. 2).

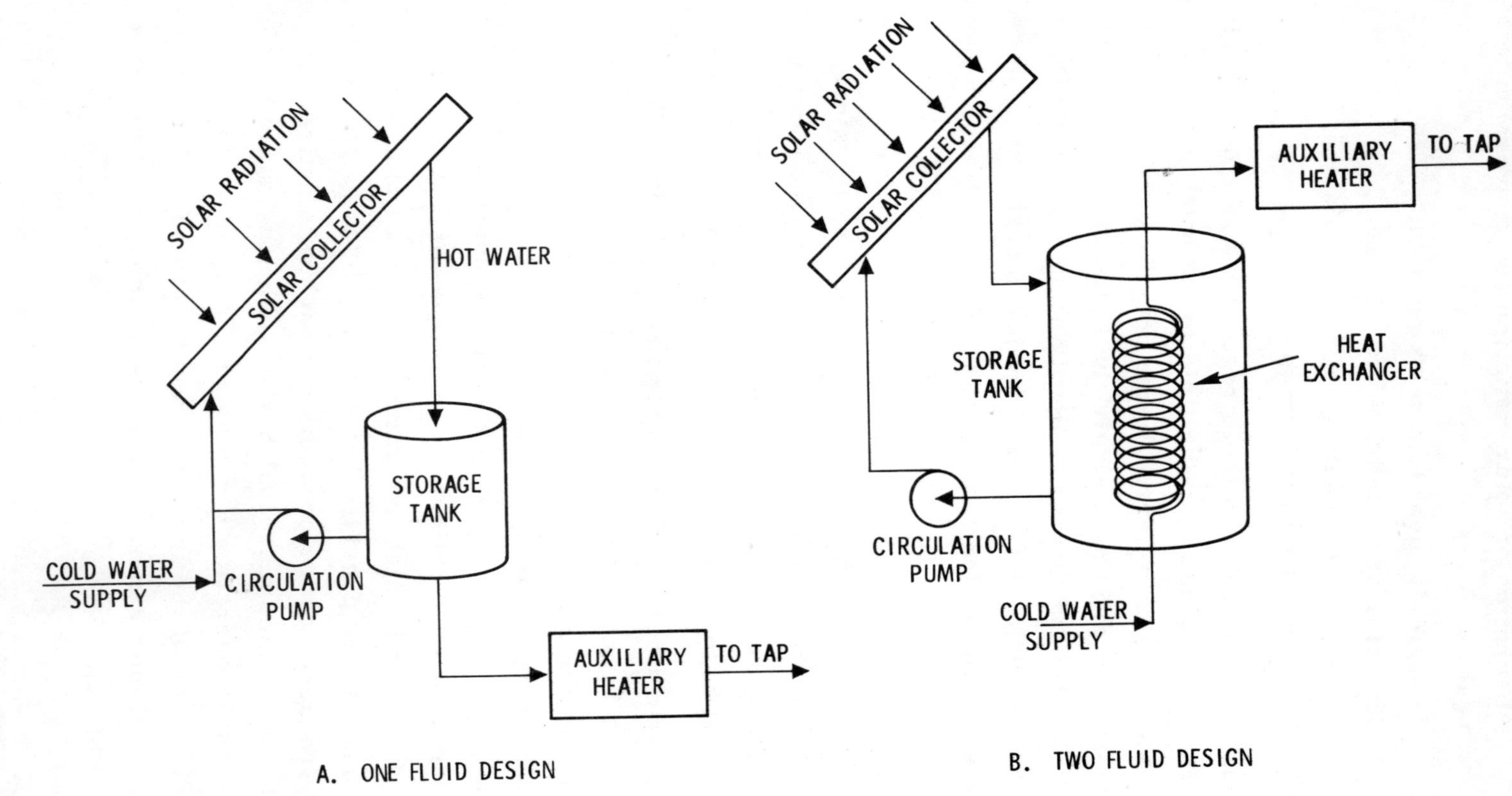

Fig. 1   Solar Water Heating Schematic

Another key element in solar heating and cooling technology, and one
which is totally absent in conventional designs, is the solar collector.
Figure 2, taken from Ref. 1, illustrates a few of the many concepts
that have been employed for collecting solar radiation and transferring
it to a working fluid.   Water is used as the working fluid in designs
A thru J.   In the first eight of these (A-H), solar radiation is transmitted by the glass covers and absorbed by the metal (or plastic) plate.
The heat is then transferred to the working fluid which flows to the
storage tank or to the point of use.   Heat losses are minimized by the
overlying glass covers and rear insulation.   The schemes illustrated in
I and J absorb the heat directly into the water, which, in the J case,

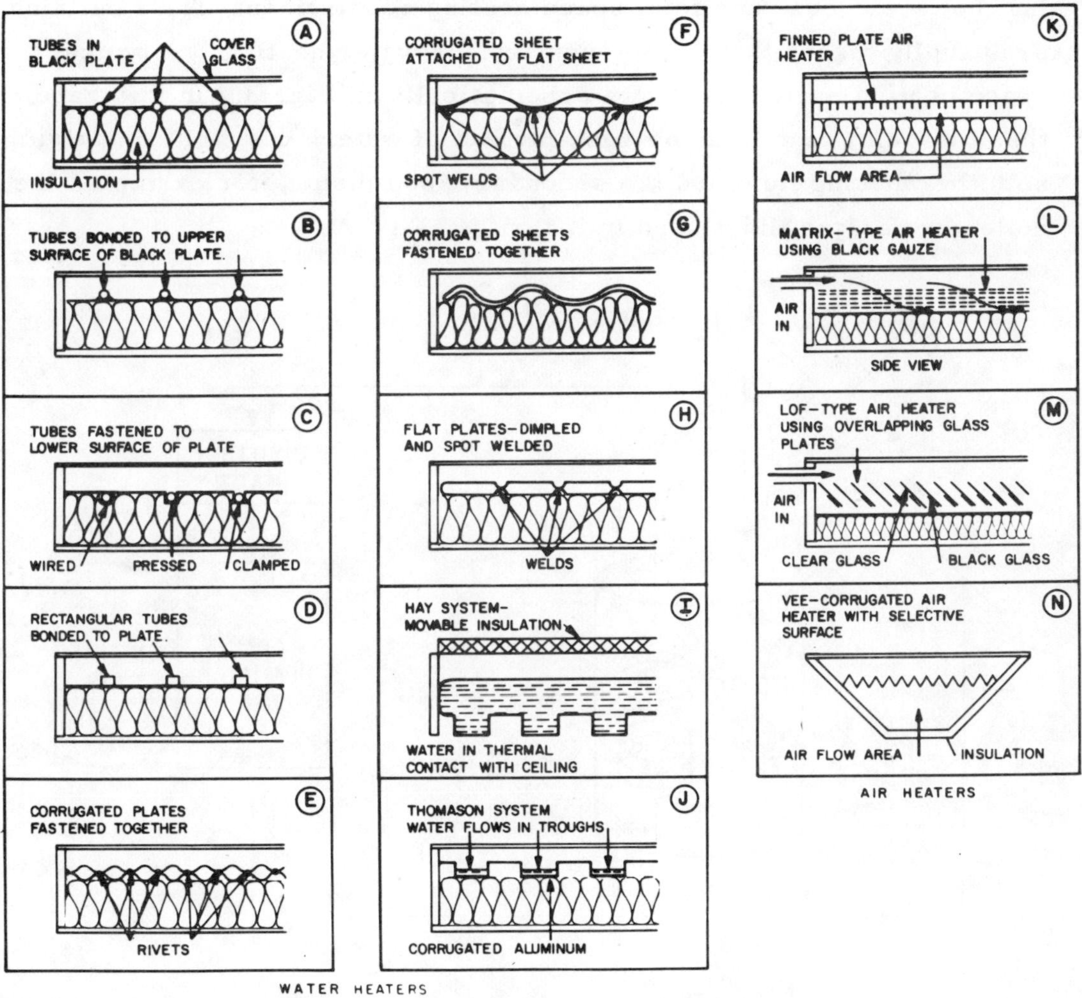

Fig. 2   Solar Collector Designs

flows to the point of use.   Temperature of the working fluid is inversely
related to efficiency of collection, with 200° F commonly achieved at
reasonable (~50%) efficiencies.   Higher temperatures for a given
efficiency are possible by the use of concentrating, tracking collectors,
but these are more complex and hence more expensive than the flat
plate design.   Furthermore, the flat plate collector gives adequate
performance for water and space heating.   Approximately 40-50 ft^2 of
flat plate collector will provide adequate hot water for a single dwelling
unit.

Space Heating

Functions for space heating are essentially the same as for water heating; however, peak and total space heating requirements for a building
are usually higher than those required for water heating.   A typical
space heating system is shown schematically in Fig. 3.   In this case,
the storage medium may be rock instead of water, with air comprising
both the working fluid and the secondary loop (using, for example, the
collector designs illustrated in K thru N, Fig. 2).

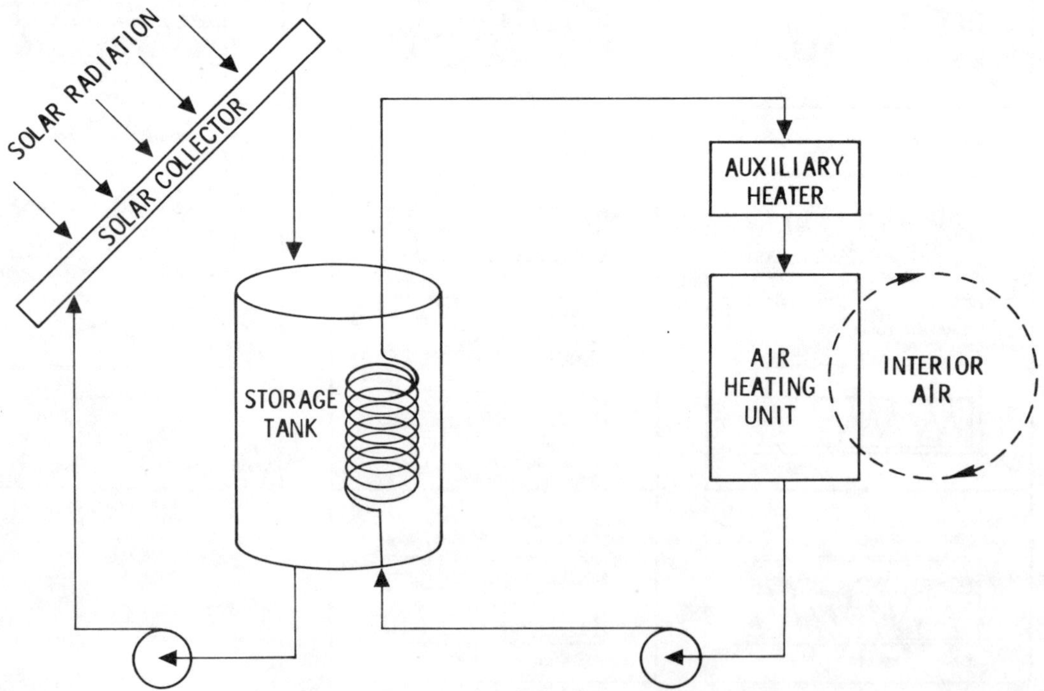

Fig. 3   Typical Space Heating System

332

It should be pointed out that a number of building heating schemes which
differ qualitatively from that shown in Fig. 3 have reached the experimental stage.   A notable example is the Skytherm residence now in
operation at Atascadero, California, which uses the collector concept
shown in I, Fig. 2.

Solar Cooling

Solar cooling is substantially more complex than solar heating, but it is
an inherently attractive application because the largest supply of energy
is available when the demand is highest.   A number of techniques have
been proposed, and a wide variety of experiments conducted, but no
clearly preferred method has yet emerged.   Among the ideas advanced
are:

1)   A solar-powered Rankine cycle engine driving a vapor
     compression refrigeration cycle.
2)   An absorption refrigeration cycle with ammonia-water or
     water-lithium bromide as working fluids.
3)   Use of dehumidifying agents with solar regeneration.
4)   Use of night sky radiation cooling and stored "cold".

Interest in the vapor compression cycle cooling has been stimulated
recently by the availability of fluorinated hydrocarbon working fluids.
A diagram of a candidate system is shown in Fig. 4.   Here the Rankine
cycle engine is on the same shaft as the conventional electric motor
(used when solar power is unavailable) and the refrigerator compressor.
This arrangement incorporates some control simplifying features in
that the ac induction motor provides the difference between the solar
engine output and the compressor requirement.   Consequently, useful
work is derived from the engine over a wide range of powers.

The absorption schemes generally suffer from the low efficiency (or
coefficient of performance) at which these units operate with temperatures easily obtained from flat plate collectors.   Development of low
temperature absorption units is now underway and their future looks
promising, particularly for large buildings.

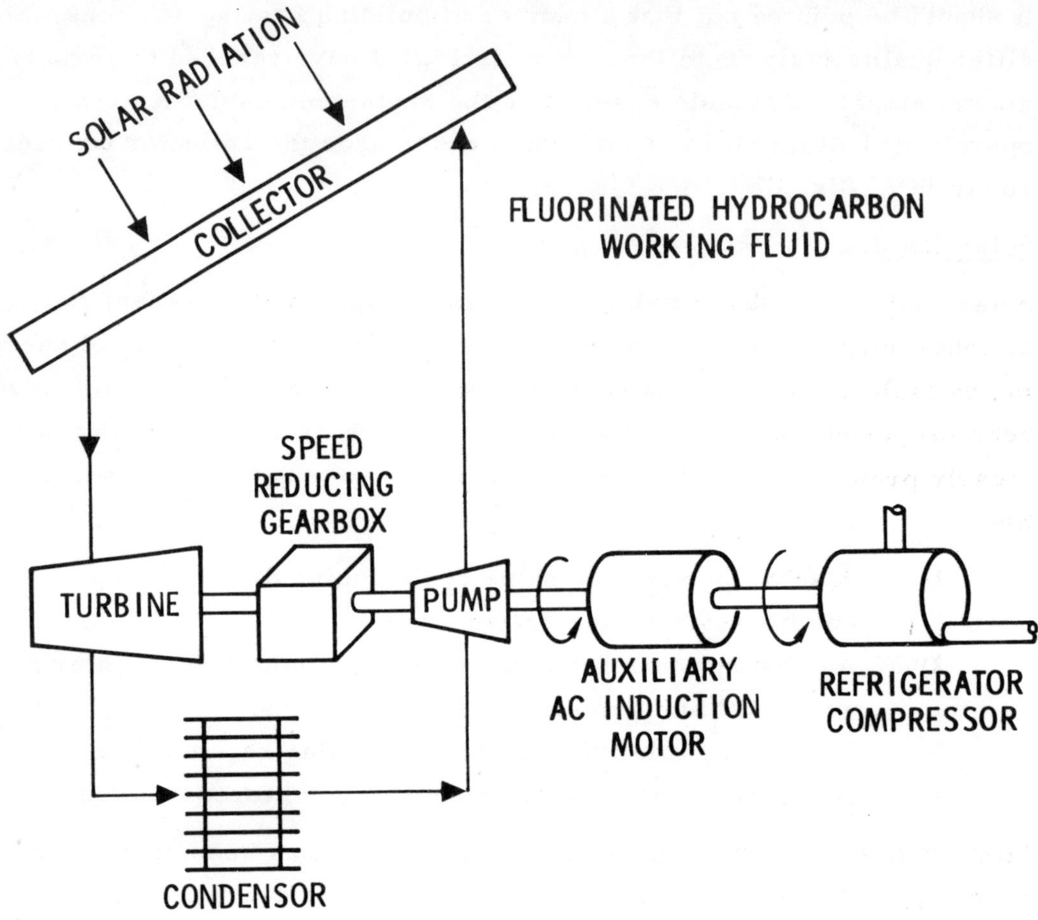

Fig. 4   Solar Powered Compression Refrigerator

Combined Systems

Water heating, space heating and cooling devices can be integrated into combined systems which provide all three functions. A schematic diagram of one alternative is shown in Fig. 5. The scheme shown combines the space heating concept shown in Fig. 3 with absorption refrigeration. The two methods are run alternately in winter and summer. Combination of other techniques are also possible. It appears that in most cases combined systems are economically more attractive than cooling alone, and they are often more attractive than heating alone (Ref. 3).

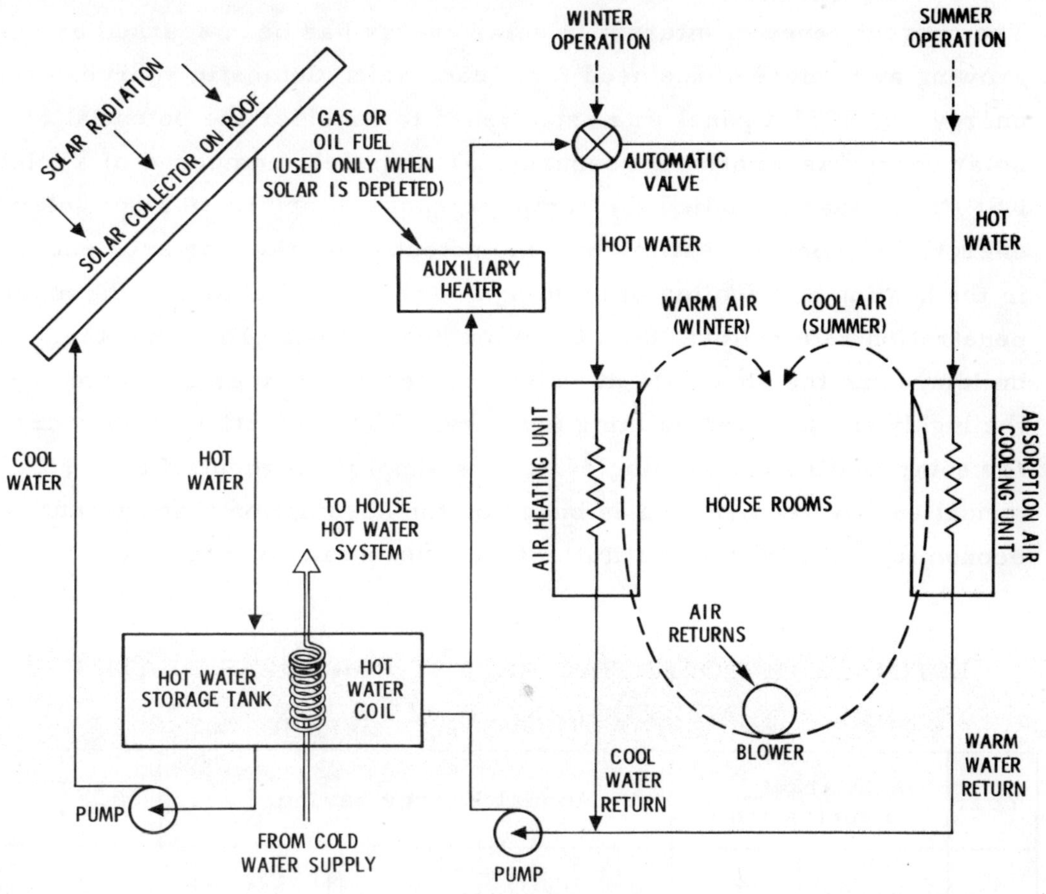

Fig. 5   Solar Heating and Cooling Combined System

POTENTIAL FOR ENERGY DISPLACEMENT

Total Market Potential

The size of market for low temperature (85°F to 220°F) thermal energy has been estimated by many sources (see, for example, Ref. 4). The energy requirement for space heating, cooling and water heating in residential, commercial and public buildings is approximately 20 percent of the total energy requirement for the U.S. The suitability of solar energy for this application has been recognized for decades. However, the relatively high cost of the use of solar energy has mitigated against broad adoption of solar energy by the building industry and little effort has been expended to develop the technology.

The current renewed interest in solar energy has been sparked by a growing awareness of the need for clean, safe, domestic sources of energy. In 1972 a panel was established to consider the potential of solar energy as a national resource. This panel, composed of a joint NSF/NASA team, studied the complete range of applications of solar energy. The panel's estimate of the potential market for such energy in the heating and cooling of buildings is shown in Table 1. The market penetration rate reflects both the more than 40 year life expectancy of buildings and the slow rate at which any new technology is adopted by the highly fractionated building industry. The projection of 30 years for solar heating and cooling system to supply 1 percent of the U. S. annual energy requirement is based on the assumption that current economic, cultural and institutional barriers can be overcome.

Table 1

ESTIMATE OF POTENTIAL MARKET FOR SOLAR HEATING
AND COOLING OF BUILDINGS

Year	Market Penetration		Annual Energy Savings*			Equipment Sales
	% of new	% of all	10^{15} BTU/yr.	%**	10^9\$/yr***	10^9 \$/yr.
1985	10	1	0. 12	0. 1%	0. 180	0. 15
2000	50	12	2. 1	1. 1%	3. 6	4. 5
2020	85	31	10. 5	3. 6%	16. 3	9. 0

*at 80 percent displacement by solar

**of U. S. requirement

***at \$1. 50/10^6 BTU

SOURCE: NSF/NASA Solar Energy Panel "An Assessment of Solar Energy as a National Resource" December 1972 (Ref. 4)

<u>Factors Affecting the Rate of Acceptance and Diffusion</u>

The diffusion throughout the building industry of the use of solar energy for heating and cooling from 0 percent of current U.S. energy requirements to 3.6 percent of projected requirements for the year 2020 will occur only if there are fundamental changes in hardware and institutions. Privately and federally supported research, development, and applications engineering is focused on reducing the cost and improving the performance of solar energy hardware and finding the most attractive applications for solar energy. The outcome of this effort will be important to successful commercialization of solar energy.

<u>Requirements on the Hardware.</u> The primary requirement for successful commercialization of solar heating and cooling is cost-effective hardware. The fundamental barriers to commercialization of solar heating and cooling has been the widespread availability of relatively inexpensive fossil fuel. For the first time in our history there is broad concern today over the availability of natural gas and petroleum. Again, new sources of supply are available only at greatly increased cost.

Although the estimates of the cost of solar energy are based on very limited experience, the future cost of the least cost alternatives for heating and cooling (i.e., fossil fuels) has now become equally uncertain. To be accepted, solar energy must be very close to the cost of the least cost alternative, i.e., natural gas; or it must offer some other feature attractive to the buyer, such as the low first cost and convenience of electricity.

In addition to meeting the primary requirement of being cost-effective, solar heating and cooling hardware must meet consumer requirements for convenience, durability, and maintainability. This has led solar water heating manufacturers to incorporate gas and electric auxiliary energy systems into their products, such as that shown in Fig. 1. Almost all systems which are considered for widespread application in this country require auxiliary energy to carry the heating or cooling requirement during peak periods. A unique effort toward an environmental control system for dwellings, which is totally independent of

the conventional energy sources, is Robert Reines' experimental house
near Albuquerque, New Mexico. Mr. Reines constructed the energy
self-sufficient dwelling by taking radical departures from conventional
dwelling design. A heavily insulated dome with limited window area
and an airlock entrance reduced the energy requirement for heating by
nearly a factor of ten. Reines' effort has demonstrated the degree of
change required if current standards of comfort are to be met without
relying on the conventional energy utility system. In the near term, it
seems clear that reliance on auxiliary fuel is required to make solar
energy both convenient and economic.

Durability of solar energy collection and conversion equipment is
extremely important. Because of overriding concern with the cost of
solar energy hardware there is a danger of sacrificing durability. At
the two ends of the cost/durability tradeoff are the solar collectors for
water heaters produced in Australia for approximately $5.00 per square
foot, and a complete solar water heating system produced in Japan and
available at a cost of about $1.00 per square foot of collector. The
Australian system is made of copper and should produce hot water for
approximately 20 years, whereas the Japanese system is made of a
clear vinyl plastic and must be replaced every year or two. Such a
maintenance problem does not appear to be acceptable to the U.S. water
heating market. In order to match existing heating, water heating and
air conditioning standards, nominal system life should exceed ten years,
and the system should continue to operate without maintenance for the
order of five years.

Changes in Construction Industry Practice

Specification of equipment. In the traditional construction process for
residential housing, heating, air conditioning and water heating, components often are not specified until the working drawings for the job
are completed. The residential developer will leave system sizing and
selection up to the mechanical subcontractor who sizes the system to
handle peak loads. The simplicity of the interface between the mechanical equipment and the building, along with the small cost penalty of
oversizing, combine to make this practice acceptable. With respect

to solar heating and cooling equipment, the penalty for oversizing will
be greater and the cost of sizing the equipment to handle peak loads
unacceptable. If solar heating and cooling are to successfully invade
the residential market system, sizing must occur earlier in the design
cycle and be based upon new criteria which have yet to be established.
Since the solar collector area approximates one-third to one-half of
the roof area for a heating and cooling system for a single family
dwelling or a two-to-three-story multiple family dwelling, it becomes
a significant spatial element which must be considered in project
planning and design.

In commercial buildings, the mechanical equipment specification
occurs earlier in the construction phase than it does in residential
practice. Professional mechanical engineers are typically involved in
detailed load calculations and the specification of equipment before
working drawings for the project are initiated. The routine involvement of professionals in the construction of commercial building should
facilitate the introduction of solar energy in this market sector.

<u>Design Criteria</u>. Equipment selection and building design are strongly
influenced by first cost. The popularity of the glass curtain wall in
high-rise commercial construction of the 1960's was a result of minimizing first cost. The result, however, has been the construction of
a large building population with a very high operating cost. If solar
heating and cooling are to have even a chance of being accepted, they
must be compared with conventional alternatives on a life cycle cost
basis, hence life cycle cost must be adopted as a design criterion.

Comparing systems on a life cycle cost basis is not as straightforward
as the phrase implies. Life cycle cost analysis requires that the
designer be very specific about the financial operations of the client/
owner of the equipment. The influence of differences in financial
structure is illustrated by the comparison of four studies of solar
energy application listed in Table 2. The key parameter is the capital
recovery factor, CRF. The CRF is the ratio of annual cost to first
cost for the solar heating and cooling equipment. The range of this

Table 2

Comparison of Parameters for Life Cycle Cost Analysis

Source	Interest on Debt	R. O. R.[2] on Equity	System Life Years	Insurance	Taxes	Maintenance	Misc.	C. R. F.[1]
Lof & Tybout Solar Heating 1972	6%	6% implied	20	0	0	0	0	0.087
Lof & Tybout Solar Heating & Cooling 1973	8%	8% implied	20	0	0	0	0	0.102
Karl Boer "Solar One"	8%	8% implied	2.5% per year dep.	0.5%	3% Real Estate	1.5%	0.5%	0.16
EQL/JPL Project SAGE	8%	13%	10	0	Note 3.	Replace at 10 years.	0	0.21[4]

NOTES

1. C. R. F. = Capital recovery factor = $/yr per $first cost = $i \dfrac{(1+i)^n}{(1+i)^n - 1}$

2. R. O. R. = Rate of return = effective interest rate on equity capital

3. 52.7 percent income tax + 2.5 percent Ad Valorem tax on capital investment

4. Ratio of equity to debt = 1.0

ratio from 0.087 to 0.21 is almost a factor of 2.5. This range
reflects different assumptions about who ultimately owns the hardware.

For government and commercial projects, accurate life cycle cost
studies, using appropriate parameters, may be practical on a project
by project basis. In the residential building market, life cycle costing
must be reduced to tabular methods which can be conveniently applied
by the specification writer.

Changes in Government

First cost influences local government in establishing the assessment
value of property. Since real estate taxes are in the neighborhood of
3 percent of market value, the preoccupation of the building industry
with first cost is not without justification. To put a 3 percent real
estate tax into perspective, consider the influence of adding a real
estate tax to the cost of solar heating estimated in Ref. 2:

$$\begin{aligned}
\text{Annual cost} &= 0.096 \text{ x first cost} \\
+3\% \text{ real estate tax} &= 0.03 \text{ x first cost} \\
\hline
\text{Annual cost} &= 0.126 \text{ x first cost}
\end{aligned}$$

While the amount of energy collected remains the same, the cost of the
energy per 10^6 BTU's is increased more than 30 percent by the real
estate tax. As long as uncertainty exists concerning the size and
manner of local government assessment of buildings which include solar
heating and cooling systems, it will be difficult to determine the life-
cycle cost of solar heating and cooling in private buildings. It follows
that an established local government policy toward the tax status of
solar conversion equipment is a necessary precursor to wide spread
use of solar energy.

There does not appear to be any technical problem in designing a solar
heating and cooling system that will meet the requirements of the
building, the health and safety code, the plumbing code, the electrical
code, etc. However, the system is not likely to be the lowest possible
cost system. Because the solar collector occupies a significant area
of the roof, many people have suggested combining the roof and the

collector to save total cost. This is the kind of potentially attractive
idea, however, which could be inhibited by current specific building
codes.

Legal issues surrounding the question of urban sun rights need clarifi-
cation. If local governments continue current practices of zoning using
two dimensional maps, buildings with solar collectors may be shaded
by neighboring buildings constructed later in time. City planning and
zoning that consider the orientation of the sun will be important if legal
disputes over sun rights are to be minimized. For example, guarantees
may be required by building owners that expensive investment in solar
collectors will not sit in the shade of a newly constructed high-rise to
the south.

Changes in Real Estate Financing Practice

In the single family dwelling market, only FHA guaranteed loans
include an estimate of operating cost when establishing the home buyer's
ability to repay a loan. Although the FHA has considered operating
cost for 20 years, insurance companies, banks, and savings and loan
companies typically do not explicitly consider operating cost. Lending
institutions will need to consider these costs on conventional loans if
the size of the market for solar heating and cooling systems is to be
significantly expanded.

With both residential or commercial income property, operating costs
are much more visible, especially when utilities are included in the
rental price. In the case of new construction, utilities companies will
provide estimates of operating cost and first cost on alternative
systems to developers. In the resale market, utility bills as well as
maintenance cost histories are routinely considered in establishing the
rate of return on investment for the property. When solar energy can
demonstrate that it is the least cost option, recognition will come
quickly from the real estate investment community.

Appraisals are used by lending institutions to establish the loan value
of owner occupied property. Appraisals made by the so-called "market
method" rely on experienced professionals to establish a fair market

value for a property. In assessing the market, appraisers find that
buyers generally do not make accurate analyses of operating and first
cost tradeoffs. However, if a property is perceived to have high operating cost because of electric heating, the value of the property is dis-
counted accordingly. On the other hand, contrary though it may seem,
if a property is perceived to have low operating cost, buyers are unwill-
ing to pay a fair premium for the property.

Inasmuch as the turnover rate for single family property is on the order
of once in five years, the average home owner will have difficulty
justifying a choice of solar heating and cooling. The real estate
financing institution could play an important role in educating the owner-
occupied residential market to make acceptance possible.

WATER HEATING FOR NEW APARTMENTS – A PLACE TO START

Background of Project SAGE*

During the year 1972 there was a growing awareness in the nation of the
need for alternate sources of primary energy. In this pre-energy crisis
period, the solar energy team at the Caltech Environmental Quality
Laboratory focused attention on applications of solar energy having the
potential for commercialization within a five-year time frame. This
search rapidly narrowed to the heating and cooling of buildings, and
the residential market place, with the emphasis, as we shall see, upon
water heating.

A brief study of the patterns of energy use in the residential sector of
Southern California led to the surprising result that solar water heating
could result in a significant reduction in the growth in demand for conventional energy supplies. The importance of water heating to reducing
the growth in energy demand is generally overlooked because only
3 percent of primary energy in the U.S. goes to residential water
heating, compared to 12 percent for space heating. In Southern
California, however, residential water heating accounts for 6 percent
of primary energy use. We suspect that similar energy use patterns

---

*Ref. 5 is a complete report on Project SAGE

dominate the whole southern United States, where conditions are best suited for use of solar energy.

Water heating for the multiple-unit residential market in Southern California was chosen over space heating or cooling as an initial enterprise for two reasons. First, space heating demand peaks during the winter and is very low in the summer, whereas the demand for hot water is much more constant. Therefore, the collector area required to supply a fixed amount of energy for water heating is about one-fourth of the collector area required to supply the same annual energy for space heating. Consequently, the unit cost of solar energy for water heating is inherently less than the unit cost of solar energy for space heating. Second, the primary energy demand for space heating is easily reduced by (1) not using electrical resistance heating, (2) improved insulation, and (3) energy conserving design. The alternatives for reducing energy used for water heating are more limited.

Multiple unit applications of solar water heaters are favored for several additional reasons: (1) Of all the applications for solar energy, multiple unit water heating appears to have the best chance to become economically competitive in this decade. (2) In this decade, 70 percent of new Southern California construction is expected to be in multiple units. (3) In an average multiple unit located in Southern California, gas water heating consumes 40 percent more energy than gas space heating and 60 percent more energy than electric air conditioning with current design and insulation standards. Trends toward improved insulation standards will make water heating even more important in the future. Thus, multiple unit water heating is likely to become eventually the largest residential consumer of energy in Southern California. (4) The capital investment required for a multiple unit solar water heating system is large enough to allow consideration of business arrangements, such as the lease service contract, which help to overcome the "first cost" barrier. (5) Compared with a single home, the multiple unit application gives greater flexibility in the technical approach, (e.g., pumped circulation, sophisticated controls, use of a two-loop system with a sealed, noncorrosive fluid in a low pressure

collector thermally exchanged with the domestic hot water at 40 psi).
(6) Sun obscuration problems are less severe for multiple units. (7)
Operating costs are more visible to multiple unit owners and investors
than to buyers of single family homes.

Because of the overwhelming concern of the housing industry with first
cost, the solar energy team was attracted to an implementation method
that would involve a leasing company or a utility. In theory, utilities
could install, own, and service solar energy collection and conversion
equipment. Monthly billing based on energy used or capacity available
would amortize the utilities capital investment in the hardware. The
best method of billing and managing such a venture easily could be
determined if the basic concept were accepted by the utility. The utility
implementation approach provides an aggregated market for the manufacturer of solar energy conversion and storage equipment. Marketing,
distribution, and service of solar energy equipment by a utility are
logical extensions of their current business operation. Also, utility
ownership of the equipment provides the real estate investment community with a guarantee that fundamental property values will not be
undermined by failure of "innovative" equipment or a failure of a new
business venture.

Analysis of Economic Feasibility

An analysis of the economic feasibility of solar water heating was made
from the viewpoint of a gas utility company. Although the exact nature
of the solar energy business operation by a gas utility was not investigated, two options were considered wherein the utility retains ownership of the equipment: (1) Offering the equipment on a lease service
contract, and (2) expansion of the existing natural gas energy business
to include the supply of solar energy, i. e. the "energy service" option.
In either case, the capital investment made by the utility in solar
energy equipment must be paid by revenues collected or by reductions in costs. For both options the basic criteria for investment is
the same.

A solar-assisted gas water heater requires full back-up capacity from gas. Therefore, there are no reductions in the cost of the distribution system. If the consumer is to pay the same for his energy services, the total investment in installing solar water heating equipment must be recouped from savings in natural gas. If a gas utility is making this investment, approximately 78 percent of the savings of natural gas must be valued at the cost of new baseload supplies of natural gas, while the remaining savings is valued at the price of gas to interruptible customers. However, because this interruptible factor is small, it can be neglected in evaluating the feasibility of solar water heating.

In order to make a preliminary estimate of economic feasibility, an interest rate and an estimate of the system life are needed. In this study, capital is amortized in 10 years with an interest rate of 16.2 percent.* This interest rate permits a utility company to make disbursements to: (1) creditors at 8 percent interest, (2) government at an income tax rate of 52.7 percent of profits and an ad valorem tax rate of 2.5 percent on capital investment, and (3) investors at a target rate of return of 13 percent on equity capital. A ratio of debt-to-investor equity of 1.0 is assumed. Although the system as presently designed should last longer than 10 years, the 10-year life is assumed for two reasons: (1) lacking sufficient information on maintenance requirements, we assumed no maintenance, with complete system failure after 10 years, and (2) the assumption of a 10-year life is typical for similar equipment, and is in accord with conservative investment practice.

The value of fuel for which the investment in solar water heating for a 32-unit apartment is amortized by the fuel saving is shown in Fig. 6. Two curves are shown, depending on whether or not solar energy is supplied to the circulation line loss. From Fig. 6 it is clear that the baseline system would be an attractive investment for a gas utility if the value of the gas saved is \$5 per 10^6 BTUs. Based on current

---

*This amortization schedule results in a ratio of $\dfrac{\text{(annual cost)}}{\text{(capital cost)}} = 0.21$.

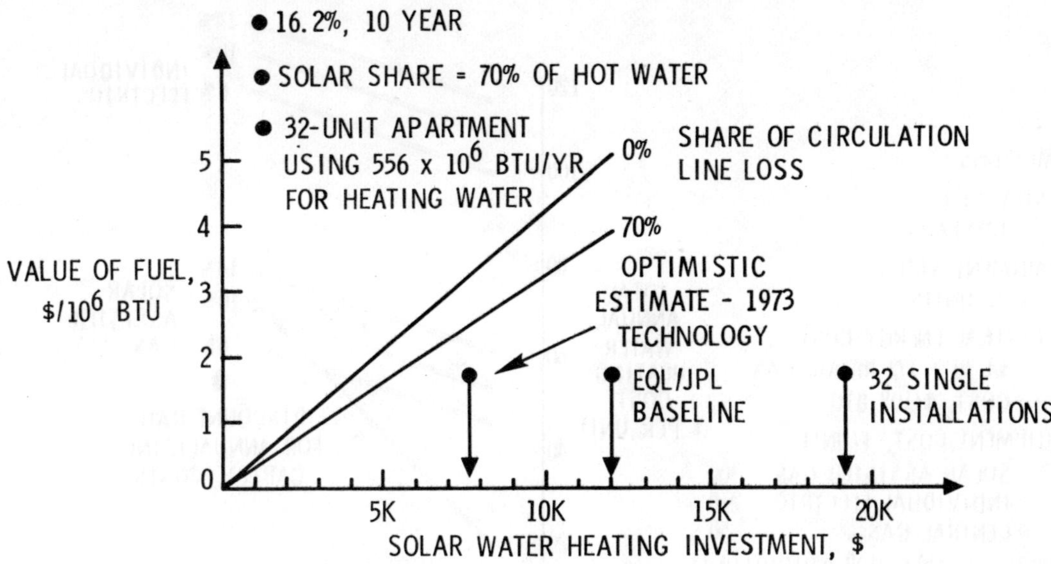

Fig. 6   Value of Fuel to Amortize an Investment
in Solar Water Heating

estimates of solar system costs, solar water heating could be
attractive if the value of natural gas were in the range of $2.50 - $3.00
per 10^6 BTUs.

The foregoing discussion has focused on the attractiveness of solar
water heating as an investment for a gas utility where the utility retains
ownership. If the solar water heating systems were sold to apartment
owners, different interest rates would be appropriate. In Fig. 7 the
annual cost of heating water by solar energy is compared to the two
dominant alternatives in Southern California — individual electric and
central gas. Although solar water heating is currently much less expensive than individual electric water heating, it would be more expensive
than central gas water heating at current gas prices of $.76 per 10^6 BTUs.
In order for solar water heating to be less expensive than central gas,
the retail price of gas would need to be approximately $3 per 10^6 BTUs.

Baseline System

In the first phase of project SAGE, feasibility of the application of gas-
supplemented solar waterheating to new apartments in southern
California was investigated. This study had three main objectives:

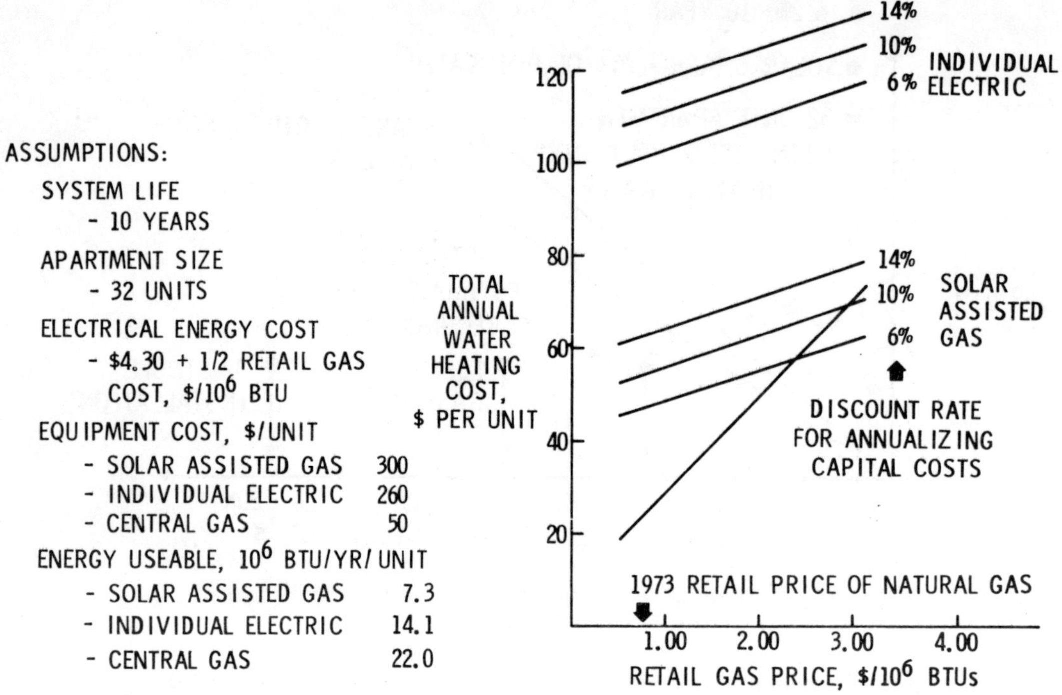

Fig. 7   Cost of Alternative Methods for Water Heating

(1) definition of a baseline system, specifying plumbing configuration, materials, components, and collector design concept, (2) estimation of system cost and performance, and (3) identification of alternate approaches to the system and component design, enabling solar water heating to become commercially viable.

After a brief examination of a wide variety of system configurations for a gas-supplemented solar water heater, one was chosen for a preliminary design study. This "baseline system" can be built completely with existing technology.

In order to estimate the cost of installing this system in an apartment, we chose a recently completed apartment in Pasadena for a detailed study of the problems involved in designing a system for an actual installation. The study building has three stories and contains 32 apartment units, a recreation room, and subterranean parking. Living density of the total complex is 40 units per acre. This building is

typical of apartment dwellings currently under construction in Pasadena
and other high land-value areas in the Los Angeles basin.   A sketch of
the study building is shown in Fig. 8.

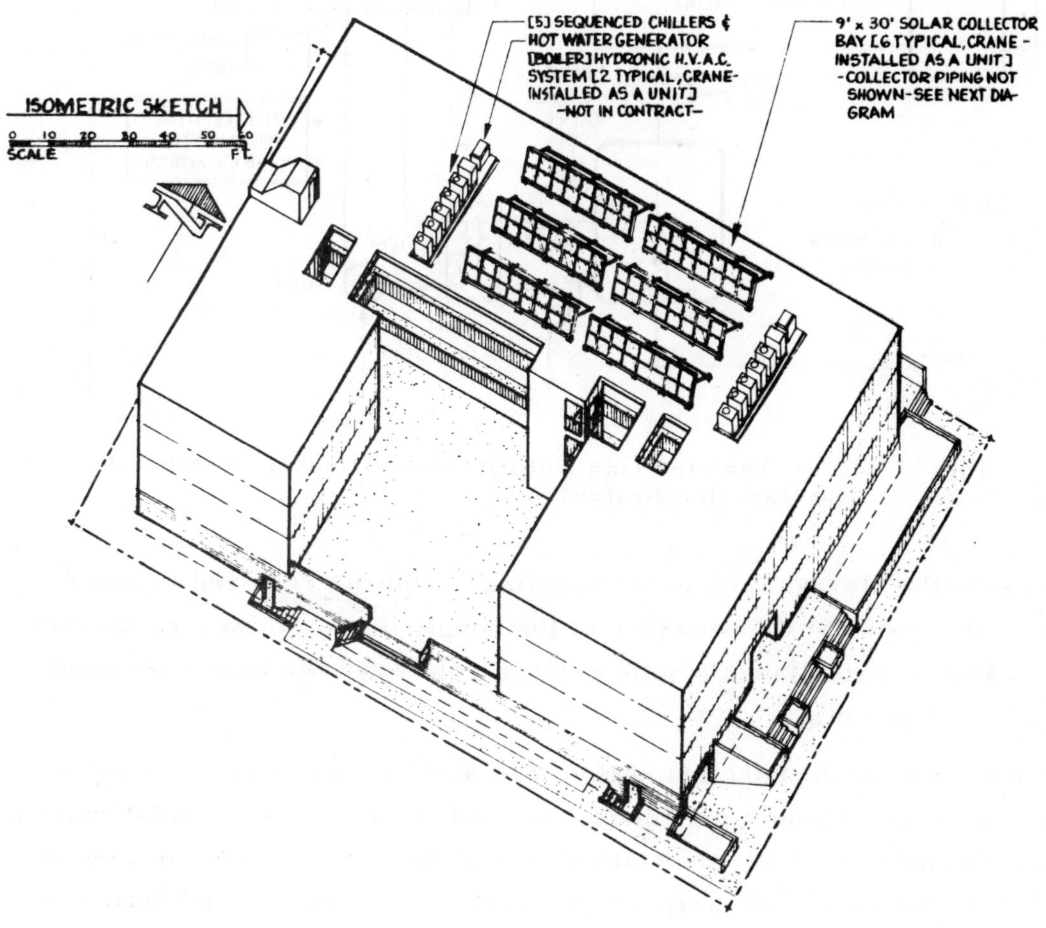

Fig. 8   Baseline Design Study Bldg. Project SAGE

The baseline system (S1), shown in Fig. 9 was chosen for two principal reasons: (1) the performance of this system is predictable; and
(2) the two-fluid design permitted exploration of low-cost options for
the collector absorber plate because of isolation from the high **pressure**,
corrosion, and scaling of the domestic water supply.   The baseline
system was initially judged to have a higher cost than alternate systems

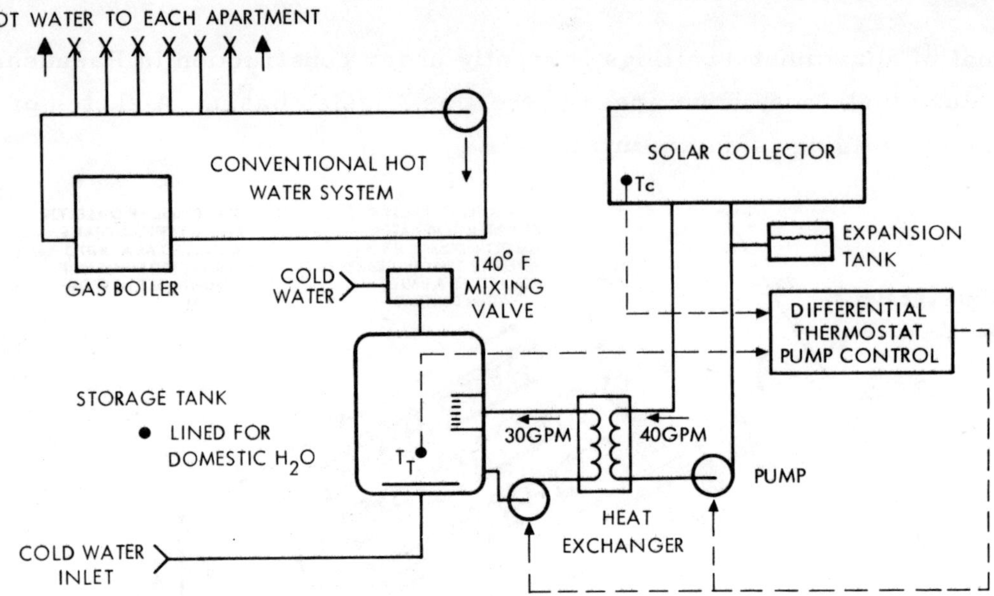

Fig. 9   Solar Assisted Gas Energy Water Heater System S1.
The Baseline System

because the storage tank in the baseline system is required to store domestic water at the pressure of the supply main. However, the baseline system establishes a solid point of reference for future research and development.

Components of the baseline system are sized to minimize the cost of hot water in a 32-unit apartment. In the baseline system, solar energy is collected by a fixed-orientation, flat plate collector with an area of 1400 ft^2 (or 43 ft^2 per unit). Water containing corrosion inhibitors is circulated between the collector and the heat exchanger at a rate of 40 gal by a fractional horsepower pump. In the heat exchanger, thermal energy is transferred to the domestic water which is stored in a 1200-gal tank (37.5 gal per unit). A second pump is used to connect the heat exchanger to the tank. The temperature difference across the heat exchanger is approximately 18° F. On a clear day, enough energy is added to the tank to raise it to an average temperature of 137° F.

As hot water is consumed by the occupants of the apartments, water at the temperature of the tank is delivered to the circulation loops.

Because this water is already heated, use of the gas boiler is minimized. Other researchers have shown that by placing the gas boiler in series with the solar preheater system, fuel consumption is effectively reduced.

It is important to point out that the baseline system does not restore that energy lost in the circulation line. We estimate that the circulation line loss for a well-insulated system accounts for 20 percent of the total energy demand for water heating (see Table 3). Although the base-line system cannot supply energy to the circulation line loss, there are minor variations of the baseline system which can do this.

Table 3

CENTRAL GAS WATER HEATING SYSTEM ENERGY BUDGET

	Annual Energy per Apartment Unit	
	Energy Supplied	Gas Energy Required*
Hot water — (58 gal/day/unit 80°F Rise)	14.1	17.6
Circulation Line Loss** (12.5 BTU/hr/ft)	3.5	4.4
Total	17.6	22.0×10^6 BTU/yr/unit

*Gas boiler efficiency assumed to be 80 percent.

**Calculated for a 32-unit apartment dwelling using insulation representing good conventional practice. Poor practice could result in a loss of 19 BTU/hr/ft, while use of foil covered ISOFOAM insulation could reduce the loss to 8 BTU/hr/ft.

The collector and tank for the baseline solar water heating system are sized to minimize the total annualized cost of supplying hot water to the 32-unit study apartment. In minimizing the total annualized cost of hot water, capital investment in solar energy equipment is balanced by reduced fuel cost over the life of the system. In Fig. 10, the annualized

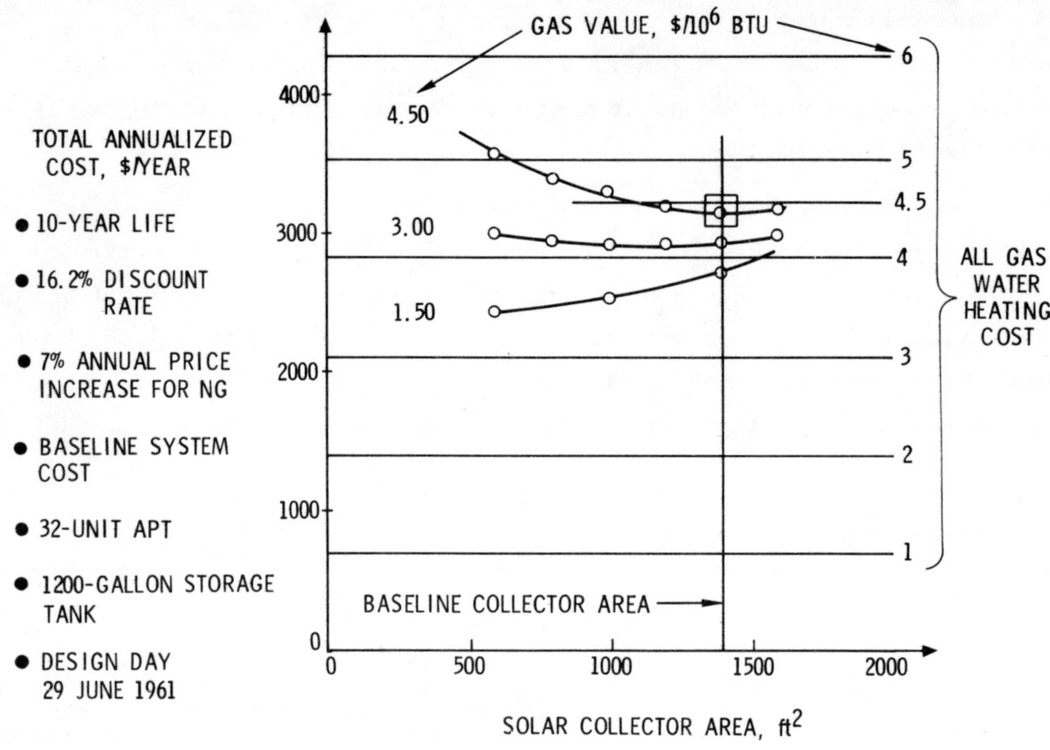

Fig. 10   Solar Collector Sizing Study

cost of the baseline system is compared to the annualized cost of all-gas water heating as a function of gas price and baseline system collector area. At a gas value of \$4.50 per 10^6 BTUs, the minimum cost of the solar-assisted system is less than the cost of the all-gas system. It can be seen from Fig. 10 that the collector area for minimum cost is 1400 ft^2. The optimum tank volume has been determined in a similar manner. This water heating system uses solar energy to provide approximately 67 percent of the annual water heating load.

To a gas utility, the seasonal and hourly variation in demand for auxiliary fuel is important. A preliminary evaluation of the demand for auxiliary fuel has been made using a simulation model of the system. Using hourly weather observations from the weather station at Burbank, California, for the year 1961 (a year with a space heating requirement typical of the long term average for the Southern California Gas Company territory), performance of the baseline system was simulated with a time interval of 20 minutes. For this simulation, the daily demand was assumed to be 58 gal per apartment unit, and the cold water inlet was assumed to be constant at 60°F. A single, typical average hourly demand profile was used for· all days of the simulation. The results of this simulation are presented in Fig. 11, and show that the monthly share of water heating carried by solar energy varied between a low of 52 percent in December to a high of 77 percent in April[*]. Although this simulation was for one year only, it is clear that variability of the weather from month to month has minor influence on the consumption of natural gas as an auxiliary fuel.

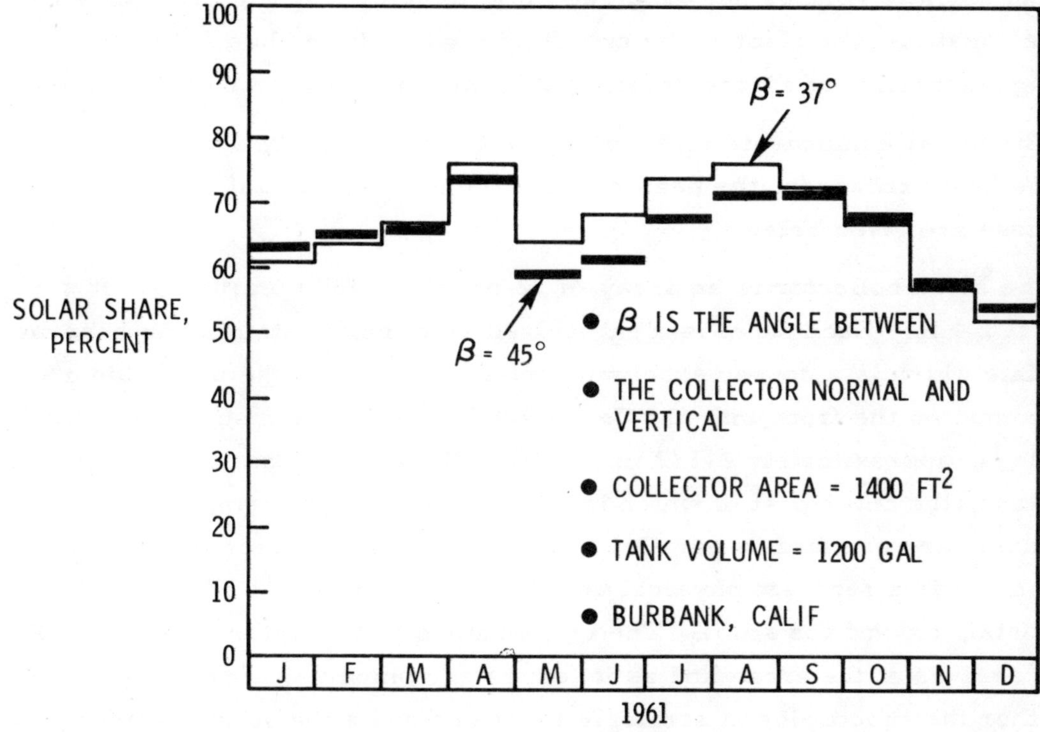

Fig. 11   Share of Water Heating Load Carried by Solar Energy

The relatively steady nature of the supply of solar energy when averaged over one month is very important. On the order of 78 percent of

---

[*]Based on Australian experience, long-term performance degradation due dust, dirt, etc. on the solar collector is not expected to exceed 5 percent.

solar energy collected by the system in 1961 could have directly
displaced production and transmission capacity in the gas utility system.
Thus the "steady supply" component of solar energy is more significant
than the "when available" component in the water heating application.

<u>SAGE Pilot Plant</u>

To verify the performance calculations of the SAGE study and to test
various components and system configurations, a solar-assisted water
heater is under construction at JPL and is expected to be operating
within weeks. A schematic of the plant is shown in Fig. 12. The plant
can emulate various one- and two-fluid loop modes; Fig. 12 shows a
schematic of the plant in the one-fluid mode. It has been sized to be
representative of a water heating system serving a ten unit apartment.

The major components of the plant are the collectors, the storage tank,
the heat exchanger, the heaters, and the pumps. Details of each of
these are given below.

The solar collector is an array of 32 panels, each measuring 2 ft x
7 ft x 4 in. Each panel is a galvanized steel box containing an absorber
plate which is a copper sheet with tubes soldered to the back side and
painted on the front with 3M Co. Black Velvet. The insulation is fibrous
glass, approximately 2-1/2 in. thick. Glazing consists of two layers of
glass, the outermost of which is double strength and tempered. Four
panels are plumbed in parallel to comprise the collector array. Each
module is a separate physical unit and all are elevated 37° above horizontal, toward the south. The temperature of the water is measured as
it arrives at the array and as it leaves each module. One panel has six
other thermocouples at strategic locations and a thermistor attached to
the absorber plate for system control.

The storage tank is of stainless steel, 42 in. in diameter by 82 in. high,
with elliptical end bells. The total volume is 580 gal of which 450 gal
is filled with water, the balance serving as expansion tank. On the
south side is a vertical line of eight penetrations for thermocouples,
reaching to the center of the tank. At the bottom center is the outlet for
water to be sent to the solar collector. The return inlet for this water

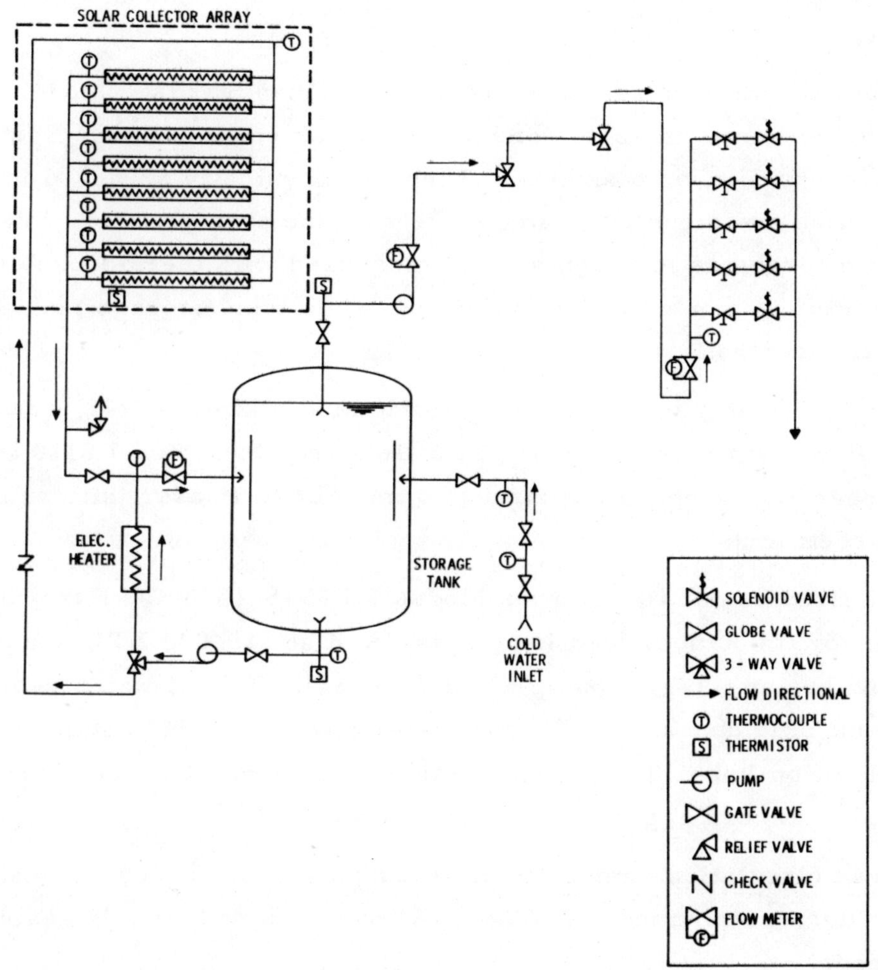

Fig. 12   Pilot Plant Schematic

is on the west side at the 250 gal level. Heated water is taken from an outlet on the north side at the 430 gal level, a few inches below the water line. This water is returned through an inlet on the east side at the 250 gal level. Water entering through either the east or west inlets impinges on the center of 3 ft^2 baffle plates placed 2 in. inside the tank wall.

The heat exchanger is composed of six American Standard Model 302 tube and shell units in series and arranged for counter flow. Each unit is 2 ft long by 3 in. in diameter. They are physically arranged in two sets of three and mounted from brackets on the storage tank. Inlet and outlet temperatures for each set are measured with thermocouples, and a thermistor is located at the outlet end of the tube (storage tank) side for system control.

The electric heater is a General Electric Model 2D421-G141, rated at 24kW. Power input is controlled by a Research Inc. Model 5110 Data Trak programmer operating through a model 646 Phaser (silicon controlled rectifier).

The gas fired heater is a Raypak Model TIP-315-16/30GG Hurricane suitable for its outdoor location. It is rated at 315,000 BTU per hr input and has an integral pump which insures sufficient water velocity to prevent lime deposits. The gas meter is a Model 310 obtained from the Utilization Laboratory of the Southern California Gas Co., and modified to record gas used in increments of 2 ft^3.

The three circulation pumps are Bell and Gosset Model 60-11S with one-quarter horsepower motors. They will pump 25 gal per min against a 25 ft head.

CONCLUSIONS

The critical state of the energy situation in this country calls for new solutions to the problems of both short supply and excessive demand. Solar energy has the potential for partially alleviating these problems. It can reduce the demand for conventional fuels and associated capital investment while maintaining current life styles. It appears that the

nearest term application of solar energy is the heating and cooling of
buildings and that today water heating is on the verge of acceptance.
In the opinion of the authors, the economic conditions for a resurgence
of a solar energy industry in the U.S. are nearly with us and the next
few years may well see water heating by solar energy again becoming
common practice.

REFERENCES

1. J. I. Yellott, "Utilization of Sun and Sky Radiation for Heating and
    Cooling of Buildings," _ASHRAE Journal_, December 1973, pp 31-43.

2. G. O. G. Lof and R. A. Tybout, "Cost of House Heating with Solar
    Energy," _Solar Energy_, 1973, Vol. 14, pp 253-278.

3. G. O. G. Lof, "Solar Energy for Heating and Cooling," Hearings
    before the Subcommittee on Energy of the Committee on Science and
    Astronautics, U. S. House of Representatives, June 7, 1973, pp 3-11.

4. NSF/NASA Solar Energy Panel, "An Assessment of Solar Energy
    as a National Resource," December 1972.

5. E. S. Davis, "Project SAGE Phase I Report," _EQL Memorandum_,
    No. 11, California Institute of Technology, 1973.

ROLES FOR SOLAR THERMAL CONVERSION
SYSTEMS IN OUR ENERGY ECONOMY

Dwain F. Spencer[*]
Arthur B. Greenberg[**]

I.   INTRODUCTION

Concentration of solar energy to produce heat dates back to 212 B.C. when Archimedes reputedly set fire to an attacking Roman fleet. This approach utilizing heliostat reflectors is the basis for the French solar furnaces in Odeillo and the central receiver concept described later in this paper.

An American, John Ericsson,[1] constructed a rectangular parabolic cylinder measuring eleven by sixteen feet in about 1883. It was claimed to have delivered one horsepower for each hundred feet of collector. In 1912, Frank Shuman initiated a monumental project in Egypt to produce approximately 55 hp from a large solar array. He employed 13,000 square feet of cylindrical troughs to produce this power. In one five-hour test, the brake horsepower never dropped below 52 hp. The economics of the plant, assuming it at 50 hp, indicated that it would have paid for itself in three years. But Frank Shuman could not make solar energy stick. This system is the forerunner of the parabolic cylindrical trough concept being explored by a number of organizations today. Will we now be more successful?

---

 * Program Manager - Advanced Energy Research and Technology
   Division, National Science Foundation
** General Manager - Civil Programs Division, The Aerospace Corp.

1. Halacy, D. S., Jr., <u>The Coming Age of Solar Energy</u>, Harper
   and Row, Publishers, New York 1973.

A Frenchman, Charles Tellier, is given credit for first building a
"flat Plate" collector in the late 1880's. Willsie and Boyle extended
the work in the early 1900's. Their collectors used water flowing
through shallow glass-covered basins, the direct forerunner of our
present day solar energy hot water heaters.

In 1695, two Italians, Targioni and Averoni, used a lens to decompose
diamonds. Lavoisier in the late 1700's developed powerful lenses to
focus sunlight and burn samples. These, of course, are the forerunners
of the Fresnel lens.

Finally, A.G. Eneas built a large reflector "resembling a hugh inverted
umbrella" to produce steam in about 1901. In the 1920's and 1930's,
both Abbot and Goddard designed parabolic mirrors. Goddard, in fact,
designed an early version of the "cavity fluxtrap." These were the
predecessors of the paraboloidal collectors being considered today.

Since it is clear from this discussion that solar collector technology is
not conceptually new, why do we think that solar thermal systems have
a role to play in our energy economy?

Certainly, the present recognition of the limitations in other fuel
sources is a dominant component. Projections have been made which
indicate that liquid and gaseous fossil fuel supplies will be depleted in
this country within 25 to 30 years. Concerns with safety of nuclear
powerplants, limitations in the availability of uranium, and problems
attendant with fast reactor breeding times and radioactive waste disposal have delayed the "age of nuclear power." Environmental concerns
with coal fired powerplants have curtailed new installations and limited
usage of this source. Clearly there is a need for a nondepletable, clean
source of energy.

Secondly, the space program has created a new interest in solar energy
as an energy source. Nearly all unmanned satellites and spacecraft,
as well as the latest Skylab manned system depend on solar energy.

These solar systems are expensive and of relatively low power levels
(0.1 to 20 kw); however, they have precipitated new technology developments which may have applications on earth. Although no solar thermal
power systems have been flown, a large number of research and development projects sponsored by NASA and DoD have been conducted.
These programs have led to the development of selective absorber
coatings to enhance photothermal conversion, fabrication of highly reflective materials and accurately controlled surfaces, and accurate
tracking and control systems for ground antennas.

Thirdly, other related technologies have been developed. Liquid-metal
heat-transfer technology, the heat pipe, closed cycle gas systems, and
a greatly increased understanding of water heat transfer have stemmed
from the nuclear program. Larger turbines and higher pressure steam
cycles have been developed for fossil fuel plants. In addition, we have
become much more sophisticated in our analytical and experimental
capabilities in addressing problems associated with the optics, transient thermal performance, and control of large systems.

These arguments certainly indicate that we should at least "give it
another try." Whether we will be successful depends in large part on
our ingenuity and ability to apply this technology base. This paper will
discuss the limited efforts performed to date in terrestrial solar thermal
conversion systems and establish some objectives which should allow us
to assess "the role of solar thermal conversion systems in our energy
economy."

II. NSF SOLAR THERMAL CONVERSION PROGRAM

Background

Table 1 shows the total federal funding for solar thermal conversion
through FY 1974. It is obvious that the solar thermal conversion program is in a conceptual analysis and exploratory research stage.
However, based on the analyses and research performed to date, there
do not appear to be any fundamental technical limitations which would
prevent application of solar thermal systems. The major problem is to

Table 1

SOLAR THERMAL CONVERSION - FEDERAL FUNDING

(Thousands of Dollars)

	FY 71	FY 72	FY 73	FY 74
Funding Source				
NSF	60	550	1430	2100
NASA				220
AEC				600

define meaningful roles (missions) for solar thermal conversion systems which can be shown to be economically competitive with alternative systems.

Program Structure

The present program has been structured to address critical technical and economic problems at various program levels; namely, (1) mission, (2) system, (3) subsystem, and (4) components and materials levels. Solar thermal conversion missions; i.e., powerplant scale and function are shown in Figure 1. Also included in the figure are the ongoing NSF sponsored projects and other related projects being conducted at either the mission or systems analysis levels. Solar thermal systems may provide electrical service only, thermal service only, or combined electrical and thermal service (powerplant function). Since solar energy is a distributed source, these services can be provided from systems with various levels of energy aggregation (powerplant scale). For simplicity, powerplant scales have been defined into qualitative energy demand categories; i.e., systems providing energy to (a) individual demand points, (b) communities, military bases, or substations, (c) municipalities, or (d) large central station powerplants. Of the twelve elements of the matrix, only two elements are not considered potential solar thermal applications due to energy distribution problems; these are central station powerplants providing thermal service or providing

362

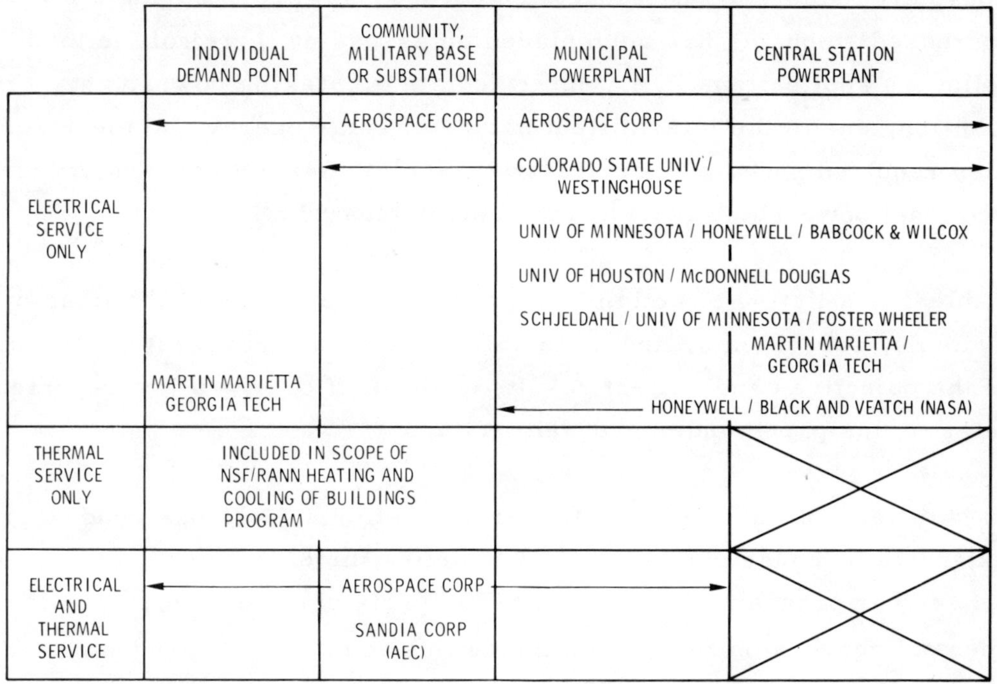

Fig. 1 Solar Thermal Conversion Missions

both electrical and thermal service. In addition, systems providing thermal service only at individual demand points or community systems have been included in the NSF/RANN solar heating and cooling of buildings program; therefore, they are outside the scope of the solar thermal conversion program.

As is evident from the matrix, to date major emphasis has been placed on the evaluation of solar thermal conversion systems for municipal or central station powerplant applications. During FY 1974 an increased emphasis has been placed on analysis of smaller scale electrical generation missions as well as combined electrical and thermal service applications. It is planned that an increasing emphasis will be placed on these lower level applications in the future.

Another important dimension to municipal or central station powerplant applications is related to the classification of the powerplant from a load-duration standpoint. Solar thermal systems had previously been

considered solely for baseload applications; however, during FY 1974
increased emphasis has been placed on assessing their role in load
following applications; i. e. , intermediate or peaking powerplants. In
addition, due to the intermittent nature of solar energy and the associated requirements for energy storage, solar thermal conversion/con-
ventional powerplant hybrids are being evaluated.

Table 2 is a listing of active NSF grants and contracts at the mission,
systems, subsystems, and materials and components levels. A summary
of the objectives and salient results of these efforts will be presented
later in this paper under program status.

In parallel with these contractor/grantee efforts, NSF has conducted an
extensive in-house planning effort to define the objectives and scope of
a five year program from FY 1975-79. This planning effort in solar
thermal conversion focused on the definition of overall program objectives and proof-of-concept experiments which could be initiated in
FY 1975.

Program Objectives and Scope

The overall objectives of the NSF Solar Thermal Conversion program
are to prove the technical and economic feasibility of solar conversion
systems, and to provide a demonstrated technology base by the mid-
1980's for load following electric energy generation applications and
electrical and thermal service applications for various types of communities including federal installations. The five year program includes a
integrated approach of analysis and research and development on key
technical, economic, environmental, social, and institutional issues
associated with the introduction of solar thermal conversion systems.

The scope of this program includes major proof-of-concept experiments
and an advanced research and technology subprogram. Specific planned
efforts include:

- Design, fabrication, integration, and test of a 10 Mwe central
  receiver pilot plant with the output from this plant integrated
  into a private utility power grid.

Table 2

NSF SOLAR THERMAL CONVERSION ACTIVE
GRANTS AND CONTRACTS

Program Level	Grantee/Contractor	Project Title	Period of Performance
Mission	Aerospace Corporation	Solar Thermal Conversion Mission	4/15/73 to 7/31/74
System 1) Parametric Analysis	Colorado State Univ. / Westinghouse	Analysis of Solar Thermal Conversion Power Systems	5/01/73 to 10/31/74
	Univ. of Minnesota/ Honeywell/Babcock & Wilcox	Research Applied to Solar Thermal Systems	7/01/72 to 7/01/74
	Univ. of Houston/ McDonnell Douglas	Feasibility Study of a Solar Thermal Power Sys. Based Upon Optical Transmission	6/15/73 to 6/15/74
2) Point Designs	Schjeldahl Corp. / Univ. of Minn. / Foster Wheeler	Solar Power Array for the Concentration of Energy	11/01/73 to 11/01/74
	Martin Marietta/ Georgia Tech.	Solar Power Sys. & Component Research	01/15/73 to 11/15/74
Subsystem	Univ. of Minnesota/ Honeywell/Babcock & Wilcox	(Same Grant as Above)	
	Itek Corporation	Solar Power Collector Breadboard Test	4/01/74 to 2/01/75
Materials/ Components	Univ. of Arizona	Chemical Vapor Deposition Research for Fabrication of Solar Energy Converters	1/01/73 to 1/01/75
	McDonnell Douglas	Elimination or Control of Material Problems in Water Heat Pipes	1/15/74 to 1/15/75
	Helio Associates	Air Stable Selective Surfaces for Solar Energy Collectors	2/01/74 to 8/01/75

- Design, fabrication, integration, and test of a 200 kwe, 2 Mwth solar total energy system providing electrical and thermal service to a new civilian community or military base installation.

- An Advanced Research and Technology subprogram encompassing (1) development of alternative concentrator/collectors, thermal transfer, and storage subsystems, (2) development and deployment of instrumentation to monitor the direct solar and circumsolar radiation in the Southwestern United States, (3) research and development on selective absorber and reflector coatings, (4) environmental, social, and institutional studies on the impact of solar thermal conversion systems, and (5) second generation systems application analysis.

Program Status

The Solar Thermal Conversion program presently includes ten active projects. The objectives and salient results from these projects are presented below.

1. Solar Thermal Conversion Mission Analysis: Aerospace Corporation

The objective of this analysis is to provide a basis for selecting the most promising missions for solar thermal conversion systems in the Southwestern United States for the period 1980-2000. The study approach includes: (a) collection and correlation of regional solar insolation and climatological data over the Southwestern United States, (b) review, analysis and projection of electrical and thermal demands in the region for the time period 1980-2000, (c) application of previously developed methodology to evaluate the potential of municipal or central station solar thermal conversion powerplants, (d) extension of the methodology to combine electrical/thermal service systems for industrial community, military base, or municipal applications, (e) development of system and subsystem requirements, (f) evaluation of siting opportunities and constraints in the Southwestern United States, and (g) estimates of the market capture potential including consideration of environmental impact and materials resource requirements.

The flow diagram for the analysis is shown in Fig. 2. The first phase of this effort focused on the development of a mission analysis methodology and the application of this methodology to Southern California. The

Solar Thermal Conversion Mission Analysis
Study Flow Diagram

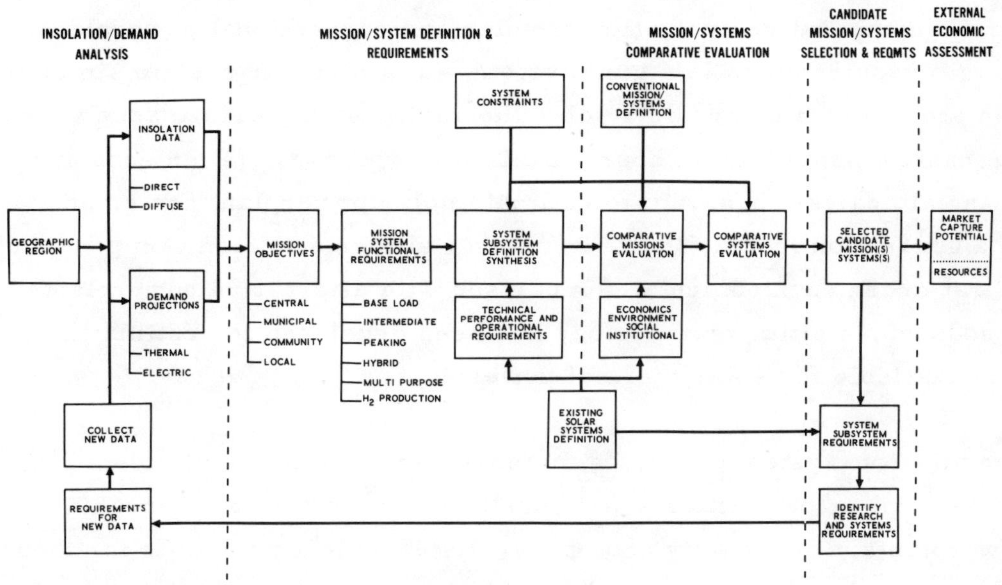

Figure 2

approach is based on a comparison of available insolation versus projected demands for a specific geographic region or subregion. These control variables are utilized to define the requirements for solar thermal conversion systems and subsystems. A computer analysis is performed to estimate the technical/economic performance of alternative solar thermal conversion systems to meet these requirements. The performance of the solar thermal conversion systems is then compared with alternative systems, including environmental factors, to assess the market capture potential of solar thermal systems. In addition, system/subsystem requirements and required research and development programs are defined.

The first phase of this effort has resulted in the formulation of a mission analysis methodology which has been successfully tested for electric service delivery in Southern California. The basic computer methodologies include: (a) insolation models for eight subregions of Southern California and Albuquerque, New Mexico which can be utilized to assess

system/subsystem performance on a consistent basis, (b) an hourly
demand forecasting model for the period 1980-2000 based on previous
Southern California Edison data, (c) margin analysis routines which
permit comparative evaluation of solar and conventional powerplant
margin requirements in a grid system, (d) a modular system simulation program which predicts solar thermal system performance versus
forecasted demand on an hourly basis for two years, (e) an economic/
financial analysis program to estimate solar powerplant costs, and
(f) a generation model for the Southern California Edison Company and
a first order environmental comparison with alternative powerplants.
In addition, a siting methodology has been developed for estimating the
land available for solar thermal powerplants.

The mission/systems analysis methods can parametrically identify
(a) powerplant performance for baseload, intermediate, or peaking
powerplants, (b) capacity and energy displacement potentials for these
plants, and (c) plant capacity factors.

Specific results from the study indicate, that with optimum siting in
Southern California, the estimated busbar energy cost in 1990 for solar
thermal baseload powerplants is two to three times conventional source
cost projections. The estimated busbar energy cost in 1990 for solar
thermal intermediate load powerplants is only 25-50% greater than projected costs for conventional sources, and the estimated busbar energy
cost in 1990 for solar thermal peaking powerplants is comparable to
that projected for conventional sources. Further, a siting analysis for
the Southern California region indicates that sufficient siting opportunities exist to supply all the additional electrical power needs of Southern
California for the year 2000. Finally, based on a postulated Southern
California Edison generation model for the period 1980-2000, the market capture potential for solar thermal conversion systems is 16,000 to
22,000 Mwe assuming commercial availability of systems by 1990.
These results are presented in a five volume report prepared by The
Aerospace Corporation, ATR-74 (7417-05)-1, January 15, 1974.

This analysis is now being extended to a seven-state region of the Southwestern United States to begin to assess the broader geographic potential of solar thermal conversion systems.

2. <u>Analysis of Solar Thermal Electric Power Systems: Colorado State University/Westinghouse Electric Corp.</u>

The objective of this project is the development of design parameters of a system or systems for thermal/mechanical conversion of solar energy to electric power at minimum cost per kilowatt-hour generated (for electric utility systems). The specific approach includes: (a) performance analysis of various types and designs of solar collectors under real solar radiation and atmospheric weather conditions for a wide range of operating temperatures, (b) evaluation of the performance of several heat engines and working fluids suitable for conversion of thermal to mechanical and electrical energy, (c) assessing suitable and cost effective energy transport and storage subsystems which will meet the requirements defined in a and b above, (d) evaluation of capital and operating costs of individual components and assessment of the changes in these costs as design and operating conditions are modified, (e) determination of the capital and operating costs of complete solar-electric generating systems, optimization of designs and operating parameters for least-cost electricity from each system, and overall performance and economic comparison analysis, and (f) evaluation of the needs for research and development for construction of the prototype solar power system(s) having the most attractive potential.

Specific progress to date is summarized below and contained in the Annual Progress Report prepared by CSU/Westinghouse - NSF/RANN/ SE/GI-37815/PR/73/4, January 1974.

Preliminary designs of specific solar powerplants are being investigated to provide a basis for performance and cost estimates and to identify critical aspects of component and system designs. A 10 Mwe powerplant was designed for Albuquerque, New Mexico, using polar-mounted parabolic cylinders in a rectangular field configuration. The collectors product 225°C steam which is transported to the turbine-generator subsystem from which the steam discharges at 50°C. Heat storage in steam accumulators was provided for one-to-four hours, and a wet cooling tower was assumed for purposes of estimating costs. The estimated energy costs of this option are 4.6¢ kwhr for one hour of storage and 3.8¢ kwhr for four hours of storage (assuming $40/m^2 for collectors). These cost estimates are based on 16% amortization rate and are 1973 dollars.

Flat-plate collectors have been studied using a "design and costing" analysis technique which provides for design decisions on the basis of performance and cost per unit of output. The outputs for 14 collector design configurations have been calculated and the cost of thermal energy has been estimated. The heat costs vary from 1.3¢ kwhr (thermal) to 1.6¢ kwhr (thermal) for 11 different collector configurations using a tube-in-plate absorber. Estimates for an evacuated plate collector yielded the least cost of heat at 0.8¢ kwhr (thermal). Based on this preliminary analysis of the per-unit-area costs of flat-plate collectors, it appears that flat-plate collectors are not likely to be as cost effective as concentrating systems for power production (at least in high direct radiation locations such as Albuquerque).

Performance equations have also been derived for concentrating collectors. The equations apply to single and double curvature concentrators as well as the central receiver (tower) system. Performance equations for absorber heat exchangers have been derived for basic geometrical configurations. A generalized method has been developed and tested for use in selecting concentrator-absorber-heat exchanger (collector) combinations.

Equations to describe the thermodynamics and fluid mechanics of heat transport have been developed for water or steam working fluids. A preliminary comparison between steam and water transport shows that steam has the least combined costs at temperatures above 150°C. Steam accumulators were selected for detailed study of heat storage.

A computer program has been developed to provide design and performance data on 3, 10, 30, and 100 Mwe steam turbine-generators as a
function of initial steam conditions, turbine capacity, and exhaust pressure. A basic turbine building block for the 10 to 15 Mwe range suitable
for use in a solar plant has been defined. The unit is capable of operating with $210^{\circ}C$ dry and saturated steam and a condenser temperature
of $40^{\circ}C$. At current market value, the cost per kw rated capacity generated is between $70 and $100.

A dynamic system simulation model is being developed to analyze specific
designs of optimal power systems to account for time varying effects.
The input overlay will be adaptable for use of real solar data or solar
insolation models and the output can be in graphical or tabular form.

3. <u>Research Applied to Solar Thermal Systems: University of
    Minnesota/Honeywell Corporation/Babcock and Wilcox</u>

The objective of this project is to perform analyses to define system/
subsystem performance for a parabolic trough solarthermal conversion
system and to initiate an integrated research and development effort
focused on key technology problems. Specific tasks include: (a) a test
program to obtain lifetime data on reflector coatings applicable to parabolic trough concentrating mirrors, (b) development of a test technique
to assess the lifetime of Al_2O_3-Mo selective coatings applied to stainless steel substrates, (c) development of a computer analysis program
to characterize the thermal performance of parabolic troughs, (d)
design, fabrication, and test of a 12 meter heatpipe, (e) design, fabrication and test of a scale model parabolic trough, (f) analysis of system
thermal performance including heatpipes, transfer, storage, and boiler
interfaces, and (g) system cost estimates.

Progress under this grant is reported in an NSF semi-annual program
report covering the period July 1, 1972 to December 31, 1972 (NSF/
RANN/SE/GI-34871/PR/72/4) and semi-annual progress report covering the period January 1, 1973 to June 30, 1973 (NSF/RANN/SE/GI-34871/
PR/73/2). A summary of the salient results is presented below.

a. <u>Reflector Coatings</u>

First surface aluminized fiberglass and anodized aluminum reflector sheeting (Alzak) and second surface aluminized acrylics, aluminized glass, aluminized Teflon, silvered Teflon, and aluminized acrylic plexiglass have been exposed on identical test racks in Arizona, Minnesota, and Florida. In general, first surface mirrors performed better in Arizona than in Florida or Minnesota. Second surface mirrors showed little change in performance after being cleaned from the original reflectivity measurement. Cleaning did increase reflectivity of the sample returned directly from the field. These tests as well as accelerated solar exposure tests are continuing.

b. <u>Lifetime Analysis of AL_2O_3-Mo Selective Absorber Coatings</u>

This phase of the effort concerns the life expectancy of solar absorber coatings and in particular of the Honeywell interference coating, consisting of 600 $\overset{o}{A}$ of Al_2O_3 on 200 $\overset{o}{A}$ of MoO_x on 600 $\overset{o}{A}$ of Al_2O_3 (AMA coating) on molybdenum or stainless steel substrates. Changes in reflectance over the wavelength range 0.2 to 10 microns after baking at various temperatures are correlated with composition versus depth profiles obtained by an in situ surface analysis technique combined with sputter etching. Auger Electron Spectroscopy (AES) was selected as the surface analysis technique because of its simplicity, high speed, quantitative capability and applicability to insulation. The composition profile is obtained in an ultra-high vacuum chamber while sputter etching of the sample is in progress, i.e., one obtains the profile as a function of sputtering time.

Relevant results include:

(1) AMA coatings on Mo substrates hold up well optically and show no composition profile changes after baking up to one hour at 950°C.

(2) The same AMA coatings on a thick (3000$\overset{o}{A}$) Mo interlayer on stainless steel show serious optical degradation. Auger analysis reveals that the stainless steel constituents Fe, Ni, and especially Cr have diffused through the thick Mo layer and entered into the AMA coating.

(3) If the same coating (2 above) was baked for 100 hours at 600°C, difficulties arose with peeling of the film from the stainless steel substrate. In unpeeled regions, no change in the diffusion profile was detectable.

(4) Compositional depth profiling of the AMA coating showed that at 400°C in vacuum, the interdiffusion of constituents is negligibly small. The life expectancy of well prepared coatings at 300°C should readily meet the 20 year goal.

(5) Studies on simple-film-substrate combinations (Al_2O_3 on Mo, Al_2O_3 on 304SS, SiO_x on 304SS, and Mo on SiO_x on 304SS) indicate that diffusion problems begin to become serious at temperatures over 600°C.

c. <u>Parabolic Trough Collector Analysis</u>

A Monte Carlo computer program has been developed which traces three-dimensionally the solar rays incident on the mirror through all absorption and reflection processes and through conversion to thermal energy on the absorber surface. The solar flux absorbed over the daily cycle is computed taking account of shadowing by structures and adjacent collectors, of end effects, and of collector orientations.

A parametric investigation was performed to define optimal collector performance. These investigations have lead to the following general results:

(1) The equatorial or polar mount is the best orientation on the basis of the total solar flux intercepted over a year.

(2) When the rotating tracking axis is aligned east-west, the required collector spacing is minimal and equal to a one-collector width gap between collectors.

(3) The optimal concentration ratio (aperature area divided by absorber coating was found to be 35. Optical imperfections and misalignments lower the optimal value to 12.

(4) The optimal rim angle (i. e., the half-angle subtended by the collector from rim to rim as viewed from the focal point) was found to be in the interval from 90 to 130°. A rim angle of 115° was selected because the collector is mass-balanced about the focus line, assuming a uniformly thin mirror.

e.   <u>Design, Fabrication, and Test of a Scale Model Collector</u>

The major objectives of this test are to:

(1)   Validate the steady state optical and thermal performance pre-
      dictions for a parabolic trough-heatpipe collector assembly
      operating at 300°C. Collectors with and without selective
      absorber coatings will be tested.

(2)   Assess start-up and shut-down characteristics of the collector,
      particularly the heatpipe.

(3)   Operate the system only under "average" environmental con-
      ditions; no evaluation of operation under adverse environmental
      conditions.

The test program will be conducted in Phoenix, Arizona and should be
initiated in mid April, 1974.

A summary of progress on the scale model includes:

(1)   Design and fabrication of a two section parabolic concentrator
      (each section 1.2M diam x 2.1m long with a rim angle of 115°)
      has been completed by General Dynamics. The concentrator
      structure is aluminum flexcore with epoxy skins. The front
      surface has an acrylic coating on which a GD-1 aluminized
      coating is deposited. The measured performance of the first
      section of the collector indicated a geometric focusing effi-
      ciency of 98%. Test samples prepared with the surface had
      average reflectances of 87%.

(2)   A water heatpipe is being fabricated by Dynatherm Corporation.
      Pipe material will be 347SS and the pipe dimensions are 2.54
      cm ox x 518 cm long.

(3)   Honeywell Corporation has installed and checked out the vacuum
      evaporation facility which will be utilized to apply the AMA
      selective coating to the heat pipe.

f.   <u>Analysis of System Thermal Performance</u>

System calculations have been performed for a transfer loop configuration in which the heat exchangers, which interface the loop with the
heatpipes; are arranged in parallel. The working fluid in the transfer
loop is water, with saturated steam being generated in each of the heat
exchangers. The calculations were focused on the determination of
heat losses and pumping energy for various piping configurations for a
200 Mwth powerplant. The results of the calculations indicate that

insulation materials with thermal conductivities comparable to air are
necessary to limit heat losses to 5 to 10%. A test fixture has been
fabricated to measure the thermal conductivity of candidate insulation
materials such as Kaowool blanket, Santocel, Thermosil, etc.

Babcock and Wilcox has performed a parametric analysis of the flow
distribution in the feeder manifolds operating in a parallel flow arrangement to determine the flow distribution and pressure drop characteristics of these systems. Designs were considered employing either uniform flow resistance coefficients or flow resistances producing uniform
flow. The designs operating with uniform flow had 50 percent less
pressure drop than the uniform resistance designs.

The transient behavior of the loop during long and short periods of no
sunshine were examined. Based on low thermal conductivity insulation
materials, morning warmup times of 1 to 2 hours are necessary prior
to deriving net power. The estimated daily energy loss is approximately
5% of the total thermal output of the collectors. Calculations were also
performed to determine whether the energy stored in the loop piping
insulation, and fluid would be sufficient to enable the system to produce
steam and continue operating during short periods in which there is no
sun. These calculations indicate that the system thermal inertia would
sustain operation for only a few minutes.

g.   System Cost Estimates

To date, there has been no reported progress in this area.

4.   Feasibility Study of a Solar Thermal Power System Based upon
      Optical Transmission: University of Houston/McDonnell Douglas

The objective of this project is to perform a preliminary technical and
economic feasibility study (including conceptual design) of a solar
thermal power system utilizing a field of heliostats (i. e. , steered
mirrors) and a central thermal collection system. The approach includes completion of tasks on (a) the solar flux concentration subsystem,
including system analysis of the optical characteristics for various

field configurations, guidance and control system studies, and reflector designs, (b) the receiver and energy transfer subsystem including conceptual designs of the tower, and (c) system definition and evaluation with particular attention to economic analysis of the collection, energy transfer, and energy conversion subsystems.

Results of this program have been documented in the semi-annual program report for the period July 1, 1973, NSF/RANN/SE/Gi394561 PR/73/3.

a. <u>Solar Flux Concentration Subsystem</u>

An analytical computer program has been developed which will provide the basis for defining the best array pattern for a field of flat, square heliostats. The program determines the optimum spacing of heliostats including effects of shading and blocking and can be utilized to determine the extent of the mirror field to minimize the cost/unit energy for a fixed receiver size. Outputs from the program will include: (1) contours of annual redirected energy per heliostat for each heliostat in the field, and (2) heliostat spacing over the field.

Preliminary analysis of reflector orientation accuracy requirements indicate that pointing and surface waviness of 1 to 2 milliradians are required to obtain concentration ratios of approximately 1000. A related factor which may limit the geographic applicability of central receiver systems is in degree of atmospheric scattering of the direct solar radiation. A literature search is underway to define the extent of these limitations.

Preliminary design of a rigid heliostat has been completed including definition of a fabrication flow plan. The preferred candidate is a hexagonal steel frame design, 15' across the flats with a second surface silvered mirror. The mirror substrate material is float glass. Preliminary cost estimates indicate that the reflector assembly costs will be in the range of $1.25 to $1.50. ft^2.

In parallel, a number of guidance and tracking concepts are being assessed. The preferred heliostat mount has the primary heliostat axis boresighted with the receiver center. With this orientation, dc motors driving a two stage worm gear in the first axis and a differential gear in the second axis appear capable of meeting the 1 milliradian pointing accuracy. Costs estimates for the tracking and control systems are in progress.

b. <u>Receiver and Energy Transfer Subsystems</u>

A number of conceptual designs for the receiver are being analyzed. Most of these designs assume a hemispherical receiver with tubes on the exterior of the structure. Configurations being considered include:

(1) "parallel" tube direct absorber, (2) a fluted absorber, (3) a staggered tube absorber, and (4) a helical wrap pattern. Another alternative configuration is a cruciform receiver. Preliminary longitudinal and circumferential heat flux distributions have been estimated for various times of the year and a specific heliostat field pattern. Incident heat fluxes are in the range 100,000 to 500,000 Btu/hr ft^2.

Five candidate cooling system alternatives are being considered; (1) 700 psi saturated steam (outlet temperature 503°F), (2) 1500 psi superheated steam (outlet temperature 900°F), (3) 1700 psi pressurized water (outlet temperature 600°F), (4) 50 psi NaK (outlet temperature 1100°F), and (5) open cycle air (outlet temperature - 1800°F).

In parallel, slip cast concrete and guyed steel tower designs are being analyzed for tower heights from 300 to 2000 feet. The major design criterion is a wind velocity of 90 mph at 50 foot elevation. Cost estimates for guyed towers range from approximately $400,000 for a 600 foot tower height to $1,000,000 for a 1500 foot tower. Cost estimates for slip cast concrete structures are significantly higher than guyed towers over this entire tower height range.

c. <u>System Definition and Evaluation</u>

Only preliminary system performance estimates have been generated.
These indicate that equinox thermal power generated varies from 33
Mwth for a 330 foot tower to 1172 Mwth for a 2000 foot tower (for a 63°
rim angle system). Optimum mirror sizes range from 10 to 20 feet on
a side and the total number of mirrors in the field range from 2000 to
150,000. More detailed system performance estimates will be developed as the project progresses.

<u>System Point Designs</u>

5. <u>Solar Power Array for Concentration of Energy: Schjeldahl
Corporation/University of Minnesota/Foster Wheeler</u>

The objective of this project is to estimate the technical and economic
performance of a heliostat (steerable reflector) array - central receiver solar thermal conversion system operating in a hybrid configuration with a conventional fossil fuel generating system. This approach
is one wherein solar energy is utilized when available in a so-called
"energy displacement" mode; thus minimizing or eliminating the requirements for energy storage and simultaneously reducing fossil fuel usage.
The scope of the effort includes: (a) Definition of hybrid powerplant
functional, performance, and operating requirements, (b) Conceptual
design of alternative systems, (c) Developments of analytical techniques
for estimating the incident flux on the receiver and the heat transfer to
the receiver working fluid, (d) Defining a baseline system design, and
(e) Conducting materials analysis and specimen testing to define the
applicability of metallized thin film reflective materials to heliostat
reflectors.

The project is being carried out in conjunction with the Northern States
Power Company. Analysis will be focused on a solar-assisted-coal
fired plant wherein approximately 500 Mw of thermal energy will be
derived from the solar system. System requirements will be defined
for central receivers providing thermal energy for the feedwater
heaters, the evaporator, superheater, or reheat stages. Since the
project was initiated recently, no documented progress is available.

6. <u>Solar Power System and Component Research: Martin Marietta
Corporation/Georgia Tech</u>

The objective of this project is to evaluate the potential of solar thermal
conversion central receiver - hydroelectric hybrid power systems, and
define requirements for a bench model central receiver test program
consistent with the CNRS Solar Energy Laboratory 1 Mwth solar furnace
in France. The specific efforts to be conducted include: (a) systems
analysis of 100 Mwe central receiver solar thermal hydroelectric
powerplant applications, (b) thermal design and analysis of a bench
model boiler and superheater consistent with the 1 Mwth solar furnace
in France, (c) definition of CNRS Solar Energy Laboratory Solar furnace modifications (if any) necessary to test boiler and superheater
components, and (d) development of specific test plans for subsequent
performance testing of boiler and superheater components.

This project is being carried out with representatives of the Salt River
Project, Phoenix, Arizona and the Bonneville Power Administration,
Portland, Oregon. The analysis will consider (a) a direct solar hydroelectric integrated system, (b) a tie-line integrated solar-hydroelectric
system, and (c) a solar powerplant providing power for an aluminum
plant in the Pacific Northwest. Since the project was initiated recently,
no documented progress is available.

<u>Subsystem Level</u>

7. <u>Solar Power Collector Breadboard Test</u>

The objectives of this project are to experimentally verify the operation
and performance of a low cost Fresnel lens collector at operating temperatures of 500-1000°F for incident fluxes of 100-1000 watts per square
meter and to estimate the cost of these units in production. The specific
approach includes: (a) Collector and test equipment concept definition,
(b) 1 meter scale concentrator - collector breadboard design, fabrication and checkout, (c) Test program planning and implementation
utilizing an MIT solar simulator with 0.5 degrees collimation, (d) Data
reduction and test data analysis, and (e) Cost estimates for collectors
in production. Since this project was initiated recently, no documented
progress is available.

8. Chemical Vapor Deposition Research for Fabrication of Solar Energy Converters: University of Arizona

The objectives of this project are to demonstrate the feasibility of fabricated multi-layered surfaces that meet the following criteria:

(a) The optical properties of these surfaces provides the visible absorption/infrared reflectance profile required for efficient photo-thermal conversion of solar energy.

(b) The surfaces retain this spectral profile for extended periods of operation at elevated temperatures in vacuum - typically over 10-20 years at 500°C.

(c) The fabrication techniques lend themselves to economical large-scale manufacture.

Specific efforts include: (a) Deposition of silicon coatings with thicknesses of 2 to 10 microns to increase absorption, (b) Development of stabilized silver infrared reflecting coatings on pyrex or stainless steel substrates, (c) Tailoring the silicon dielectric film index of refraction to decrease reflection, (d) Development of vacuum deposition coating facilities, and (e) Performance of life tests by heating samples in vacuum at temperatures of 300° - 500°C.

Results of this program have been documented in the semi-annual progress report for the period January 1, 1973 to June 30, 1973, NSF/RANN/SE/GI-36731x/PR/73/1 and in the annual progress report for the period January 1, 1973 to December 31, 1973, NSF/RANN/SE/GI-36731x/PR/73/3.

A summary of the salient results of the project to date include:

(1) It has been determined that a set of chemical vapor deposition parameters exist that permit the deposition of a 2 micron thick silicon film of satisfactory optical quality onto a metallized pyrex substrate. Dependence of the optical character of the silicon film was obtained through systematic variation of temperature, flow rate, and composition of the gas mixture performed under in-process emittance control.

(2)   Deterioration of the thin silver reflector through agglomeration at
       temperatures above 350° was eliminated through the use of a
       Cr_2O_3 stabilization layer. Deposition of the stabilizing layer on
       the reflector prevented agglomeration for temperatures in excess
       of $850^{\circ}C$. The infrared reflectance of a silicon film deposited on
       the stabilized silver reflector shows no degradation. Auger
       spectroscopic studies performed at the University of Minnesota
       indicate there was no migration of the silicon through the silver.

(3)   Theoretical analysis has been performed of the infrared reflectance
       to identify the processes that degrade the optical performance of
       an absorber - reflector stack. Comparisons of calculated and mea-
       sured spectra have guided the development to the point where these
       degradation processes could be minimized, and calculated and
       measured reflectance closely agreed.

(4)   Preliminary results of annealing studies show that no degradation
       in the optical performance of the absorber-reflector stack occurred
       after 150 hours anneal at $540^{\circ}C$.

(5)   Anti-reflection coatings of silicon dielectrics have been fabricated
       with indices of refraction $1.45 < n < 2.0$. Critical process control
       variables have been identified and shown to be controllable.

(6)   The basic feasibility of the silver-silicon coating has been demon-
       strated on pyrex substrates.

Specific efforts planned for the next year include:

(1)   Deposition of increasing thicknesses of the silicon absorber to 5
       microns and subsequently to 10 microns.

(2)   Establishment of a baseline chemical vapor deposition process to
       tailor the refractive index of silicon dielectric material over the
       range $1.46 < n < 2.0$.

(3)   Lay down of silver-buffer layer-silicon-silicon dielectric coatings
       on planar stainless steel substrates.

(4)   Development and construction of chemical vapor deposition
       facilities for coating planar stainless steel substrates.

(5)   Measurement of the reflectance of flat samples over the spectral
       range from 0.4 to 15 microns, and at temperatures up to $500^{\circ}C$.

9.   <u>Elimination or Control of Material Problems in Water Heat Pipes:
     McDonnell Douglas Astronautics Co.</u>

The objective of this research project is to determine an effective combination of materials, fabrication methods, and operating procedures
for a water heatpipe that will provide satisfactory long life operating

performance for temperatures up to $310^{\circ}C$. Specific research efforts
include: (a) Definition of the most effective combination of design,
materials, fabrication methods, and operating procedures that eliminates or minimizes noncondensible gases in water heatpipes, (b) Fabrication and test of a 15-20 meter long water heatpipe that utilizes the
procedures established in (a) and is projected to be cost effective and
compatible with full-scale production and preparation methods, and
(c) Establishment of material and fabrication process specifications
applicable to a full-scale solar collector heatpipe life test. Since this
project was initiated recently, no documented progress is available.

10. <u>Air-Stable Selective Surfaces for Solar Energy Collectors: Helio
    Associates</u>

The objectives of this research project are to develop selective absorber
coatings suitable for various substrate materials and verify the performance of the coatings at $150^{\circ}C$ in the presence of air for one year and to
define specific collector configurations most conductive to the use of
these coatings. The specific approach includes: (a) Screening of candidate coatings and selection of those coatings deemed most appropriate
for air stable operation at $150^{\circ}C$, (b) An optical measurements program
for characterizing the optical properties of the coatings, (c) Coating
lifetime analysis and one year test program, (d) Collector performance
analysis for selected coatings, (e) Collector test module design and
test plan definition, and (f) Cost estimates for producing coated collectors in production. Since this project was initiated recently, no documented progress is available.

III. OTHER RELATED PROGRAMS

The two other major projects underway in solar thermal conversion
are: Dynamic Conversion of Solar Generated Heat to Electricity being
performed by the Honeywell Corporation under NASA contract and the
Solar Community and the Cascaded Energy Concept Applied to a Single
House and a Small Subdivision being performed by the Sandia Corporation under AEC sponsorship and in-house funding. As shown in Figure
 1 the   Honeywell project parallels the work being performed by

Colorado State University and Westinghouse. The Sandia system study addresses one potential solar total energy application.

1. Dynamic Conversion of Solar Generated Heat to Electricity: Honeywell Corporation/Black and Veatch

This summary is of interim results based on monthly progress reports 1 through 4. The project was initiated September 6, 1973 and is for a 10 month duration.

The objective of this project is to identify concepts for large-scale exploitation of solar energy for the generation of electricity and to evaluate those concepts technically and economically and recommend the most promising for further investigation. This investigation shall result in the conceptual design, estimation of costs and the identification of additional research and/or development required to achieve realization of the design concept and/or cost goals.

Some of the major conclusions of the study, to date, include:

(a) The optimal siting location for concentrating collectors and low power level central receivers in Inyokern, California. The optimum location for large central receiver towers is an area near Blythe, California.

(b) Based on optimum flat plate collector temperatures, the overall efficiency of a 100 Mwe powerplant is estimated to be less than 3 percent. Comparison of this system with a solar parabolic trough system indicates that 40 times as many flat plate collectors and 1.5 times as much land area is required to generate 100 Mwe of electric power. Based on this analysis, flat plate electrical generating systems are marginal, at best.

(c) The efficiency for a dual cycle conversion system (closed cycle helium followed by steam bottoming cycle) with a maximum temperature of 1529°F has been estimated to be 43.6% assuming dry cooling tower heat rejection. This conversion system is applicable to the central receiver.

(d) Analysis of the performance of parabolic dish collectors indicates that although temperatures of 1500°F can be obtained, it may not be practical to employ parabolic collectors because of difficulties with heat transport fluids, the piping network, and the requirement for a secondary heat exchanger between the transport fluid and the turbine working fluid.

(e) Polar mounted parabolic trough collectors produce the maximum
net energy over a year compared to E-W or N-S oriented systems.
For the same aperture area, the net energy collected by a trough
system is greater than that of a parabolic dish; however, achievable temperatures with the trough are approximately 600°F compared to 1000°F to 1500°F for the parabolic dish collectors.

2. <u>The Solar Community and the Cascaded Energy Concept Applied to
a Single House and a Small Subdivision - A Status Report: Sandia
Corporation SLA 73-0357, May 1973</u>

Sandia Laboratories has recently been studying methods in which solar

energy can be utilized for residential energy needs. A solar community

concept has resulted wherein solar energy is used to provide electricity,

heating, cooling, and domestic hot water. The solar community may

consist of a single house, a small subdivision, or a large complex of

1000 to 2000 houses. The system considered involves the central

collection and storage of solar energy at temperatures of about 450°F

to 500°F and power generation with an organic Rankine cycle turbine

which has been derated to allow the exhaust energy to be used for heating,

cooling, and hot water.

To date, the analysis has been limited to an isolation input which could

be expected in Albuquerque, New Mexico. Results of the preliminary

analyses indicates: a) Solar communities utilizing the Cascaded energy

concept with 450°F to 500°F collection temperatures could be cost

competitive between 1985 and 2020. b) The solar community could pro-

vide about 60 to 70 percent fuel savings. c) Central collection and

distribution of the electrical and thermal energy is more cost effective

than collection at individual residences. d) Cost estimates for a para-

bolic trough collector which attains temperature of 500°F are estimated

to be between $2.00 and $4.00 per square foot. e) Energy storage is

accomplished with the intrinsic heat capacity of water in large, pressur-

ized tanks. f) Turbine performance is based on the use of n-pentane as

the working fluid with minimum inlet temperature of 360°F.

Although other programs in solar thermal conversion system analysis

and technology development are undoubtedly underway, the preceding

summary of NSF sponsored and other agency sponsored efforts is representative of today's state of knowledge on solar thermal systems within the United States.

IV. SYSTEM ALTERNATIVES AND MAJOR TECHNICAL PROBLEMS

Although there are a large number of solar thermal system alternatives which can be identified, the major system factor affecting performance is the maximum expected collector temperature. Table 3 summarizes the major alternative systems by projected operating temperature and collector approach. Comments relative to the performance of these systems are also presented. All focusing systems require climates which have a high fraction of direct solar radiation available during daylight hours. Therefore, the geographic applicability of these systems w ill be limited primarily to the Southwestern United States. The flat plate or low concentration systems appear to have marginal performance for electrical power generation; however, they appear to have a potential role in solar total energy applications. These systems will have broader geographic applicability.

A major technical problem associated with all solar thermal conversion systems stems from the intermittent nature of solar energy. These systems must be capable of accepting diurnal thermal transients and rapid transients during daily operation. This fundamental problem will probably be the major factor in selecting the most preferred system alternatives. The other limiting factor will probably be the achievable lifetimes of components and subsystems. Since solar thermal conversion systems are capital intensive, component lifetimes of 15 to 20 years are necessary to obtain competing energy costs. Accelerated test programs which can be utilized to project component lifetimes are essential.

A summary of other major technical problems which must be considered in the design of alternative solar thermal conversion systems is given in Table 4. As is well known, the major problem is obtaining efficient collectors with costs of \$2. 00 to \$4. 00 per square foot. In the case of the central receiver, this is the required cost of the heliostats.

Table 3

MAJOR ALTERNATIVE SOLAR THERMAL SYSTEMS

Projected Operating Temperatures	Collector Type	Comments
1000-1500°F	Central Receiver (2 axis tracking)	Operating temperatures competitive with fossil fuel sources. Thermal conversion in relatively small region. Performance not well defined.
1000-1500°F	Parabolic Dish (2 axis tracking)	Operating temperatures competitive with fossil fuel sources. Thermal energy extraction from focal zone and transfer to energy conversion equipment, major limiting factor.
600-1000°F	Fresnel Lens (1 or 2 axis tracking)	Operating temperatures between nuclear and fossil fuel plants. Cost of refractive optics and thermal losses major problems.
400-600°F	Parabolic Trough (Single axis tracking)	Operating temperatures competitive with pressurized water nuclear reactors. Performance well characterized. Requires selective surfaces and vacuum envelopes.
300-400°F	Boosted Flat Plates, Low Concentration Collectors (Fixed)	Operating temperatures marginal for electrical power production. Probably limited to solar total energy system applications.
200-300°F	Flat Plate, Planar Collectors (Fixed)	Operating temperatures probably too low for electrical power production. Most likely will be limited to individual building heating and cooling.

Table 4

MAJOR TECHNICAL PROBLEMS FOR
SOLAR THERMAL SYSTEMS

<u>Solar Collector Type</u>	<u>Major Technical Problems</u>
Central Receiver	Transient Thermal Performance of Receiver
	Low-Cost High Surface Accuracy Flat Mirrors (Heliostats)
	High Temperature Thermal Storage
	Gas Turbine Conversion Technology
Parabolic Dish	Thermal Energy Extraction From Focal Zone
	High Temperature Energy Transport to Storage or Conversion Equipments
	Highly Efficient Insulation Materials
	Low-Cost High Surface Accuracy Parabolic Mirrors
	High Temperature Thermal Storage
Fresnel Lens	Low-Cost, Refractive Optics
	High Temperature Energy Transport to Storage or Conversion Equipment
	Highly Efficient Insulation Materials
	High Temperature Thermal Storage
Parabolic Trough	Low-Cost, High Surface Accuracy Parabolic Cylinder Mirrors
	Vacuum Jacketed Receiver Heat Pipe
	Intermediate Temperature Thermal Transport to Storage or Conversion Equipment
	Efficient Thermal Insulation
	Intermediate Temperature Thermal Storage
Boosted Flat Plates or Low Concentration Systems	Low-Cost Collector Designs
	Absorber Selective Coatings Which Are Air Compatible
	Organic Rankine Cycle Conversion Technology
Flat Plate or Planar	Low-Cost Collector Designs
	Absorber Selective Coatings Which are Air Compatible
	Organic Rankine Cycle Conversion Technology

Another major requirement for all of the systems operating at temperatures above 400°F is the need for efficient thermal storage approaches. Steam conversion technology can be utilized for most applications; however, there appears to be a meaningful role for high temperature closed cycle gas turbines in conjunction with the central receiver and organic Rankine cycle conversion systems in conjunction with low-concentration systems.

V.   CONCLUSIONS

Based on the analyses and limited technology developments performed to date, the following general conclusions relative to the roles for solar thermal conversion systems are drawn:

1. Solar thermal conversion electrical powerplants appear to be most competitive with fossil fuels powerplants for intermediate or peaking power applications.

2. Solar total energy systems servicing industrial centers, communities, military bases, and rural needs are promising alternative applications.

3. Central receiver solar thermal systems appear to offer the greatest potential for competitive economic performance for electrical generation applications. Domestic use of these systems may be limited to the Southwestern United States.

4. Non-concentrating or low-concentration systems offer the greatest potential for solar total energy systems and appear to have a broad geographic applicability.

5. No fundamental technical breakthroughs are required to develop solar thermal conversion systems. A focused research and development program is necessary to assess the potential of these systems. Subsequently, pilot plant operations of 5 to 10 Mwe are necessary to assess system performance, particularly transient thermal performance.

THE SOLAR COMMUNITY - ENERGY FOR RESIDENTIAL HEATING, COOLING, AND ELECTRICAL POWER[*]

W. H. McCulloch
D. O. Lee
W. P. Schimmel, Jr.[**]

A series of systems studies on the potential uses of solar energy has been conducted at Sandia Laboratories. The outcome of these studies is a new concept, the Solar Total Energy Community. This is a residential community which could significantly reduce its fossil fuel energy consumption by using the sun as the source for most of the community's energy needs.

A system computer program, developed for the study, has been used to examine several candidate systems and to optimize the operation of the interrelated components which provide space heating, air conditioning, water heating, and electricity for residences and light commercial buildings. An experimental program has been initiated to investigate various technological areas relative to the concept.

The study has shown that the Solar Community is technologically feasible and that the projected costs warrant the further investigation of solar energy as an alternative residential energy source. This paper reviews the previous work, reports recent findings and improvements, and presents the current status of the continuing analytical and experimental efforts.

INTRODUCTION

Recently, the industrialized nations of the world, particularly the United States, have become acutely aware of the set of circumstances which have become known as the energy crisis. One consequence has been consideration, or more appropriately, reconsideration, of several

---

[*] This work was supported by the United States Atomic Energy Commission.

[**] Criteria and Heat Transfer Division, Sandia Laboratories, Albuquerque, New Mexico

possible alternative energy sources. A group of engineers and scientists
at Sandia Laboratories has been considering the possibility of effectively
utilizing solar energy as a means of reducing the drain on our fossil fuel
resources. A series of systems studies on the potential uses of solar
energy has been conducted and some conclusions have been reached in this
initial effort. First, the technology required for effective collection
and utilization of solar radiation is, for the most part, presently
available. Second, due to the increasing costs of energy obtained from
fossil fuels, the economic feasibility of solar energy may be expected
within a reasonable time. Third, the utilization of solar energy meets
with public approval, because it can be accomplished without the environmental impact of conventional, and some of the proposed new, energy
sources, and in a manner which requires no unacceptable perturbations of
our life style. Fourth, solar energy can be used to displace a significant segment of energy now derived from fossil fuels by supplying the
total energy needs of residential communities.

The specific outcome of these studies is a new concept: the Solar Total
Energy Community. This is envisioned as a residential community which
substantially reduces its fossil fuel consumption by using the sun as
the source of most of the community's energy needs. By including air
conditioning and generation of electricity, this system goes somewhat
beyond the economically and technologically feasible solar-powered space
and water heating systems currently being developed for the commercial
market. A number of configurations for the Solar Community system have
been considered.[1] The most promising system incorporates the total
energy concept, i.e., "cascading" energy requirements such that the energy
rejected by high quality energy uses, such as electrical generation, can
be used to meet lower quality needs such as space heating.

Analyses of the Solar Community and solar central power stations based
on what was regarded as presently available technology have been performed and the results show that Solar Communities utilizing the cascaded
energy concept are more cost effective than central power stations. The
advantage of residential communities over central power stations derives
from the demand for both high and low quality energy at the same location.

Transmission of low quality energy over even moderate distances, e.g., "waste" heat from a central power station to a housing subdivision large enough to make use of the energy for space heating, is very expensive. The analyses have also shown that, relative to communities using fossil fuel energy sources, the Solar Total Energy Community with a fossil fuel back-up system to provide energy during periods of inadequate insolation can provide about a 60 to 70 percent annual savings in fuel consumption.

SYSTEM DESCRIPTION

The Solar Total Energy Community concept is a residential community, perhaps with some light commercial buildings, in which solar energy is used to provide electricity, space heating, air conditioning, and domestic hot water. The energy system shown schematically in Fig. 1, consists of a central collection area, an energy storage facility, a turbine/generator, an auxiliary boiler, absorption air conditioning equipment, heat exchangers, and an energy transmission system. It should be noted that the description of the system presented here reflects the results of studies to this time. Future developments, from analysis, experimental investigations, and surveys of available components, parts, and materials, can be expected to have some impact upon the system configuration and the operational parameters for the various components.

The system collects solar energy with focused collectors to provide storage temperatures (450°-500°F) sufficient to meet the input temperature requirement for the turbine. An auxiliary fossil fuel fired boiler is included to provide energy in the event of insufficient energy in the storage system. Electrical generation for the community is powered by a turbine which operates at relatively low input temperatures, about 360°F, and which uses a high molecular weight, organic fluid as the working fluid. The turbine is derated, i.e., its output temperature is elevated so that the exhaust energy may be used for space and water heating and air conditioning. A lower temperature storage unit, about 200°F, is the buffer between the turbine output and the residential demands. A cooling tower, not shown in the schematic, rejects energy from the absorption air conditioner or excess energy when the storage units are filled.

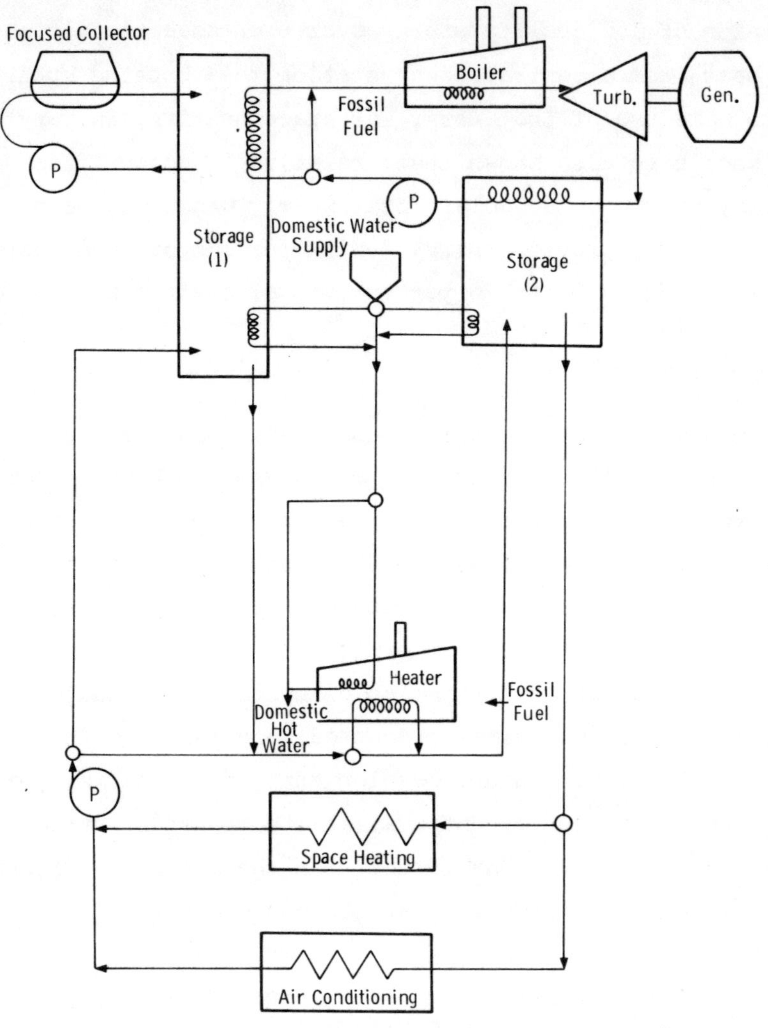

Fig. 1   Schematic Representation of Solar Community

The optimum number of residential units to be included in the community
is yet to be determined. Selecting the most advantageous size for the
community involves a trade-off between the savings in cost and convenience
realized by sharing the collection, generation, and storage facilities
and the expense of the low-temperature distribution system. One study
has shown that a community of 20 units is more cost effective than a
single house[2], but a very large community (thousands of units) requires
an extensive distribution system which proves to be prohibitively expensive.

PREVIOUS ANALYSIS

Solar Input Data

The first step in the analysis of the collection and utilization of
solar energy is the acquisition of solar input and weather data which
reasonably represents the actual situation. For this study, the data
were derived from Weather Service records for Albuquerque, New Mexico,
which spanned the period 1950-1972. To avoid being overwhelmed by hourly
data for 23 years, consideration was limited to four 5-day periods during
the year, centered about the autumnal and vernal equinoxes and the summer
and winter solstices. This gave approximately 100 data points to be used
to determine each hourly point for the four characteristic days used in
the analyses. Because weather data are not distributed normally, a
weighted mean based on a most probable value for a particular time was
used to determine meaningful values for the solar insolation. The results of this statistical determination are shown in Table 1.

Table 1

INSOLATION[*] FOR FOUR CHARACTERISTIC DAYS

Hour	March 21	June 21	September 21	December 21
1	0	0	0	0
2	0	0	0	0
3	0	0	0	0
4	0	0	0	0
5	0	0	0	0
6	0	152	0	0
7	188	236	188	0
8	273	289	261	212
9	307	311	297	257
10	328	326	310	299
11	344	335	320	323
12	341	333	317	333
13	329	331	306	325
14	321	318	302	319
15	304	295	292	301
16	292	265	254	271
17	254	232	222	224
18	191	185	150	0
19	0	128	0	0
20	0	0	0	0
21	0	0	0	0
22	0	0	0	0
23	0	0	0	0
24	0	0	0	0

[*]Solar flux (Btu/ft^2-hr) on a surface normal to incident beam.

The weather data records show cloudiness for about one-fourth of the
hourly readings. In this analysis, cloudiness is included by considering
one day in four to have no solar input. Data for the month-to-month

variation of specular radiation, i.e., that fraction of the total incident solar radiation which can be collected with a focusing collector,
are not presently available for Albuquerque. However, such data do
exist for Gilat, Israel[3], and the normal incidence records for Albuquerque indicate that the ratio of direct or specular to total radiation is
about 0.8 for cloudless summer days. This information was used to estimate the average monthly variation of specular radiation for Albuquerque.
The estimates are shown in Table 2.

Table 2
ESTIMATED AVERAGE FRACTION OF SPECULAR RADIATION
IN ALBUQUERQUE, NEW MEXICO

Month	Fraction of Specular Radiation
January	0.65
February	0.59
March	0.63
April	0.66
May	0.68
June	0.77
July	0.80
August	0.79
September	0.79
October	0.76
November	0.75
December	0.64

Collectors

The fundamental component of the Solar Community system is the solar
collector, and the systems studies have shown it to be critical to both
the cost and performance of the system. Therefore, most of the effort
has been directed toward understanding the improving the solar collector
design and performance. Any object exposed to sunlight absorbs energy
in the form of primarily short wavelength radiation at a rate dependent
upon its configuration and surface properties. The object, heated to a
temperature above that of its surroundings, then loses energy to the
surroundings by conduction, convection, and radiation. The radiation
heat loss is infrared radiation, i.e., predominatly of a longer wavelength which is characteristic of the temperature of the object, and is
also dependent upon configuration and surface properties. The radiative
properties of many surfaces are such that the surface absorbs short wavelength sunlight more effectively than it emits long wavelength radiation.
Such surfaces are said to be selective and are quite advantageous in
solar collectors.

394

A solar collector, then, is an apparatus designed to maximize the rate
at which solar energy is absorbed while minimizing the losses to the
surroundings such that there is a net rate at which energy is available
for beneficial utilization. The mechanisms by which energy is lost all
become more significant as the temperature difference between the collector and its environment increases. Therefore, the efficiency of a
solar collector, defined as the ratio of the rate at which usable energy
is delivered to the rate at which solar energy is intercepted by the
collector, is a decreasing function of the temperature at which the output energy is provided.

Several solar collector designs have been considered in this study
including flat plates and a variety of focusing collectors. The configuration shown schematically in Fig. 2 has been tentatively chosen as the
most appropriate for the Solar Community system. The reflector is a
parabolic cylinder which focuses the insolation onto a receiver tube
which is selectively coated. The thermal losses from the receiver are
minimized by a glass envelope. This glass enclosure, due to the radiation properties of glass, acts as a radiation "trap," transmitting the
short wavelength sunlight but absorbing the infrared radiation emitted

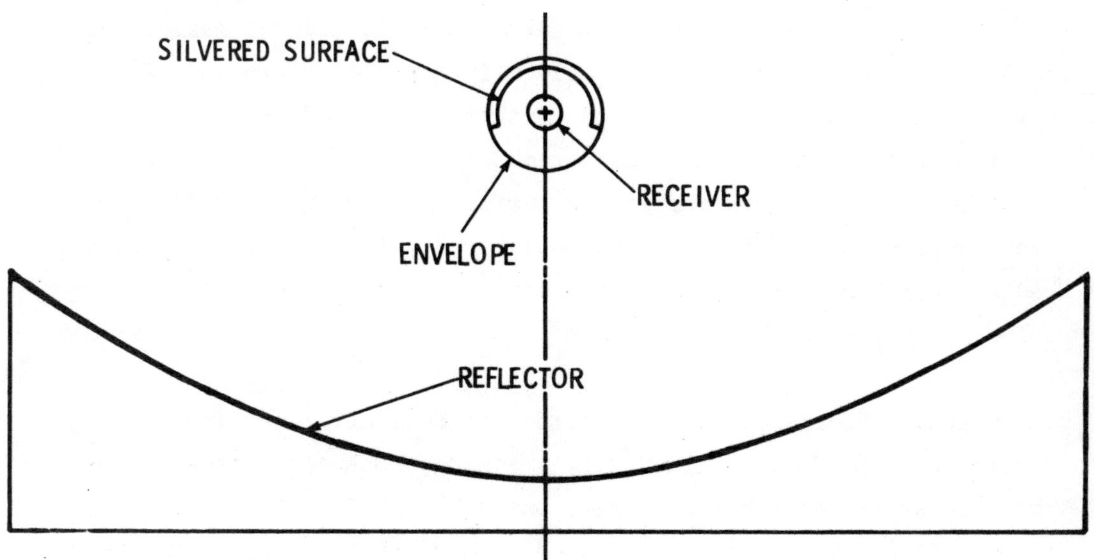

Fig. 2   Cross-Sectional View of Focused Collector

from the receiver. The envelope may be (1) given a reflective coating
on its inside surface to impede the radiation heat transfer, (2) evacuated to prevent convection, and (3) insulated on the outside to minimize
losses to the environment. The silvering and insulating are limited to
that part of the circumference which does not receive the reflected
solar flux. Analyses have indicated that collectors of this design can
provide adequate performance for the Solar Community using materials and
surface coatings which are presently available and without requiring
evacuation of the envelope. This permits the construction of a system
which is relatively inexpensive and which is independent of technological
breakthroughs.

Once the basic design of the collector has been selected, factors external
to the collector must be considered. Specifically, how are the performance and cost of a matrix of collectors affected by orientation and
spacing? Collectors mounted such that they maintain the maximum inter-
ception of solar radiation and spaced such that there is never any
shading of one collector by another obviously provide the maximum performance. However, fully tracking, unshaded collectors also require the
most land area, the most interconnecting plumbing, and the most complex
tracking hardware. Alternatives to the fully tracking system include
(1) collectors in which the longitudinal axis is located horizontally
east to west and which track rotationally north to south and (2) collectors in which the axis is located north to south and is tilted at some
angle toward the south and which track rotationally east to west. Each
of these configurations sacrifices performance because a smaller projected
area is presented to the incident solar radiation and because some of the
radiation intercepted by the reflector is not reflected onto the receiver, i.e., there are end losses.

The hourly performance of these different configurations for a clear June
day is shown in Fig. 3. The fully tracking, unshaded system provides
the highest collection efficiency, but the unshaded, N-S collector,
tilted 35° (latitude of Albuquerque) toward the south, performs almost
as well and should be significantly less expensive because of a simpler
mounting and tracking system. For these calculations, the receiver tube

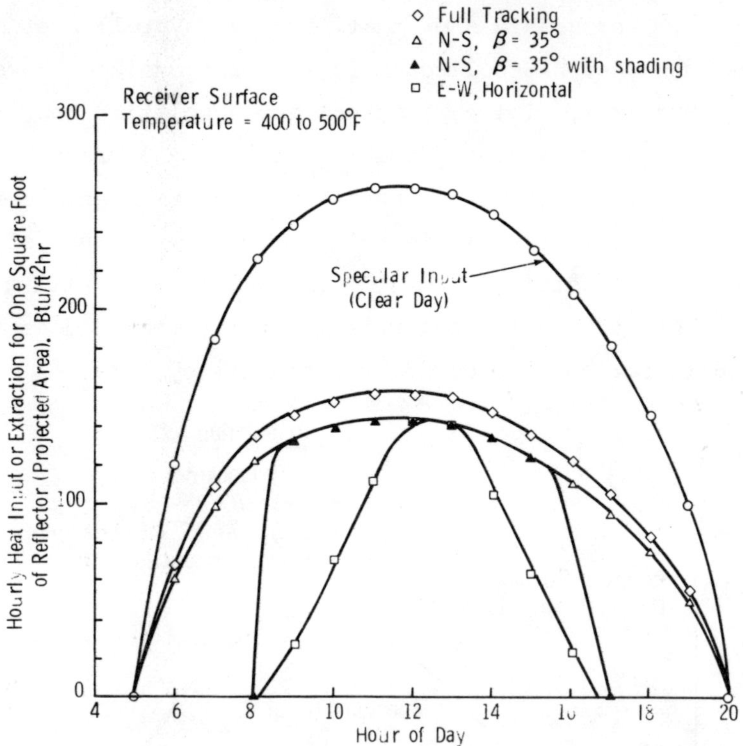

Fig. 3   Hourly Variation of Solar Flux and Extracted Energy
for Focused Collector (Albuquerque, June 21)

in the N-S collector was considered to be extended to minimize end losses.
The E-W collector, without end extensions, is substantially less efficient.

Figure 3 also shows the impact of shading on the performance of the N-S
collector. The output is shown for the N-S collectors which are spaced
such that there is no shading on a collector due to its southern nearest
neighbor and which have a space equal to the width of the reflector
between each collector and its eastern and western nearest neighbors.
The edge effect of the field of collectors, i.e., no shading on the row
of collectors nearest the sun, is ignored. Comparison of the curves for
shaded and unshaded N-S collectors show that the effect of shading is to
give the collectors rather abrupt start-up and shut-down times so that
early morning and late evening insolation is lost. The same impact
would obtain for similarly spaced fully tracking collectors.

The annual input and extraction performance of the above configurations is
shown in Fig. 4. These data show that, from an overall system performance
viewpoint, an unshaded N-S collector is the best choice. However, the
realities of land area, piping, pumping, and insulation requirements make
the consideration of very widely spaced collectors impractical. Similarly,
from an architectural viewpoint, the E-W collectors might be the most
desirable, but a severe penalty is paid in the performance of the system.
Therefore, a N-S collector tilted 35° toward the south with a spacing
equal to the width of the reflector is generally regarded as the most
likely candidate for commercial Solar Community systems.

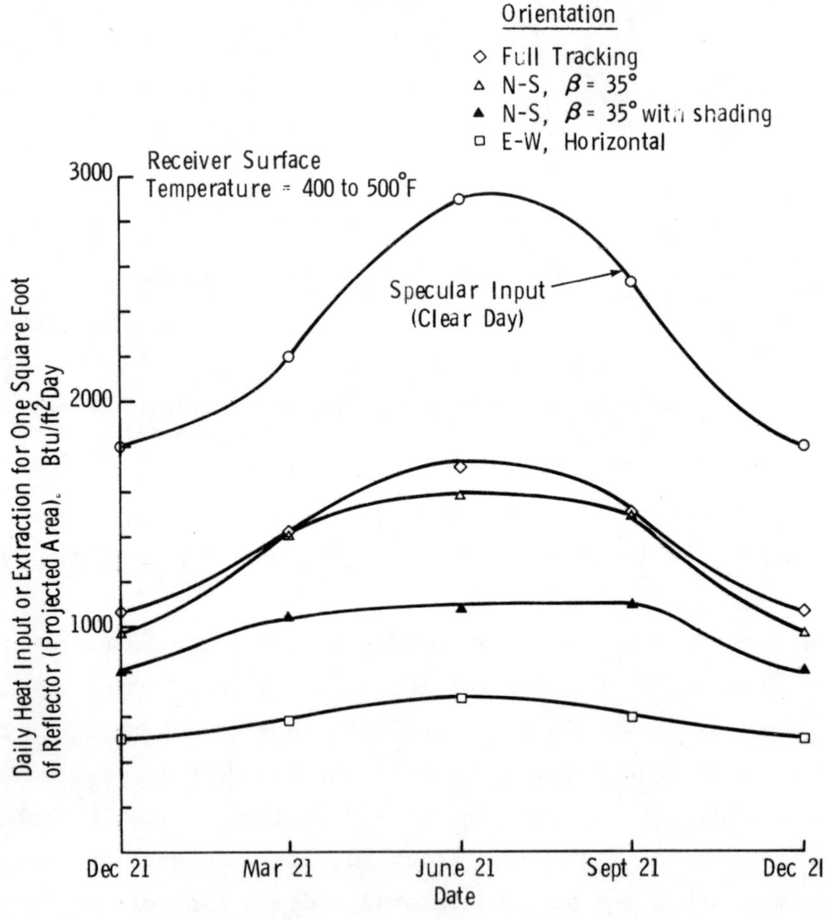

Fig. 4   Approximate Daily Variation of Solar Flux and
Extracted Energy for Focused Collector (Albuquerque)

Thermal Energy Storage

The storage of energy is accomplished with the intrinsic heat capacity
of water in large, pressurized tanks which are well insulated to prevent
excessive losses. Storage Unit 1 provides relatively high-quality energy
for the turbine. When the water temperature is between about $365^{o}F$ and
$450^{o}F$, energy can be extracted for the turbine. The space heating, air
conditioning, and hot water systems can demand energy from Storage Unit
1 if energy is not available in Storage Unit 2. For the systems
analysis to date, the losses from the storage systems have been neglected.

Turbine

As a result of the low working temperatures available for electrical
generation in the Solar Community system, Rankine turbine cycles using
organic working fluids have been considered the best means of driving the
generator. Of the various organic fluids considered, n-pentane provides
the highest cycle efficiency for the specific temperatures considered.
Other factors, such as toxicity and corrosiveness, should also be weighed
in the final selection of the working fluid for commercial applications.
The turbine must operate between the temperatures of the storage units,
$400^{o}F$ and $200^{o}F$. For the cascaded system analysis, the turbine was
assumed to operate between $360^{o}F$ and $254^{o}F$ with the inlet and outlet
pressures equal to 391 psia and 74 psia, respectively. The overall
efficiency for such conditions is approximately 11 percent.

System Loads and Size

Representative values for the energy demands for electricity, space
heating, air conditioning, and water heating have been developed and are
reported in some detail in Reference 2. Both the annual electricity
consumption and the average hourly load variation have been estimated
from data from individual homeowners and the Public Service Company of
New Mexico. The hourly heating and cooling requirements are determined
for each analysis using average ambient temperatures obtained from
Weather Service data and effective heat transfer coefficients for residences insulated about 20 percent better than present FHA standards.
The heating is assumed to be provided by a circulating hot water system

with 100 percent efficiency within the residence. The air conditioning
is assumed to be provided by an absorptive refrigeration system with a
generator temperature of $150^{O}F$ and a coefficient of performance of 0.4.
The demand for domestic hot water is assumed to be 120 gallon per residence per day.

The analytical model of the system and the required inputs as described
here have been used to predict the annual performance of the Solar
Community in terms of fuel consumed and effective cost. These calculations do not, however, indicate the performance which could be expected
during extreme weather conditions, and it is the extreme conditions
which determine the capacities or sizes of the system components.
Therefore, data have been obtained for the extreme cold and hot periods
which have been recorded in Albuquerque and are used for sizing the system.

Cost Analysis

In the final analysis, the Solar Community (as well as any other alternative energy source) must compete economically to be a significant part
of the energy picture of the future. The analyses performed to date have
attempted to include in a realistic way consideration of the costs of
utilizing "free" solar energy. Some of the costs can be reasonably
estimated, but two areas exist in which there is extreme difficulty in
reliably predicting costs. The cost of mass-produced solar collectors
is, at best, an educated guess because so few have been built, and estimates range from less than one dollar per square foot to tens of dollars
per square foot as reasonable costs. The second major uncertainty in
the cost comparison is the standard to which the Solar Community is
compared, the "normal" community. The normal community is defined as
having the same energy demands as the Solar Community currently under
consideration. However, the electricity is provided by a public utility.
Air conditioning is provided by an absorption air conditioner, heating
by a forced air furnace, and hot water by a conventional water heater,
all fueled into the future. Early in 1973, a rather extensive literature
search provided estimates of the costs of natural gas and electricity
projected to the year 2020. The projection has already proven to be
inadequate, but, because of the rapidly changing fossil fuel market and

the wide variation in prices geographically, updated estimates are impossible to make with any degree of certainty. Perhaps the greatest uncertainty in projections of energy consumption and pricing is the inability to predict the changes in governmental policy and procedures which can be expected over the next few years.

One other factor which has a strong influence upon the competitive position of the Solar Community is the rate used to capitalize the required initial investment. These studies have used two values, 12 percent, which is thought to represent standard capitalization, and 8 percent, which is the applicable rate for government project type of financing.

With these uncertainties in mind, comparative cost analyses have been made; and the most recent estimate is that energy supplied by a Solar Community system will be about twice as expensive as energy presently derived from fossil fuels. Therefore, it would appear quite possible that, by the time the concept could be made available on a commercial basis, the Solar Community will be economically competitive. On this basis, it has been concluded that further investigation of the Solar Community as an alternative residential energy source is warranted.

To help in understanding the costs of a Solar Community, the relative estimated costs of the major components for one specific system, shown in Fig. 5, are presented in Fig. 6. This Solar Community is a twenty-house subdivision with 2000 ft^2 of floor space in each residence. There is central collection (20,000 ft^2 of collector area), electrical generation, and thermal storage; and it is assumed the capitalization rate is 12 percent and that the collectors can be built for $4 per square foot. As is the case for all solar systems, the Solar Community cost is dominated by the collector system.

Further details concerning the costs of components and their operation and other information relative to the cost analysis are reported in References 1 and 2.

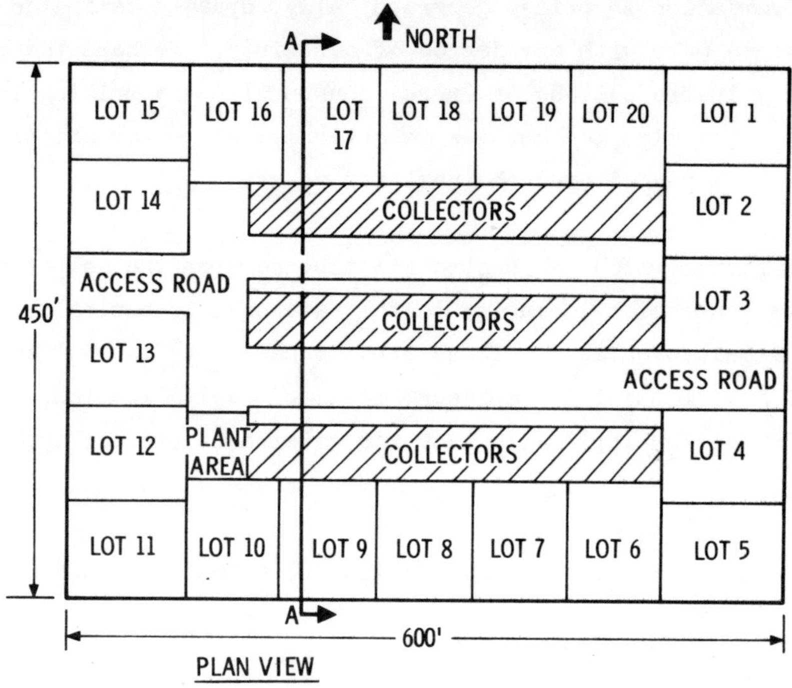

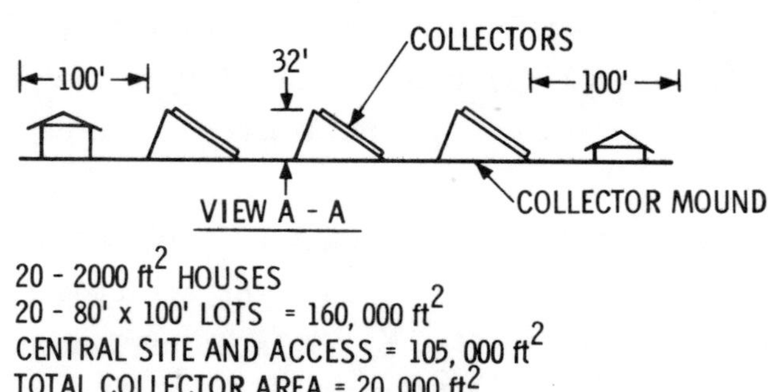

20 - 2000 ft^2 HOUSES
20 - 80' x 100' LOTS = 160,000 ft^2
CENTRAL SITE AND ACCESS = 105,000 ft^2
TOTAL COLLECTOR AREA = 20,000 ft^2

Fig. 5   Schematic of a Typical Solar Community Subdivision Consisting of Twenty 2000-Square-Foot Houses and Central Collection and Processing

Conclusions to Date

The analyses which have been performed have yielded a number of general conclusions, as well as much detailed information.

1. A Solar Community which saves from 60 to 70 percent on fuel consumption could be built on the basis of present or near future

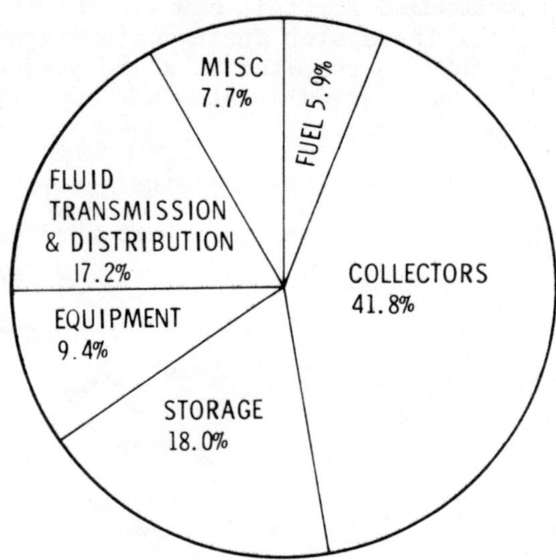

Fig. 6   Annual Total Energy Cost Percentage for a Solar Community Subdivision
(20 houses in subdivision, 20,000 square feet of collectors, 12%
capitalization rate, $4 per square foot - collector cost)

technology. The system would utilize relatively low temperature
(400°F-500°F) energy collection, storage in water, a vapor turbine,
and cascaded energy for space heating, air conditioning, and domestic hot water.

2. Both a large residential area, about 2000 residential units, and a
single solar house have been shown to be less advantageous than a
small solar subdivision of twenty houses. Some savings is possible
by sharing the collection, electrical generation, and thermal storage facilities, but, as the size of the community increases, the
cost of the low-temperature energy distribution system becomes prohibitive. Also, the subdivision concept allows more architectural
flexibility than a single house.

3. While cost parity of the Solar Community cannot be assured because
of the many uncertainties in projecting energy supply and demand
into the future, neither can it be stated conclusively that the
Solar Community will be unable to compete economically. Therefore,
this effort should be pursued vigorously, at least to the point at
which a definitive assessment can be made.

4. The size of a collector field for a particular application can be
reduced to provide a more economical system without seriously de-
grading the fuel savings. In one study, a 38 percent reduction in
collector area resulted in a 30 percent reduction in cost and the
fuel savings relative to a normal community decreased only from 67
percent savings to 58 percent.

5. The air conditioner coefficient of performance needs to be improved
    over the value 0.4 assumed for this study. The extreme weather con-
    ditions proved to tax the system during extended hot periods.
    Increasing the COP to 0.75 resulted in a 480 percent increase in fuel
    savings for an 8-day hot period and the air conditioning pump require-
    ment was reduced by 47 percent.

It should be emphasized that the above conclusions represent a status
report, presenting the results of analyses performed primarily before
July 1973, and, as such, they are subject to further investigation, both
analytical and empirical. As the effort proceeds toward the design of
specific systems, some modification of the configuration and operational
parameters should be expected.

CURRENT STATUS AND ACTIVITIES

This effort has been and continues to be made up of activities in four
broad categories. First, much time is required in understanding and
communicating the relationships between Sandia's Solar Community Program
and the general area of solar energy. Second, a large number of systems
analyses have been made to evaluate the performance of a variety of ways
to utilize effectively the solar energy incident on the earth's surface,
and these analyses continue as the system design is developed. Third,
existing technologies are continually studied and evaluated to determine
what hardware is available or could reasonably be made available for use
in the Solar Community. Fourth, an experimental program is underway with
two primary goals: to develop an efficient solar collector for this
application and to demonstrate the feasibility of the concept by operating a representative model of the Solar Community system.

Establishing a Perspective

The energy crisis is a problem which is much more than simply a technical
challenge, and anyone endeavoring to determine solutions must concern
himself with both the technical difficulties in providing equipment to do
the job and the socio-economic implications of any proposed activity. To
accomplish this task, we have studied the energy problem in general to
better understand what impact the Solar Community might have on future
energy consumption patterns. Because the proposal[4] is a residential
community, a significant portion of the effort involves relating to

planning departments of cities and utility companies, builders and con-
tractors, and the public, i.e., the potential residents of a Solar Community. The concept has been met with widespread enthusiasm, encouragement,
invitations to present the idea to various groups, and inquiries as to how
the ideas might have application in other activities. Some of the more
tangible results in this area are the inclusion of communities which could
be solar-powered in the long range planning for the City of Albuquerque
and the interest expressed by a local utility company in participating in
the construction of a demonstration Solar Community in the late 1970's.

Systems Analysis

The data for the performance of various systems which have been presented
to date have been the product of systems computer programs developed
early in the effort. These programs have been modified as needs have
dictated. Presently near completion is the first version of a more com-
prehensive and useful systems program which gives the analyst a more
flexible and powerful capability. In the new program, effort is being
made to model each component more representatively, and the program is
designed such that more precise, updated, analytical representations of
the system components can be incorporated without major revisions to the
overall program.

Recently a computer analysis has been developed which treats the performance of the solar collector as a function of distance along the receiver.
Some results from this analysis are shown in Figure 7. This plot shows
the rate of increase in fluid temperature as it flows through the receiver as a function of the flow rate for various solar input levels.
These data were determined for a receiver inside diameter of 1.2 inch and
an inlet temperature of $500^{o}F$. This analysis has been used to examine a
number of different situations but has not yet been included in the overall system program. The output of this program will be particularly
useful in determining the proper series/parallel arrangement of collectors in a collector field.

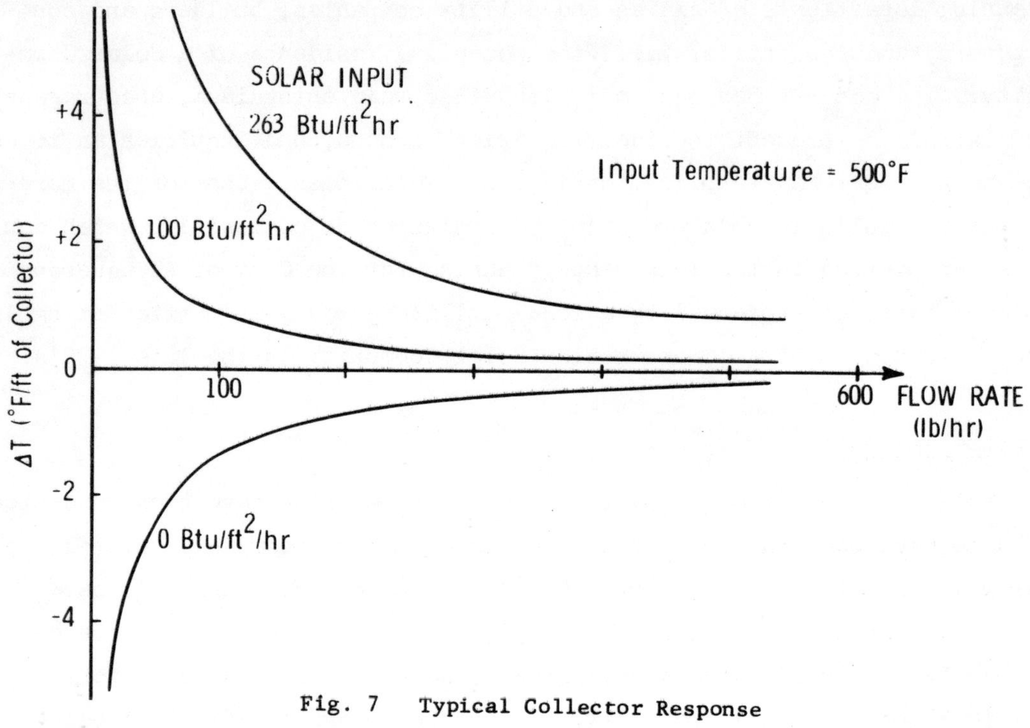

Fig. 7   Typical Collector Response

Experimental Program

The concept of the Solar Community has progressed to the point at which
it is possible to begin to procure and develop the necessary equipment
to establish a viable experimental program. The first objective of the
experimental program is to develop an efficient solar collector for the
Solar Community. To this end, the apparatus pictured in Figure 8 has
been constructed to test the performance of operating collectors. This
test facility is designed to supply water (or oil, if water proves to be
an unsuitable working fluid) to a solar collector(s) at a variety of
temperatures and flow rates and to measure the increase in the sensible
heat of the water as it flows through the receiver. The apparatus may
also be operated with no solar input to the collector to study the
mechanisms by which energy is lost from the collector. The data obtained
from tests will be very useful in conjunction with the analytical capability to evaluate the accuracy of calculational models and the validity
of conclusions drawn from the analytical studies and to consider some
aspects of collector design which are not amenable to theoretical analysis.

Fig. 8   Experimental Apparatus

The first series of tests to be carried out on this facility is designed
to evaluate several aspects of the receiver design.   Receiver tubes with
different design features have been constructed and they will be tested,
each in turn, in a common reflector.   The test program will determine
experimentally the answers to several questions pertinent to the detailed
design of the receiver.

1.   Should a plug be inserted inside the receiver tube to reduce the
     cross-sectional area available for fluid flow? Analytical consider-
     ations of this problem have not been conclusive.

2.   Two surface coatings for the receiver, iron oxide and lead sulfide,
     have been selected as inexpensive, easily deposited, and sufficiently
     stable in air. Which of these provides the better performance? The
     evaluation of other candidate coatings will probably be required also.

3.   How much does a glass envelope around the receiver improve the per-
     formance? Calculations have shown the envelope to be quite advanta-
     geous but experimental verification is needed.

4.   Should the envelope be given a reflective coating over that part of
     the inside surface not exposed to be reflected solar flux? Effects
     due to circumferential asymmetries have not been adequately treated
     analytically.

5. What is the effect of varying the size of the envelope and/or the
   receiver tube? This involves the convection heat transfer mechanism
   on both sides of the envelope and the determination of optimum tube
   size to intercept the focused image from an imperfect reflector.

6. Is evacuation of the envelope advantageous, and, if so, to what
   degree? From analysis, it is thought that the increased system
   complexity and costs required for an evacuated envelope more than
   offset the gain in thermal performance, but this should be substantiated.

7. What fluid flow rates are required for optimum performance of
   various receiver configurations?

In addition to aiding the design of the receiver, these tests will provide better information about the operation and performance of collectors
which will be useful in subsequent Solar Community systems analyses.

In January, 1974, a project was outlined which is designed to meet the
second major objective of the experimental program, i.e., to demonstrate
the feasibility of the Solar Community concept with a representative,
working model. The project is to provide the total energy needs of a
building which provides office and laboratory space for some of the
Solar Community effort at Sandia Laboratories in Albuquerque.

An existing building which has about 12,000 square feet of floor space
is to be converted to solar power by July, 1975. The electricity, space
heating and cooling, and hot water requirements will be the equivalent of
about 15 residences. About 8,000 square feet of collector area, spread
over approximately one acre of land area, will be required. The required
electricity (approximately 40 kilowatts) will be provided by a turbine/
generator system obtained as a complete unit including the control system
and condenser and evaporator heat exchangers and capable of outputs up
to 120 kilowatts. The turbine uses toluene as the working fluid and is
to be adaptable to input temperatures which range up to 550°F. Turbine
output conditions will also be variable, from 100°F-200°F.

An absorption air conditioning system capable of 70 tons of refrigeration
at a heat input, or generator, temperature of 190°F is being ordered.
The existing space heating system for the building uses both a forced air

system and a circulating hot water system. Energy for these systems, as
well as for the hot water supply, will be supplied by the solar collectors.

The thermal energy storage units for the project will be of a type which
has been under development at Sandia Laboratories in Livermore, called
stratified storage. These storage units utilize the differential in
densities of hot and cold water to "float" the hot water over the cold
water in a single, well-insulated tank as shown schematically in Fig. 9.
Calculations and preliminary experiments have indicated that a thermo-
cline can be maintained such that the heat transfer from the hot to the
cold water is only a few percent. This type of storage has the great
advantage that virtually all the energy stored in the tank is available
near the highest temperature, while, in mixed storage systems, the available temperature decreases from the maximum as energy is removed. The
storage systems for this project will be sized to allow the equivalent
of four hours' operation at the rated output, sufficient for overnight
operation. These components when combined into a working model of the

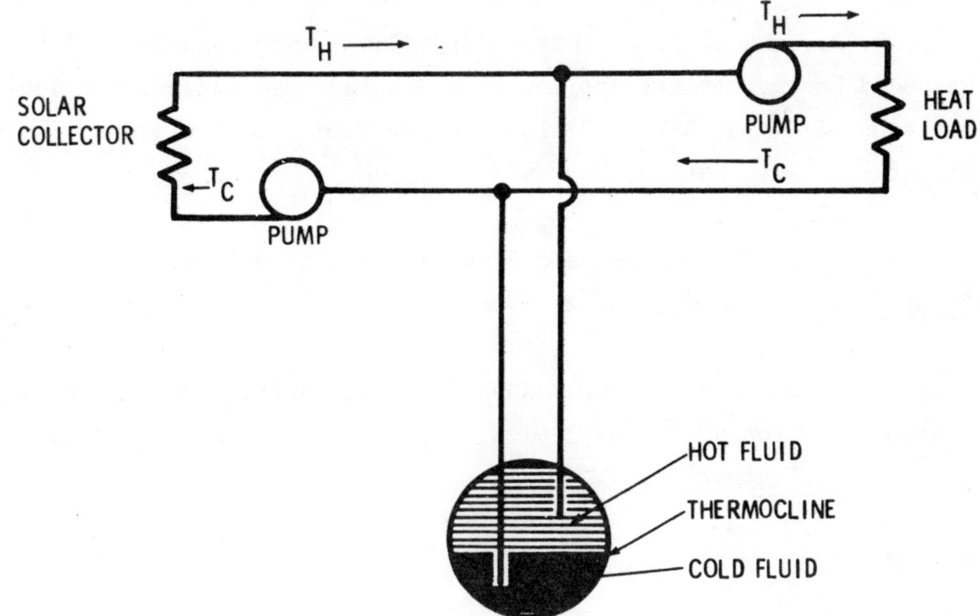

Fig. 9   Stratified Energy Storage

Solar Community will give valuable experience in the construction of the
system as well as provide actual operating data. Flexibility will be
maintained which allows experimental optimization of the system relative
to various components and subsystems and evaluation of candidates for
augmenting or replacing some parts of the system.

ACKNOWLEDGEMENT

The Solar Community project at Sandia Laboratories is a widely ranging
effort and to list all the individual contributions to the program is
impractical. The authors, therefore, gratefully acknowledge that this
report is a result of a team effort.

REFERENCES

1. Pope, R. B., W. P. Schimmel, Jr., D. O. Lee, W. H. McCulloch, and
    B. E. Bader, "A Combination of Solar Energy and the Total Energy
    Concept--The Solar Community," Eighth Intersociety Energy Conversion
    Engineering Conference, Philadelphia, Pennsylvania, August 13-17,
    1973.

2. Pope, R. B., and W. P. Schimmel, Jr., "The Solar Community and the
    Cascaded Energy Concept Applied to a Single House and a Small Sub-
    division--A Status Report," SLA-73-0357, Sandia Laboratories,
    Albuquerque, New Mexico, May 1973.

3. Stanhill, G., "Diffuse Sky and Cloud Radiation in Israel,"
    Solar Energy, Vol. 10, No. 2, 1966.

4. "Draft Proposal for a Solar Community: Feasibility and Exploratory
    Development Program," SLA-73-0181, Sandia Laboratories, Albuquerque,
    New Mexico, February 1973.

SOLAR/HYDROELECTRIC

COMBINED POWER SYSTEMS

Floyd A. Blake*

Drawing on background technology from space solar
power programs, both research and flight, Martin
Marietta has defined a terrestrial solar energy
conversion power system that can be of near-term
significance in relief of the power generation
portion of our nation's energy crisis. The system consists of equipment that, when installed in
an existing power system, augments power generation without increasing normal energy comsumption.
The most promising initial application is direct
solar hydroelectric integrated installation. During periods of sunlight availability, generation
is supplied by the solar conversion system, with
corresponding saving in water above the dam for
later use at increased rates or more prolonged
periods.

The concentrating heliostat powered solar energy
to steam conversion system is described together
with its applications associated with hydroelectric systems. Benchmark historical data in
solar power technology is included to provide
perspective and support the conclusion that near-
term development is readily attainable.

INTRODUCTION

Drawing on background technology from space solar power programs, both

research and flight, Martin Marietta has defined a terrestrial solar

energy conversion power system that can be of near-term significance in

relief of the power generation portion of our nation's energy crisis.

The system consists of solar conversion equipment that, when installed

in an existing power system, augments the system's overall power

* Program Manager, Solar Power System and Component Research (National
Science Foundation Grant GI-41305), Martin Marietta Aerospace,
Denver Division

generation without increased energy consumption. The solar/hydroelectric
integrated system is a promising initial application of the concept; it
provides a method of increasing system output without increasing the
water supply.

Under sponsorship of the National Science Foundation Research Applied to
the National Needs Program, a team from Martin Marietta Aerospace and
Georgia Institute of Technology is currently performing system analysis
and component design that will lead to an early proof-of-concept demonstration. The NSF program manager is Mr. Dwain Spencer. Mr. J. D.
Walton is leading the Georgia Tech effort. Counseling and consulting
support will be provided by the Centre National de la Recherche Scien-
tifique (CNRS) solar laboratory and by Babcock and Wilcox. Advisory
support regarding the applications within existing hydroelectric power
systems is being provided by Salt River Project and Bonneville Power
Administration. This paper is an update of preliminary design information developed during the course of proposing the current program.

SOLAR/HYDROELECTRIC POWER SYSTEM CONCEPT

Coupling a solar-energy-conversion power system to an existing stored
energy power system provides a short cut through the development cycle
for large scale solar power. Principal benefits include (1) full value
solar generated power even though it is generated on an as-available
sunlight basis, and (2) increased system capacity for hydroelectric
systems currently water limited. Figure 1 is an artist's concept
aerial view of envisioned solar power development installations at
Horse Mesa, Arizona. These interface with the hydroelectric power plant
of Horse Mesa Dam via a short low voltage transmission line. The 300
kWe plant shown matches the French solar furnace in thermal capacity and
would enable long term operation to be carried out on the components
initially performance/demonstration tested there. The 1000 kWe installation incorporates multiple field geometry vital to efficient land
usage of large solar power plants.

Augmentation of power from an existing hydroelectric installation that
operates in a round-the-clock mode is achieved by operation of a matched-
capacity solar conversion system 25% of the time. The principle is
illustrated in the bar chart of Fig. 2, an example typifying the

Figure 1. 300 kWe and 1000 kWe Solar Energy Conversion Development
Installations Envisioned for Horse Mesa, Arizona.

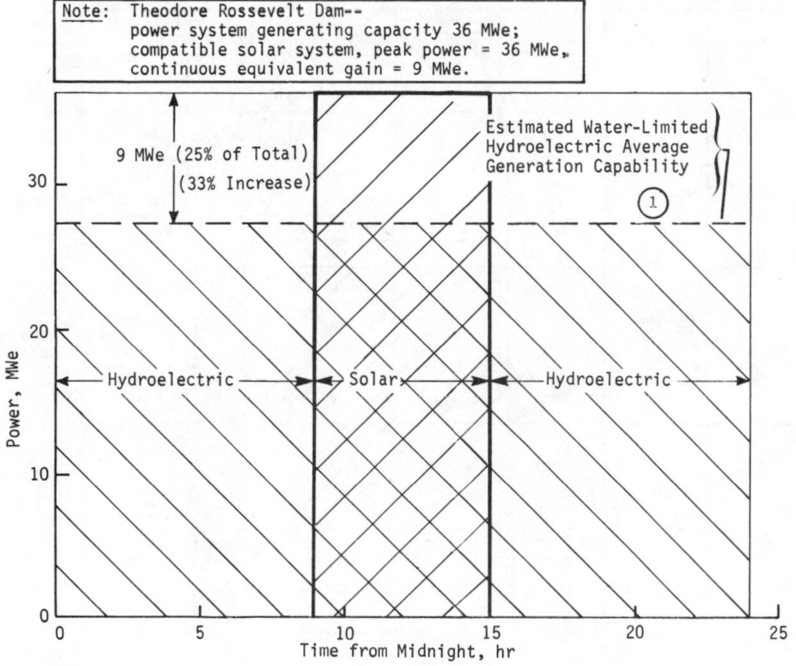

Figure 2. Average Day Solar Augmentation Operating Pattern Taken for
Theodore Roosevelt Dam of Salt River Project, Phoenix,
Arizona.

average 6-hr day. The water limited performance for the 24-hr period
of hydroelectric operation is shown as line 1. With solar augmentation, the power production for the central six hours of daylight is
taken over by the solar system, with consequent saving of water above
the dam. For this example, hydroelectric operation during the remaining hours is augmented 33% by use of water saved during solar operation.
For many installations that are currently being used only for peaking,
the augmentation fraction approaches 100% because solar power will
generate the demand for half of the days of the year. The 6-hr day is
based on annual hours of sunlight reduced to a daily average. The
occasional cloudy streak is covered by existing storage energy of installation. The relationship of the solar operating period, required
operating threshold, and direct normal solar insolation is shown in
Fig. 2 for a typical set of winter days (Phoenix, Arizona, January
30-February 2, 1962). The design-point operating threshold sunlight

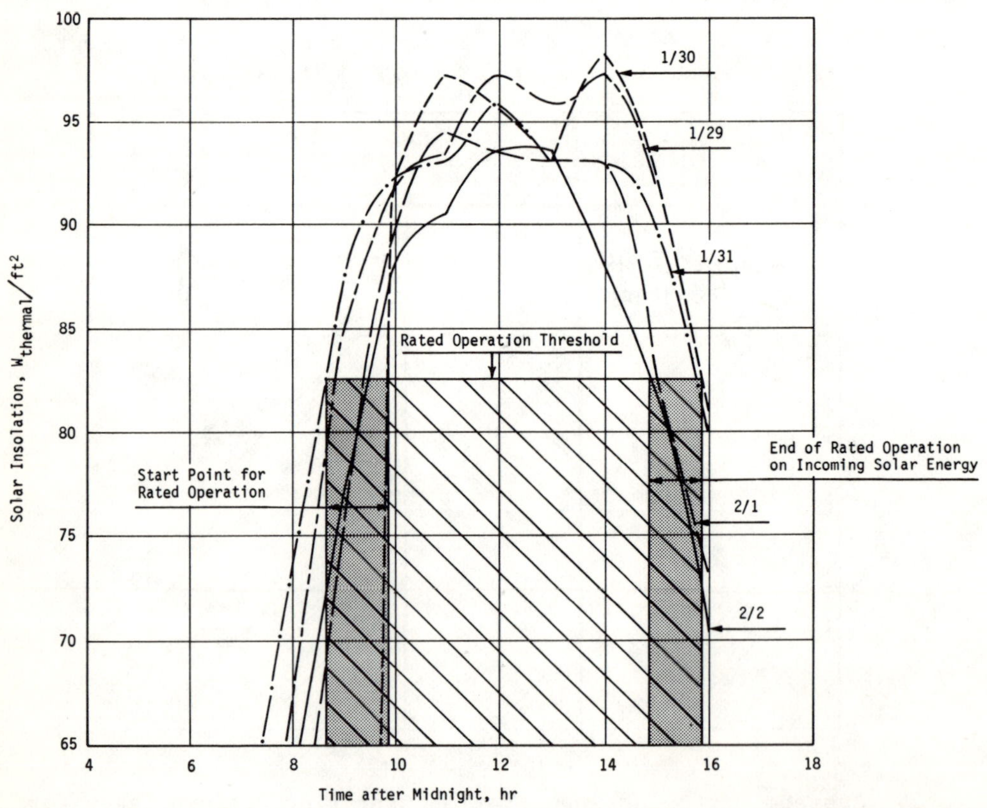

Figure 3. Solar Energy System Operational Period vs. Daily Solar
Insolation Patterns

414

energy level of 82.5 W/ft^2 was reached between 8:35 and 9:50 a.m. for the sample days. Insolation fell below threshold level between 2:55 and 3:55 p.m. Build-up operation to rated load and tailoff to no load occur during the preceding and following 1-hr periods. Normal operating periods will range up to nine hours for individual days, and must average six hours on an annual basis to achieve the preliminary design performance cited throughout this paper.

It is important to cite that direct co-location of solar and hydroelectric plants is not the only application of the concept. An off-site location with connections into network transmission lines makes solar augmentation applicable to the Bonneville Power Administration system of the Pacific Northwest. As shown in Fig. 4, solar power

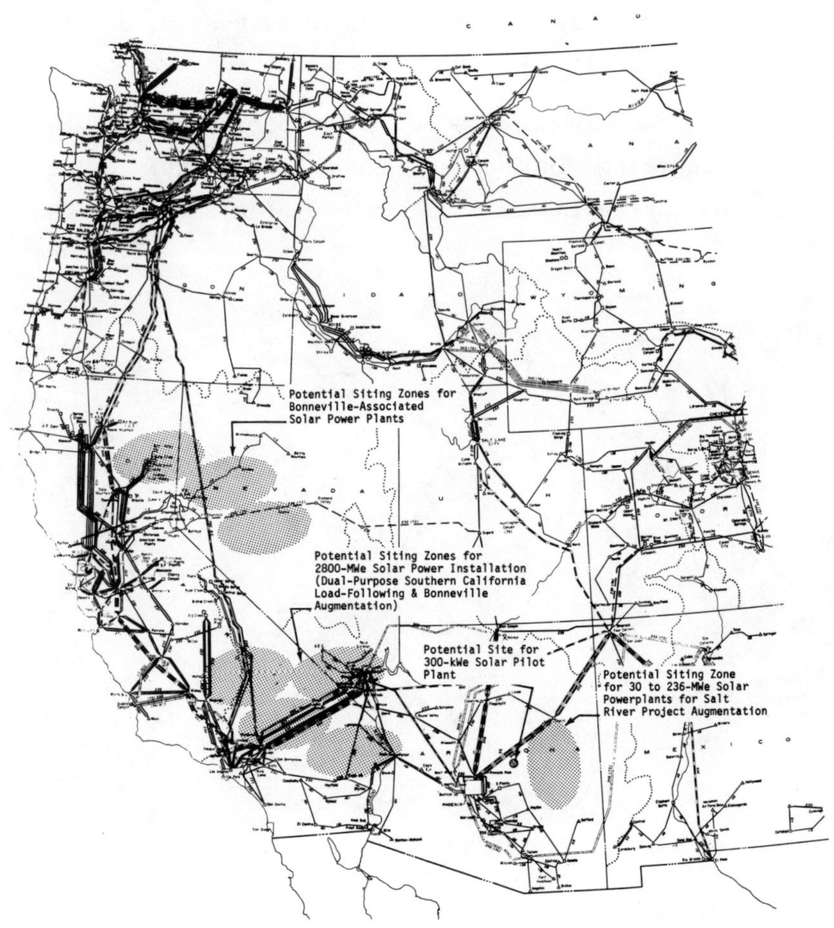

Figure 4.   Potential Siting Zones for Solar Energy Conversion Plants
            Using Existing Transmission Networks for Hydroelectric
            System Augmentation

systems in northern Nevada connecting into the network leading from
Bonneville south into California could augment the 10538+ MW system.
Sites as far south as the desert east of Los Angeles could be directly
associated with Bonneville, using the 750-kv d-c line now in use. The
southwestern United States Sun Bowl, with an expected 2200 solar system
operating hours per year, is shown on the National Climatic Center map
of annual sunlight hours (Fig. 5). With a slight system size increase,
it is likely that the sunlight islands of eastern Oregon and Washington
would be suitable for solar power systems.

SOLAR ENERGY CONVERSION SYSTEM DESCRIPTION

The solar-energy-conversion power system consists of three major subsystems: (1) a concentrating-heliostat energy-collection/concentration
subsystem, (2) a cavity-type heat-exchanger subsystem, and (3) a steam-
turbine-driven electrical generation subsystem, as shown in Fig. 6.

Sunlight striking the mirrored faces of the heliostat modules is reflected and concentrated in the heat-exchanger cavity. Solar energy
retention in the cavity is maximized by use of a secondary concentra-

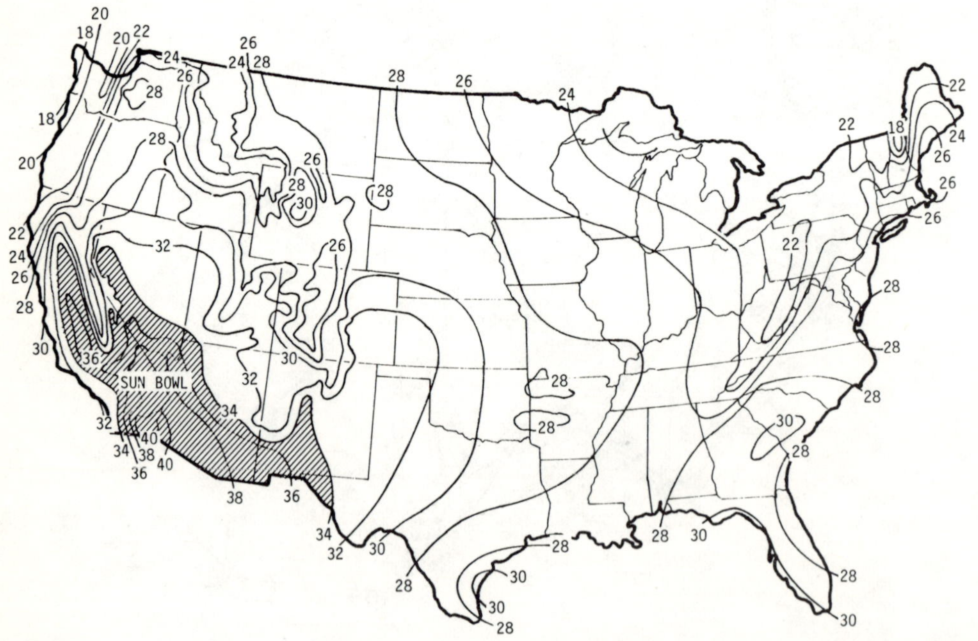

Figure 5. National Climatic Center Annual Sunshine Hours Map with
"Sun Bowl" Accented (Hours in Hundreds)

Figure 6. North-Side Field of 1000-kWe Solar Power System

tor and a quartz window in the focal plane. Energy conversion to both steam and superheated steam is accomplished in passages lining the cavity. Final conversion to electricity takes place in a conventional steam turbine typical of those used in cycling loaded industrial applications.

The concentrating-heliostat, which combines the large area solar energy collection function and the 500 to 2500 concentration function is the major investment element of the system. For one design concept, a 100 MWe plant contains 20,000 heliostat mirrors to 12 boiler installations, 12 superheater installations, and 4 turbogenerator installations. Use of existing mirror manufacturing capacity embodied into the functioning optical system is key to attainment of economic feasibility. Flat glass second surface mirrors are currently in production in excess of 100,000 ft^2 per day in individual plants. Curvature to obtain concentrating performance is obtained by dishing the mirrors in a manner shown in Fig. 7. This type of concentration permits the individual

mirror sizes to be raised somewhat independently of the boiler size, enabling
high flux, minimum sized heat exchanger
components to be designed.

The general plan view of the concentrating-heliostat sections is that of
a 45 degree right triangle to enable
approach angles of reflected rays to
be effectively handled by the secondary
concentrator. Both north- and south-
side fields of tracking heliostat modules are used to increase the total
module energy collection size and
thereby permit use of larger conversion equipment (Fig. 8).

The north-side field must be on a slop-
ing terrace to maximize packing of
reflected area in close proximity to
the heat receiver. This is required
by the generally vertical attitude of
the heliostats and eliminates shadowing of the mirrors by nearby mirrors.

Figure 7. Concentrating-
Heliostat Mirror
Segment Formed by
Dishing Flat Glass
Mirror, Focal
Length = 65 feet

The south-side field can be on a level
foundation because the generally horizontal attitude of the mirrors
does not present as severe a shadowing problem.

Durvature of each mirror in the system is unique and is based on its
distance from the focal plane center point.

A precedent for a large field of heliostats has been set by the CNRS
Solar Energy Laboratory in Odeillo, France (Fig. 9). Individual units
of its heliostat have areas comparable to those being proposed in the
Martin Marietta system, 484 vs 400 ft^2. Control is by individual sensors mounted in the reflected beam.

The 63 mirrors shown, together with the parabolic concentrator of the
furnace, collect and concentrate sufficient solar energy to operate
solar-power experimental equipment up to 300 kW.

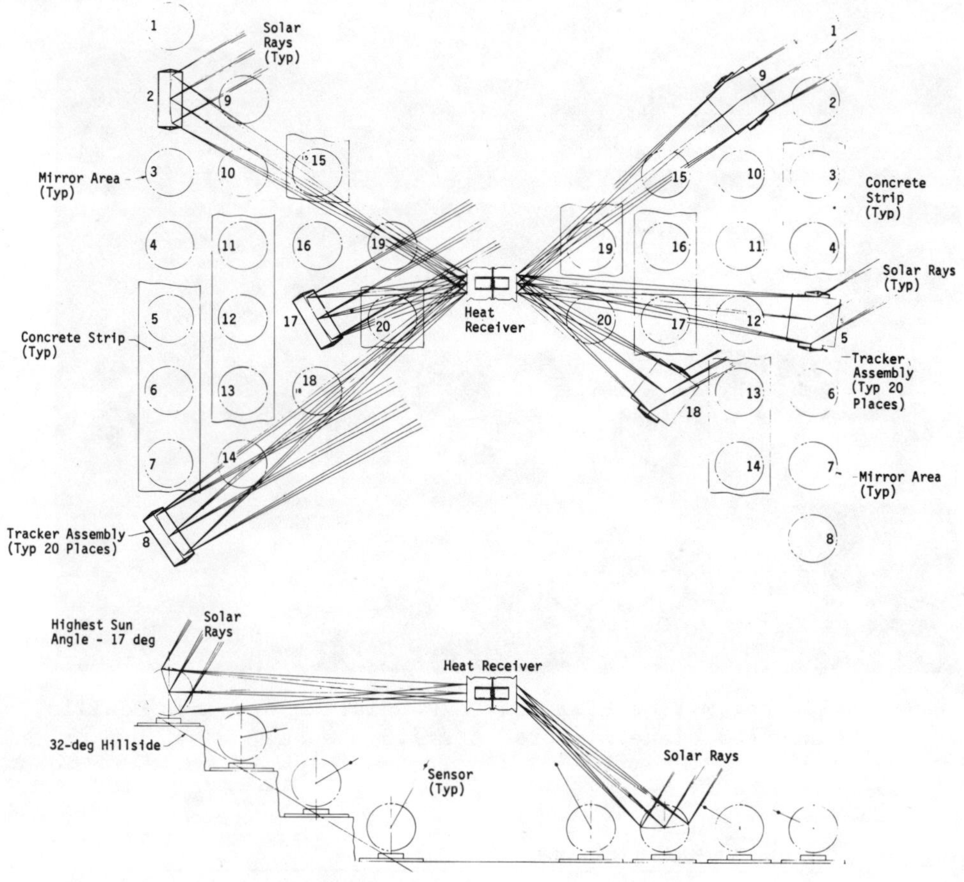

Figure 8. Concentrating-Heliostat Layout for 200 kWe

Features of the solar-energy-to-steam conversion component and its
supporting flow subsystem are shown for the double-cavity heat receiver
that would be associated with the 1000-kWe power-system module (Fig.
10).

Each cavity has identical aperture configurations, using the same secondary concentrators and quartz windows. The boiling sections in each
cavity are also identical, and the two are in parallel in the fluid
flow circuit. Saturated steam from the boiling sections is separated
and fed to the superheater in one cavity and then to the turbine. Return from the high-pressure turbine is reheated in the second cavity
and directed to the low-pressure turbine. The close proximity of the
turbine and heat exchanger in the installation is of paramount importance to minimize flow losses in the low-pressure steam system.

Figure 9. Heliostat Mirror Field of CNRS Solar Laboratory, Odeillo,
France (63 plane mirrors, 6 x 7.5 m each).

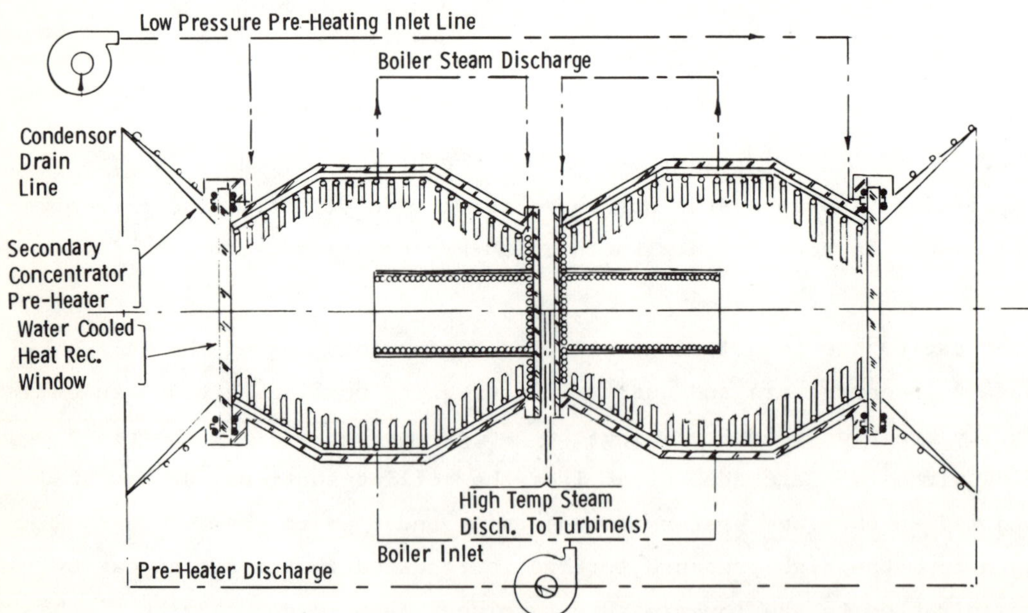

Figure 10. Double Cavity Boiler/Superheater for 1000 kWe Solar-
Energy-Conversion System.

In larger systems such as the 100 MWe being studied, separate boilers
and superheaters are more logical since the mirror fields are being
raised to practical maximums limited by sun image size. The balance
between the preheat/boiling energy and the superheating energy ranges
between 86/14 and 80/20 with corresponding effects on the field size.
Separation of the two functions into two unique devices adds considerable design flexibility in meeting specific system size and cycle
requirements.

TECHNICAL PERSPECTIVE

An overall perspective of the range of solar concentration and result-
ant operating temperatures achieved by associated conversion systems
(Fig. 11) illustrates the state-of-the-art boundary for various concentration types.

Concentration levels from 1 to 64,000 have been used with corresponding
temperature levels from 100 to 3170°C. Devices have ranged from flat-
plate collectors, to parabolic cylinders, to full paraboloids of revolution.

There is a large step between the temperature levels achievable with
parabolic cylinders and with full parabolas that spans the 538°C
requirement for domestic utility steam generation. The concentrating-
heliostat with concentrations between 500 and 2500 is an optical com-
promise with full parabolic mirror performance, but meets the steam
requirement with a comfortable, but not excessive, margin.

Examples are shown of installations that serve as benchmark references
for the temperature levels of various forms of solar-energy-conversion.
The Meadi, Egypt, parabolic cylinder installation generated 55 hp for
water pumping. The solar thermionic installations at GE, Phoenix, and
Table Mountain routinely achieved temperatures above 1500°C for direct
electrical power generation. The CNRS solar furnace at Odeillo is
capable of retaining 750 kW of thermal energy at 2200°C.

Elements that contribute to solar energy collection losses are shown in
the heat balance for the cavity vapor generator (searchlight mirror/
thermionic generator) in Fig. 12. The seven front-end losses have
varying degrees of applicability to all solar collector/concentration
systems. These fall into two general groups--optical losses and thermal

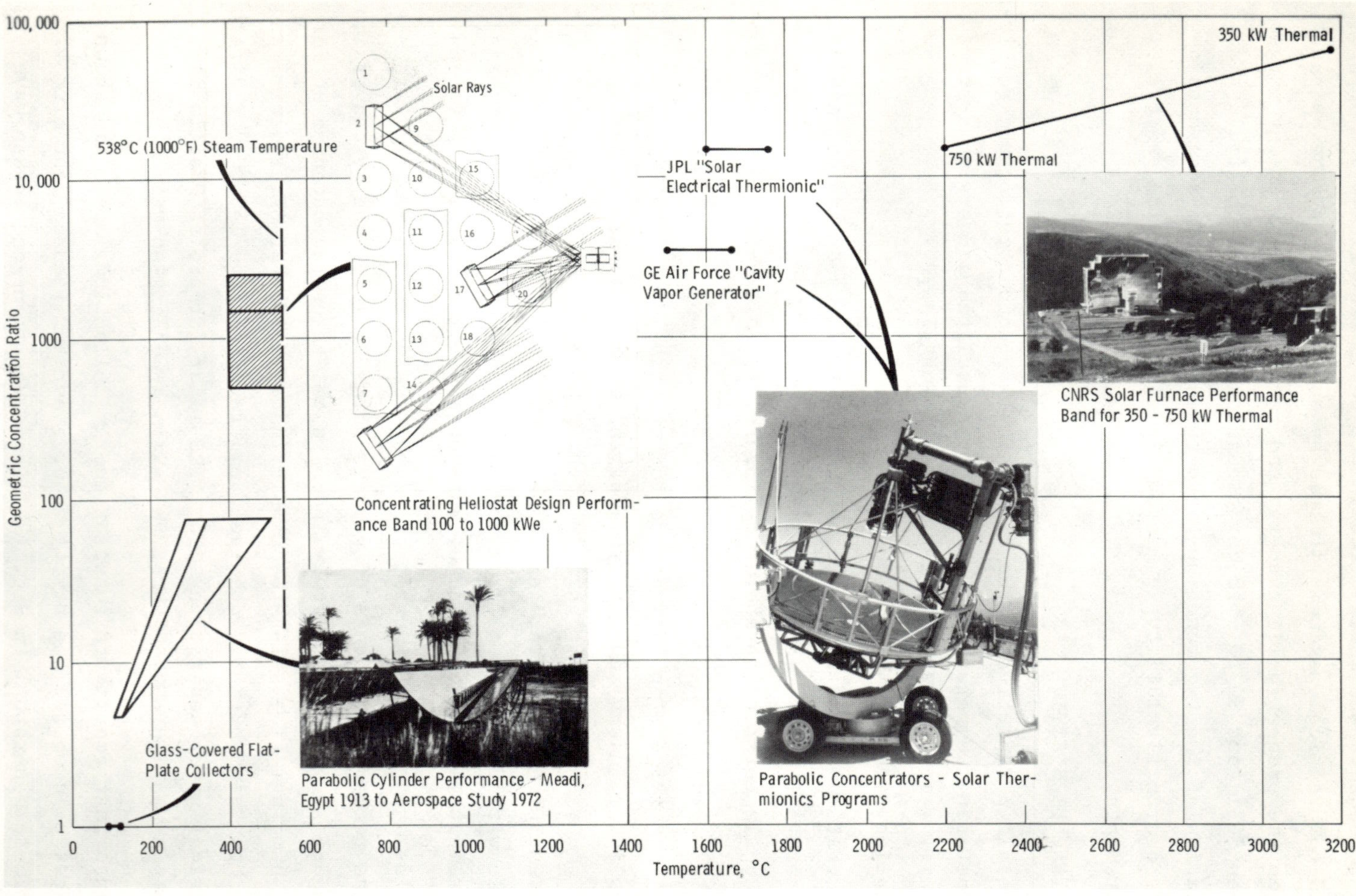

Figure 11. Solar-Energy-Concentration/Temperature Spectrum

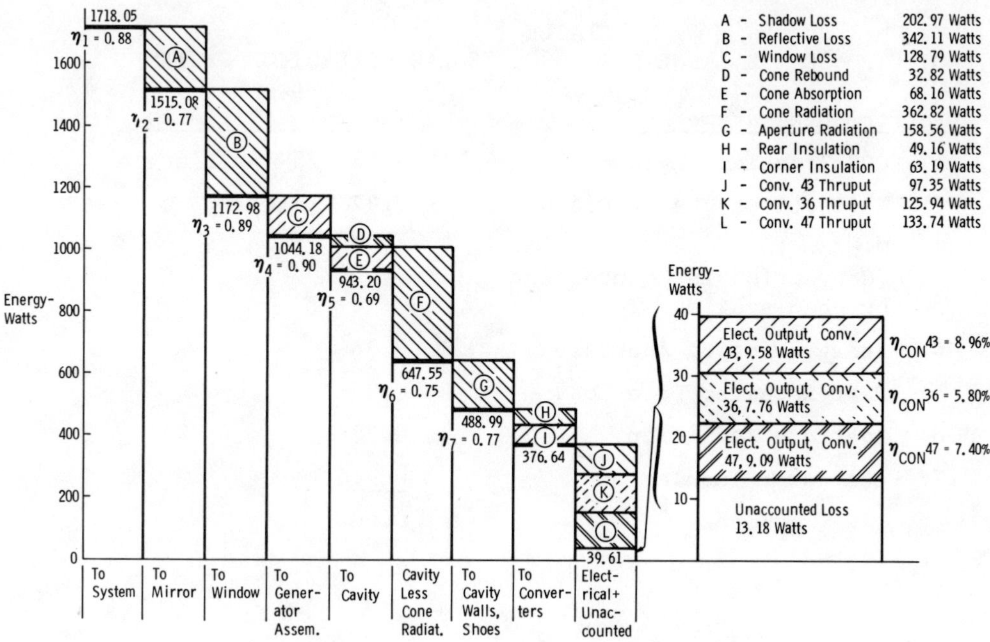

Figure 12.   Energy Balance-Solar Thermionic Generator Operating With 1632°C Cavity Temperature

losses.   The example shown with a 21% thermal collection efficiency is pessimistic regarding terrestrial applications due to lower temperature levels, design improvements, and potential for using glass mirrors.

An evaluation of individual solar-energy-collection efficiency factors available for a concentrator-cavity absorber generating 538°C (1000°F) superheated steam resulted in the values shown in Table 1.   The front-end thermal efficiency attainable in a straightforward development should be 60.8%.

Coupling the 60.8% efficiency with a 24%-efficiency industrial steam turbine would permit attainment of 100 kWe from a concentrating-heliostat mirror field of 8000 ft^2.

Key areas of improvement are in mirror reflectivity, thermal losses that are drastically reduced by lower temperature level, and improved insulation design.

ECONOMICS

The first point of the learning curve for cost of a solar-energy-conversion power system is illustrated by Fig. 13.   The chart's main

TABLE 1

PROJECTED HEAT BALANCE FOR 1000°F SOLAR COLLECTION SYSTEM

Performance Parameter	Thermal Efficiency
1) COS Tracking Angle	0.88 (28° half angle)
2) Reflectivity Using Glass	0.92
3) Window Loss (Convection is Counterpart on open system.)	0.89
4) Energy Missing Aperture	0.96
5) Heat Receiver Face Loss	0.986
6) Aperture Radiation	0.976
7) Generator Insulation	0.914
Total	0.608

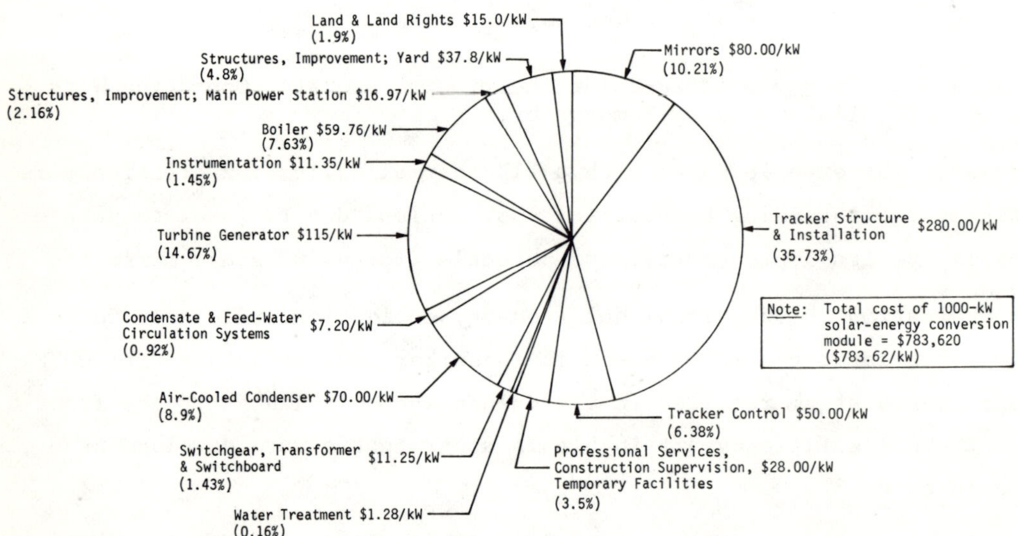

Figure 13. Solar-Energy-Conversion Power System Cost Breakdown, 1000 kWe

usefulness is identification of the many small items that contribute to system cost and its presentation of elements in perspective. The mirrors, which perform fundamental energy collection/concentration, form only 10.2% of system cost. Their associated equipment, trackers, and control, form more than 41% of the cost and represent the area of most fruitful cost reduction through design simplification and mass production.

424

Scale-down penalties shown for the turbine and boiler will also largely
disappear with design of systems larger than 1000 kWe for which this
sample study was made.

RECENT HISTORICAL BENCHMARKS

Skylab 16.5 kWe Solar Photovoltaic Power System

Solar electric power systems using both photovoltaic conversion and
concentrated-energy thermal conversion have been developed during the
United States space program. Only photovoltaic systems have been
applied to spacecraft because of the general advantages of state of
development, operating efficiency, operating temperature, and minimum
constraint on spacecraft operation. The largest of these systems to
date is Skylab, which has a beginning-of-life capability in excess of
16.5 kWe (Fig. 14). (Design power was 22 kWe; 16.5 kWe capacity resulted from loss of one Orbital Workshop solar panel array.)

Figure 14.   Skylab - Powered by 16.5 kWe Solar Photovoltaic Power
             System.

With regard to terrestrial application, the photovoltaic system has serious economic hurdles because of the number of cells required for significant area coverage. Spacecraft solar panels use 208 to 211 2x2-cm solar cells per ft^2, at an assembled cost greater than $10 per cell.

Orbital Workshop Solar Power Circuit - 1.57 kWe During Terrestrial Testing

The relationship of intercepted area to electrical output for terrestrial applications is illustrated by the Orbital Workshop solar array module group shown in Fig. 15 being tested at the Martin Marietta Denver Division solar facility. This unit generated 1.57 kW from an area of 181 ft^2 (specific power = 8.67 W/ft^2). For comparison, the specific performance of the 100-kWe solar thermal preliminary design is 12.5 W/ft^2 of the mirror field reflecting surface.

Figure 15. Orbital Workshop Solar Array Terrestrial Test

<u>Temperature-Level Benchmark for Solar Thermal Conversion</u>

Temperature performance from concentration devices is frequently quoted
to establish a meaningful figure of merit for the system. Use of this
figure has some of the same pitfalls as use of open circuit voltage in
electrical devices.

Coupling temperature level with actual power generation or steam generation is more meaningful because it indicates final operating conditions.
In this regard, the highest temperature level obtained during electric
power generation is 3189°F (1754°C) reached during testing of the JPL
solar electrical thermionic system with a 9.5-ft-dia parabolic mirror
(Fig. 16). This illustrates the excessive margin of performance inherent in full parabolic systems when referenced against the 1000°F
(538°C) steam generation requirement.

<u>Solar-Generated Superheated Steam Benchmark</u>

The best measure of heliostat performance in regard to the steam conversion requirement is the work of Dr. G. Francia of the University of

Figure 16. JPL Solar Electric Thermionic Test - 1964

Genoa, whose third pilot plant is shown in Fig. 17. These plants
efficiently generated 500°C superheated steam at 150 atmospheres
(2205 psi).

Two features of Francia's system do not lend themselves to convenient
scaling to a multikilowatt system: (1) the location of the boiler/
superheater directly above the mirror field, and (2) use of mirrors
equal in size to the boiler aperature. The latter is the most serious
and is overcome by use of concentration curvature in the heliostat
mirrors.

The primary benefit of overhead positioning is that it permits use of
an open cavity, trapping the hot convection gases within the chamber.
This may develop into the most efficient approach.

Total Solar-Energy Collection Benchmark

The 200-m^2 parabolic concentrator of
the CNRS solar energy laboratory at
Odeillo, France, leads the solar facilities of the world in total energy
collection capability. Collection of
1000 kW of thermal energy has been
achieved in this facility (Fig. 18).
The existance of this facility and
its availability for power conversion
equipment development (pending negotiations) is a significant factor in
the attainability of near-term power
system development. It permits parallel development of the optical
collection/concentration subsystem in
the United States and boiler superheater development in France, with
subsequent combination of the same
equipment into a pilot plant for
operational and endurance development testing.

Figure 17. Solar Plant No. 3,
St. Ilario-Neriv,
Genoa, Italy. Sur-
face Area = 200 m^2;
Steam Production =
150 kg/hr at 150
atm and 500°C.

Figure 18. Concentrator and Focal Zone Buildings, CNRS Solar
Laboratory, Odeillo, France

SUMMARY

All the elements required for a solar-energy-conversion system are
readily available, and applications for such a system that will deliver
full-value power to existing systems independent of dedicated "Solar
Energy Storage" have been identified. Research programs to finalize
configurations and create operational pilot plants in the near term
(2 to 3 yrs) have been proposed and are vital to attainment of the
benefits realizable from our solar energy resource. While the economics of such systems are not advantageous when referenced to oil,
fossil fuel, and nuclear plants at this time, the dynamic nature of
this picture is such that "free-fuel" solar plants should be attractive
by the time their development cycle is completed.

REFERENCES

1. J. Bruno, "Sunlight Checkout Test for OWS/SAS (Orbital Workshop/ Solar Array System." MCR-71-320. Martin Marietta Corporation, Denver, Colorado, March 9, 1972. (Contractual document, Contract 100GB1-SC).

2. F. A. Blake, "Calibration of Solar Concentrator for Power System Research." Progress in Astronautics and Aeronautics, Vol 11 (Subtitled "Power Systems for Space Flight") 1963, p 669.

3. D. L. Purdy, R. C. Keyser, F. A. Blake, and J. F. Williams, "Cavity Vapor Generator Program." ASD-TDR-62-899, October 1962.

4. F. A. Blake and R. R. Herrick, "Performance-Endurance Testing of the Cesium Vapor Thermionic Cavity Generator." IEEE Transactions on Aerospace, Vol 2, No. 2, April 1964, p 688. (Transactions Subtitled "Proceedings of International Conference on Aerospace Electro-Technology").

5. F. A. Blake, "Solar Performance Evaluation Test Program of the 9.5-foot-Diameter Electroformed Nickel Concentrator S/N 1 at Table Mountain, California." NASA Technical Memorandum 33-206, Jet Propulsion Laboratory, June 15, 1967.

6. P. Rouklove and F. A. Blake, "Performance Testing of a Solar Electric Thermionic Generator System." Solar Energy Society Conference Paper, 1965.

7. C. G. Abbot, "Utilization of Solar Energy." Annals of the Astro-physical Observatory (Smithsonian), Vol IV, Chapter IX, 1922.

8. G. Francia, "Pilot Plants of Solar Steam Generating Stations." Solar Energy, Vol 12, No. 1, September 1968, pp 51-64.

9. F. Trombe and L. Phat Vinh, "Thousand kW Solar Furnace, built by the National Center of Scientific Research in Odeillo (France)." Solar Energy, Vol 15, No. 2, June 1973.

SESSION IX

INCREASING THE SUPPLY: GEOTHERMAL SOURCES

Chairman: Bill Ogle
           Consultant to the Atomic Energy
           Commission, Anchorage, Alaska

CURRENT WORLDWIDE UTILIZATION AND ULTIMATE
POTENTIAL OF GEOTHERMAL ENERGY SYSTEMS

L. J. P. Muffler[+]

The primary use of geothermal energy to date is for the generation of electricity. Present world geothermal electrical capacity is approximately 1075 megawatts, most of which is concentrated in Italy, northern California, and New Zealand. This geothermal electrical capacity is only 0.1% of the total world generating capacity from all modes. Other important uses of geothermal energy include space heating, agricultural heating, and product processing.

The ultimate potential of geothermal energy is far greater than present utilization, and could be a significant component of the energy supply of the world. Inasmuch as geothermal energy is the natural heat of the earth, and since temperatures increase with depth, the resource base potentially available for man's use is indeed immense. Under present technology and economics most of this heat is either too diffuse or too deeply buried to be considered as a resource. But changing economics, the increasing shortages of fossil fuel, the demands for environmentally acceptable energy sources, and promising technological advances in extraction and utilization indicate that during the next few decades, geothermal energy could become an important source of heat and electricity in many parts of the United States and the world.

GEOTHERMAL ENERGY

Geothermal energy is heat produced naturally within the earth and transmitted towards the earth's surface by conduction, convection, and radiation. We know from measurements made in drillholes and mines that temperatures increase with depth in the earth. Furthermore, various geophysical data indicate that temperatures continue to increase below the deepest drillhole. Thermal models of the earth[1] indicate that at the

---

+   Coordinator, Geothermal Research Program, U. S. Geological Survey, 345 Middlefield Road, Menlo Park, California

base of the continental crust (25 to 50 km) temperatures range from 200°C
to 1,000°C, and at the center of the earth (6,371 km) reach perhaps 3,500
to 4,500°C.

Most of the earth's heat, however, is far too deeply buried to be exploit-
able under any reasonable technological assumptions. The world's deepest
drillhole reached about 9 km, at a cost of over $6,000,000. Although
drilling may someday reach two or three times this depth, such deep holes
are likely to be exceedingly expensive. Accordingly, commercial geothermal
extraction in the forseeable future is likely to be restricted to heat at
depths less than 10 km.

The amount of heat in this outer 10 km, however, is still immense. White[2]
has calculated that the heat above surface temperatures in the earth to a
depth of 10 km is 1.25×10^{27} joules. This heat is over 2,000 times the
heat that could be derived from the combustion of all the world's coal
resources (as estimated by Averitt[3]). The geothermal resource base* is
thus very large and deserves careful consideration as an energy source
for man's use.

Most of this heat in the outer 10 km of the earth is evenly distributed
and therefore is an extremely low-grade, albeit large, resource. At most
places throughout the earth, the flow of heat to the surface is dominated
by conduction, resulting in an average temperature gradient of only 30°C
per km. Consequently, temperatures sufficiently high for utilization
usually occur only at depths too great to be tapped economically.

Locally throughout the earth, however, high temperatures occur at relatively shallow depths owing to the upward transport of heat by the movement of liquid rock (magma) and water. At the margins of major crustal
plates, magma can be generated at depth, move upward in the crust, and
lodge as intrusive bodies near the earth's surface. These bodies act as
heat sources that drive overlying convective cells of meteoric water, thus
bringing the geothermal heat near enough to the surface to be tapped by
drillholes[5]. Such near-surface concentrations of geothermal energy are

---

*   "Resource base" is defined by Schurr and Netschert[4] as all of a given
      material in the earth's crust, whether its existence is known or un-
      known and regardless of cost considerations."

termed "convective hydrothermal systems". Water is essential, both to
transport heat from the deep thermal anomaly to a shallow reservoir accessible to drilling and to transport heat from the reservoir rock to the
drillhole and thence to the surface. Virtually all the geothermal energy
produced to date has come from convective hydrothermal systems.

CURRENT USE OF GEOTHERMAL ENERGY

Geothermal energy at present is used mainly to generate electricity. A
permeable reservoir at depths of 1 to 3 km and temperatures of at least
180°C is tapped by a drillhole, through which water and/or steam flows to
the surface. At the wellhead the steam is separated from any entrained
water and is passed through a convectional turbine-generator set to produce electricity. At present, electricity is generated from geothermal
resources in only seven countries (Fig. 1) and accounts for only 0.1% of
the world's generating capacity from all modes.

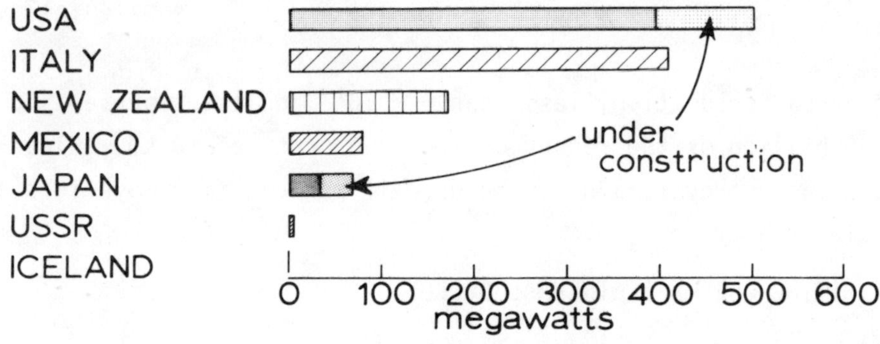

FIGURE 1

Geothermal energy is also used directly to heat buildings and hothouses.
In several countries, particularly Iceland, the USSR, and Hungary, exten-
sive district and municipal heating systems use natural warm to hot waters.
Geothermal energy is also used for product processing (paper at Kawerau,
New Zealand; diatomite at Namafjall in Iceland) and for air-conditioning

using a LiBr absorption system. Geothermal heat also has potential application in refrigeration and desalination, and some geothermal waters contain potentially valuable concentrations of metals (e.g., the Salton Sea geothermal system).

The present-day extraction of geothermal energy from the earth is summarized in table 1.

TABLE 1

ANNUAL EXTRACTION OF GEOTHERMAL ENERGY FROM THE EARTH
(from Fig. 1 and from data given in table 49, Ref. 6)

Heat extracted to generate electricity	
Vapor-dominated geothermal systems	1.57×10^{17} j
Hot-water geothermal systems	1.03×10^{17} j
Space heating	$.16 \times 10^{17}$ j
Agriculture	$.46 \times 10^{17}$ j
Product processing	$.05 \times 10^{17}$ j
TOTAL	3.27×10^{17} j

The total annual production, less than 4×10^{17} joules, represents only 3 parts in 10 billion of the resource base to a depth of 10 km. Clearly much geothermal energy remains to be used if man can find ways to exploit it commercially.

FACTORS LIMITING USE OF GEOTHERMAL ENERGY

A major limitation to the use of geothermal energy lies in the nature and distribution of the resource. Near-surface geothermal reservoirs are concentrated in regions of recent volcanism and mountain-building. Because of heat losses and pipeline costs, steam and hot water can be transported only a few kilometers, and geothermal energy has to be converted to electricity if it is to be distributed to users distant from the production site. Accordingly, geothermal energy does not have the flexibility of distribution that characterizes oil, gas and coal and cannot readily compete with these sources for many purposes, such as transportation.

Geothermal developments such as The Geysers in northern California amply demonstrate that generation of power from a vapor-dominated (dry steam) geothermal system is economically competitive with other modes of power generation. But the economics of the far more abundant hot-water-geothermal systems are far less favorable. However, in the past year the soaring cost of petroleum products has caused increased interest in the development of the hot-water resources of the United States. Clearly, the extent to which these hot-water systems are developed depends in great part on how expensive the traditional fossil fuels become.

Economic and convenience considerations also have limited the use of geothermal energy in the United States mainly to electrical generation. But increasing fuel costs and electricity prices may encourage direct use of geothermal heat in space heating and product processing. Because of the relatively low temperature and pressure of natural steam. geothermal generation has a low thermodynamic efficiency (14% at The Geysers, a relatively favorable situation). Several industrial processes use large quantities of low-grade heat and could well realize greatly reduced energy costs from direct utilization of geothermal heat.

Utilization of geothermal energy for the generation of electricity has to date been restricted almost entirely to convective hydrothermal systems having reservoir temperatures of at least 180°C. At lower temperatures, the amount of steam produced by flashing to any reasonable turbine intake pressure (e.g., 5 to 7 kg/cm^2) is insufficient for economic generation (Ref. 6, Fig. 28). However, in recent years proposals have been made to use low-temperature geothermal waters in a heat exchanger to boil a secondary fluid such as iso-butane or freon[7,8]. This low-boiling-point fluid (as a gas) would drive a turbine and generator, be condensed, and then return to the heat exchanger in a continously circulating loop. The geothermal fluid would not be allowed to vaporize and would be ultimately returned to the ground. The world's only plant using this "binary system" for generating electricity from geothermal energy is at Paratunka, Kamchatka[9,10]. In the Imperial Valley of California, San Diego Gas and Electric, Magma Power Company, and Magma Energy, Inc. are testing a modified binary system. The test results have not yet been released and the economics of this and other binary systems for generating geothermal power are not known. Some published cost analyses are very optimistic[11]; other analyses[12] are more cautionary, suggesting that under most circumstances direct steam generation would be more economic than a heat-exchanger system.

Binary conversion systems may also make economic the use of warm to hot
waters that are found at depth in rapidly subsiding sedimentary basins of
late Cenozoic age. These "geopressured systems" differ from the convective hydrothermal systems described above in that the contained water is
not meteoric but connate (water trapped in the sediments at the time of
deposition). This connate water was trapped by impermeable shale beds
during compaction of the sediments and remains in pores at pressures
approaching lithostatic[13]. The water does not circulate over a hot intrusive body. Instead, temperatures are high because the normal heat flow
is trapped by insulating water-rich sediments[14].

Geopressured systems are known primarily from the Gulf Coast of the United
States, where extensive drilling for oil has incidentally delimited large
reservoirs of geopressured water at temperatures up to 273°C (Ref. 13).
These geopressured reservoirs have not yet been tapped commercially,
owing presumably to economic constraints. The resource of geopressured
reservoirs is not only the thermal energy but also the kinetic energy of
the high-pressure water and the energy available from combustion of the
entrained methane.

Extraction of geothermal energy to date has been limited to convective
hydrothermal systems and is dependent on natural permeability. Under
present technology, rocks of low permeability do not form an economic geothermal reservoir, however hot they may be. Two groups in the United
States are currently trying to extract geothermal energy from these "hot
dry rock" systems. Battelle-Northwest, in cooperation with Rogers
Engineering and Southern Methodist University, is drilling a deep test hole
at the heat flow anomaly recently discovered at Marysville, Montana[15].
Los Alamos Scientific Laboratory is investigating an area of elevated heat
flow in the western Jemez Mountains, New Mexico[16,17]. The Los Alamos program involves drilling a hole until temperatures of approximately 300°C
are reached (at perhaps 3 to 6 km in that part of the Jemez Mountains).
The lower part of the hole will be packed off and fractured hydraulically,
producing a vertical, circular crack perhaps 1 km in radius. A second
hole will be drilled to intersect the top of the crack. Cold water will
be pumped down the first hole, heated at depth, and produced at the surface at 250°C and 80 kg/cm^2.

Should extraction of heat from hot dry rock be proved economically feasible, the size of the nation's geothermal resource would be greatly increased, for even at normal or slightly elevated geothermal gradients (30-60°C/km) temperatures approaching 300°C can be reached at 5 to 10 km. However, there are major technical problems to be solved in the hydro-fracturing, heat transfer, and crack propagation, and the economics of the proposed system are speculative.

Finally, use of geothermal energy is limited by the depth to which man can drill economically. The deepest well yet drilled for geothermal energy is just slightly over 3 km. At increasing depth costs go up exponentially, to many millions of dollars per well at depths of 7 to 10 km. Yet at these depths under only an average geothermal gradient (30°C/km) temperatures reach 225°C to 285°C. Particularly if the extraction of energy from hot dry rock becomes feasible, the ability to drill to great depths cheaply would greatly increase our use of geothermal energy.

ENVIRONMENTAL CONSTRAINTS

The environmental effects of geothermal development have been outlined in detail in the Environmental Impact Statement prepared for the Federal geothermal leasing program[18]. The summary presented here is taken with slight modification from reference 6.

It appears that generation of electricity from geothermal energy can be accomplished with only a slight effect on the environment. Geothermal plants produce neither atmospheric particulate pollutants nor radioactive wastes. Geothermal plants do produce thermal pollution; indeed, the amount of waste heat per unit of electricity generated is higher for geothermal than for either fossil-fuel or nuclear modes, owing to the low turbine efficiencies at the low geothermal steam pressures. Geothermal effluents, as well as being warm, commonly contain dissolved salts and

metals and thus present a chemical pollution hazard to surface or ground waters. Accordingly, all proposed geothermal developments in the United States plan to dispose of unwanted effluent by reinjection into the geothermal reservoir. Reinjection of water in fault zones may increase the incidence of earthquakes, by a mechanism similar to that demonstrated for

the Rocky Mountain Arsenal well in Colorado[19]. Furthermore, intensive
withdrawal of geothermal waters may cause ground subsidence[20].

Other environmental effects of geothermal development appear to be con-
trollable at reasonable cost.   These include noise (drilling, testing,
and production), gaseous emission (particularly H_2S), drilling accidents,
casing failure, and industrial scars.

Bowen[21] correctly pointed out that "To understand properly the impact of
the production of electric power on the environment, it is necessary to
evaluate more than just the power plant, whether it is geothermal, nuclear,
or fossil fueled; the entire fuel cycle from mining, processing, transportation, and disposal of spent wastes must be considered." When viewed
in this light, the environmental impact of geothermal generation appears
to be minor compared with fossil-fuel or nuclear generation. The environmental impact of geothermal generation is restricted to the generating
site, whereas much of the environmental impact of other modes of generating takes place at sites such as mines, processing plants, and disposal
sites as well as the power plant itself.

CONCLUSIONS

Although the amount of geothermal energy used today is small, the amount
of heat available in the upper part of the earth's crust is very large
and has great potential for use by man. How much of the immense geothermal resource base will ever be used depends on the interrelationship of
many technologic, economic, environmental, and governmental factors and
cannot be predicted with certainty. As other less expensive and more
flexible sources of energy are exhausted, use of geothermal energy may
increase greatly. Increased use could result also from one or more of
the following items:

1.   Utilization of the demonstrated thermal, kinetic, and combustion
     energy of the geopressured reservoirs.

2.   Perfecting techniques such as the binary system that will allow
     generation of electricity from geothermal systems at temperatures
     less than 180°C.

3.   Development of drilling techniques that will allow economic drilling
     to depths of 10 km or greater.

4. Expansion of the use of low-grade geothermal resources for product processing, space heating, agriculture and desalination.

5. Development of techniques to fracture hot dry rocks at depth and economically extract the contained heat.

REFERENCES

1. A. H. Lachenbruch, "Crustal Temperature and Heat Production: Implications of the Linear Heat-flow Relation," _J. Geophys. Res._, vol. 75, No. 17, June 1970, pp. 3291-3300.

2. D. E. White, "Geothermal Energy," _U. S. Geol. Survey Circular 519_, 1965, 17 p.

3. P. Averitt, "Coal Resources of the United States, January 1, 1967," _U. S. Geol. Survey Bull. 1275_, 1968, 116 p.

4. S. H. Schurr and B. C. Netschert, _Energy in the American Economy_, 1850-1975, The Johns Hopkins Press, Baltimore, 1960, p. 297.

5. L. J. P. Muffler and D. E. White, "Geothermal Energy," _The Science Teacher_, v. 39, March 1972, pp. 40-43.

6. L. J. P. Muffler, "Geothermal Resources," _in_ D. A. Brobst and W. P. Pratt, eds., "United States Mineral Resources," _U. S. Geol. Survey Prof. Paper 820_, 1973, pp. 251-261.

7. A. Hansen, "Thermal Cycles for Geothermal Sites and Turbine Installation at The Geysers Power Plant, California, " _United Nations Conf. on New Sources of Energy (Rome, 1961)_, vol. 3, 1964, pp. 365-379.

8. V. K. Jonsson, A. J. Taylor, and A. D. Charmichael, "Optimisation of Geothermal Power Plant by Use of Freon Vapour Cycle," _Timarit VFI_, 1.-2 heftig arg., 1969, pp. 2-18.

9. B. F. Shubin, "Experimental freon geothermal power station," _Elektricheskiye stantsii_, No. 5, 1967, pp. 20-22 (Summarized in English in Soviet Geothermal Electric Power Engineering, Rept., 2, Advanced Research Projects Agency, Dept. of Defense, 1972, p. 33-37).

10. V. N. Moskvicheva, "Geothermal power plant on the Paratunka River," _Geothermics_, Special Issue 2, vol. 2, 1970, pp. 1567-1571.

11.   J. H. Anderson, "The Vapor-turbine Cycle for Geothermal Power Genera-
      tion, <u>in</u> Paul Kruger and Carel Otte, eds., <u>Geothermal Energy,
      Resources, Production, Stimulation</u>, Stanford Univ. Press, Stanford,
      California, 1973, pp. 165-175.

12.   B. Wood, "Geothermal Power," <u>in</u> H. C. H. Armstead, <u>Geothermal Energy:
      Review of Research and Development</u>, UNESCO, Paris, 1973, pp. 109-121.

13.   P. H. Jones, "Geothermal Resources of the Northern Gulf of Mexico
      Basin," <u>Geothermics</u>, Special Issue 2, vol. 1, pt. 1, 1970, pp. 14-26.

14.   C. R. Lewis and S. C. Rose, "A Theory Relating High Temperatures and
      Overpressures," <u>Jour. Petroleum Technology</u>, vol. 22, No. 1, 1970,
      pp. 11-16.

15.   D. D. Blackwell and C. Baag, "Heat flow in a 'Blind' Geothermal Area
      near Marysville, Montana," <u>Geophysics</u>, v. 38, 1973, pp. 941-956.

16.   M. C. Smith, ed., <u>A Preliminary Study of the Nuclear Subterrene</u>, Los
      Alamos Scientific Laboratory, LA-4547, 1971, 62 p.

17.   M. Smith, R. Potter, D. Brown, and R. L. Aamodt, "Induction and Growth
      of Fractures in Hot Rock," <u>in</u> Paul Kruger and Carel Otte, eds.,
      <u>Geothermal Energy: Resources, Production, Stimulation</u>, Stanford Univ.
      Press, Stanford, California, 1973, pp. 251-268.

18.   U. S. Dept. of the Interior, <u>Final Environmental Impact Statement
      for the Geothermal Leasing Program</u>, U. S. Govt. Printing Office,
      Washington, D. C., 4 vol., 1973.

19.   J. H. Healy, W. W. Rubey, D. T. Griggs, and C. B. Raleigh, "The
      Denver Earthquakes," <u>Science</u>, v. 161, No. 3848, 1968, pp. 1301-1310.

20.   T. M. Hunt, "Net Loss from Wairakei Geothermal Field, New Zealand,"
      <u>Geothermics</u>, Special Issue 2, vol. 2, pt. 1, 1970, p. 487-491.

21.   R. G. Bowen, "Environmental Impact of Geothermal Development," <u>in</u>
      Paul Kruger and Carel Otte, eds., <u>Geothermal Energy: Resources,
      Production, Stimulation</u>, Stanford Univ. Press, Stanford, California,
      1973, pp. 197-215.

NEW TECHNOLOGY CHALLENGES IN EXPLORATION, EXPLOITATION AND ENVIRONMENTAL IMPACT OF GEOTHERMAL SYSTEMS

George V. Keller *

In recent years, the possibility of using geothermal energy as a supplement to more conventional energy sources has received widespread attention. Estimates of the importance of geothermal energy vary widely. Some have suggested that geothermal energy might account for a major fraction of the electrical energy generated at the end of this century, but a much more common view appears to be that geothermal energy is a curiosity, and in the forseeable future, will fill only a negligible fraction of our energy needs, as it does at the present time. Geothermal steam is used to generate less than one-one thousandth the electrical energy now being used in the United States. Whether this remains the case, or whether geothermal energy becomes an important energy source freeing more valuable forms of energy for types of use that require special characteristics in a fuel depends on the solution or non-solution of a wide range of problems in exploration, exploitation and environmental impact.

Problems in exploration for geothermal energy stem from the fact that we know relatively little about the geological control for the occurrence of such energy sources. We know in a qualitative way that normal heat flow from the interior of the earth is not a significant energy resource. The average heat flow from the earth is only 1.5 microcalories per square centimeter per second. In the equivalent electrical units, this amounts to 46 kilowatts per square kilometer. Considering that the population density in the United States is 22 people per square kilometer, this rate of energy flow would be just adequate to meet our current demand for electrical energy. Even ignoring economic realities, it

---

* Department of Geophysics, Colorado School of Mines, Golden, Colorado

would be physically impossible to intercept and convert any appreciable fraction of this energy flow.

One must consider stored heat energy in the earth before geothermal energy seems at all possible. The amount of heat energy stored in a rock depends on the temperature and on the heat capacity of the rock. Of the materials that comprise a rock, water has the highest heat capacity, ranging from 0.6 to 1.0 calories per cc per °C. The solid silicate minerals usually have heat capacities which are slightly smaller per unit volume. A reasonable heat capacity for a rock is typically 0.6 to 0.8 calories per cc per °C. Thus, for each degree C that the temperature of 1 cubic kilometer of rock can be lowered, the heat energy that could be extracted is $0.6 \sim 0.8 \times 10^{15}$ calories. The electrical equivalent would be $0.7 \sim 0.9 \times 10^{9}$ kilowatt hours, or considering a lifetime of 30 years for a conversion plant, $0.27 \sim 0.34 \times 10^{4}$ kilowatts. Of course, the conversion of heat energy to electrical energy is beset by many inefficiencies, so that only a small part of the heat energy in a rock can be converted to electrical energy. These figures merely point out that the stored heat in rocks is a much more interesting form of energy than is heat flow from the earth's interior, even though stored heat may not be replaced in the span of time of interest to man.

The normal rate at which temperature increases with depth in the earth is about 20°C per kilometer of depth. Abnormal rates are considered to be five times as great, and rarely, temperature gradients several tens of times this great have been observed. Intuitively one might expect that the greater the efficiency with which heat could be collected from the earth, and the higher the temperature at which it could be extracted, the better would be the probability of having a commercially viable source of energy. In all of the geothermal fields developed to date, natural steam is extracted and expanded to drive turbines for the production of electricity. In some cases, as at the Geysers, in northern California, and at Lardarello, in Italy, only steam flows from the

wells, while at other places, as at Cerro Prieto, in Mexico, and
Wairaki, in New Zealand, hot water under pressure flows from the
wells, and a portion is allowed to flash to steam to drive turbines.
The temperature and pressure of this steam is low in comparison
with that of modern high-pressure boiler systems. Because of the
low efficiency under these conditions, a large amount of steam is
required for each kilowatt-hour of electricity produced, an amount
ranging from 5 to 10 kilogram in presently operating geothermal
plants.

Inasmuch as a single kilowatt-hour of electricity is of no
great value, geothermal plants are economically feasible only if
large volumes of steam can be produced cheaply from the ground.
In areas where other energy sources are still abundant, geothermal
steam would be competitive only if it could be produced at a cost
of 2 to 4 mills per unit, with a capital investment of $50 to $100
per kilowatt capacity in wells and piping. In areas where fuels
are lacking, geothermal energy is an attractive alternative at
about five times these costs.

Drilling a well and completing it for the production of geothermal steam currently costs between $200,000 and $600,000.
Assuming 10 kilograms of steam may be required for each kilowatt-
hour of electricity production, and that a capital investment of
$100 per kilowatt capacity is allowed, an average geothermal well
must produce 20,000 to 60,000 kilograms of steam per hour. If
the wells are drilled in a field that produces hot water, the
total amount of water produced by a well must usually be 5 to 10
times greater than the amount of steam required. These casual
calculations are intended to point out that the scale of production
required from steam wells is quite different than the scale of
production required from oil wells, where a production of a mere
1000 kilograms of oil (about 7 barrels) per hour would be considered to be worth producing. In other terms, a kilogram of

steam is worth about 0.02 to 0.04¢, while a kilogram of oil is
worth about 5.0¢ at present U.S. prices.

The lower value of steam requires that it be producible at
rates which are two orders of magnitude greater than those suitable
for oil production. Such production rates are possible only if
heat can be extracted by flowing a heat-exchange medium through
the rock containing stored heat at very high rates. It is necessary
that the rock be fractured in order that the permeability be
great enough for this transfer to take place. The geothermal
reservoirs now being exploited are naturally fractured. Whether
this will always be the case is not known, but it is reasonable
to expect that development of large-scale rock fracturing techniques will be essential if extensive development of geothermal
energy is to take place.

A major difficulty being faced in the current efforts to
locate additional sources of geothermal energy is our lack of
understanding of their geologic settings. Geothermal reservoirs
appear to be intimately associated with modern volcanism or
intense tectonic activity. This has led to the supposition that
heat has been supplied to the geothermal system by an underlying
magma chamber. The nature of a magma chamber is very poorly
understood, inasmuch as no drill hole has penetrated into one,
and indirect methods, such as geophysics, have tended to indicate
that reservoirs filled with molten rock are rare, if they occur
at all. A major effort may be required to investigate the nature
of magma chambers before the geologic controls on occurrence of
geothermal systems can be understood.

Assuming that magma chambers do exist at shallow depths in
the crust, say from 3 to 10 kilometers depth, the manner in which
heat can be transferred to shallower depths is still poorly understood. If heat is transferred from the molten rock by conduction
through a frozen shell, the rate of heat flow is very slow because
of the low thermal conductivity of solid rocks. It is possible
that mass transfer of heat takes place from the magma chamber by
evolution of volatile materials, or by adsorption and release of
water from the rock into which a magma chamber has been intruded.

This latter mechanism has been suggested by several geologists
as the explanation for sealed steam reservoirs. In this model,
water from the host rock around an intrusive is converted to
steam, and in so doing, deposits its mineral content to make an
impermeable caprock around an intrusive. Inside the caprock, all
the water is converted to steam above its equilibrium temperature
with water, and held in place by the caprock. If such "dry steam"
fields exist, they would provide an attractive target for development because of their high temperatures and the efficiency of
conversion to electrical energy would be relatively high. However,
no drilling has yet penetrated into such a supercritical temperature
regime.

Beyond such an impermeable caprock, heat transfer to the
surface may be conduction, if rocks are impermeable, or by convection, if the rocks are permeable. When convection takes place,
temperature remains high as water rises through the rock, and
water containing considerable amounts of energy can be extracted
at relatively shallow depths. The most favorable circumstances
occur when permeable zones penetrate into high temperature areas,
and carry this high temperature to shallow depths. Convecting
systems are known where the temperature of the moving fluid is as
high as 350°C, as at Cerro Prieto, or as low as 100°C, as at
Kilauea Volcano, Hawaii.

In some systems, the convecting system may rise to the point
where the overburden pressure is not great enough to prevent boiling.
At this stage, boiling will occur, and the temperature will drop.
Often, the temperature _vs_ depth curve will follow the boiling point
curve. In other cases, the near-surface rocks may become filled
with steam at pressures below the equilibrium pressure for conversion from steam to water. It would appear that for most geothermal
systems now being used for power production, only the upper part
of these convection system has been tapped by drilling.

This description of the types of geothermal systems that
may develop around a shallow intrusion or magma chamber is probably
simplistic, but if even partly correct, it suggests that the best
place to look for geothermal energy is a place where there is
modern volcanism or tectonism; that is, in oceanic spreading
centers, in subduction areas, and around mantle hot spots. Unfortunately, the United States is not richly endowed with such
geologic phenomena. Best prospects appear to lie in the Western
United States, in three areas:

a.   the coastal ranges, which may represent a subduction zone
that has recently died out.

b.   the basin and range area, which may represent a northerly
extension of the east Pacific Rise spreading center, and

c.   the Rio Grande valley - Snake River valley trend, which
may represent a minor spreading center.

In the more distant future, it may become feasible to extract
heat from normally hot rock. With thermal gradients of 20° to 30°
per kilometer of depth, rocks at depths of 8 to 10 kilometers would
be hot enough to be of interest. In most areas, rocks at these
depths would be crystalline and impermeable, so that extensive
rock fracturing would have to be accomplished and heat extracted
by circulating a heat exchange fluid. In a few areas, particularly
the Gulf Coast, sedimentary rocks extend to such depths, so that
natural waters may be used as the heat exchange fluid if permeability
is present. Pressures may be great enough to cause waters to flow
naturally to the surface, which would simplify the extraction of
energy from such deep reservoirs. However, the economic constraints
on the development of normal geothermal heat are the same as those
on the development of hyperthermal systems. Each well must be
capable of producing at least 100,000 kilograms of steam per hour,
and must cost less than $1,000,000. Such accomplishments do present
a major technological challenge.

At the moment, there are sufficient technological challenges involved in the development of even the hyperthermal systems at shallow depth. Not the least of these is the problem of locating the systems by some method cheaper than random drilling. There are several hundreds of active volcanoes distributed around the world, and several thousands of associated hot springs areas, suggesting that there might be several thousand hyperthermal systems suitable for exploitation around the world. Even if each on the average is capable of sustaining a power plant of only 100 megawatts capacity, the worldwide potential for the development of geothermal power is significant -- 200,000 megawatts. In the western United States we would expect no more than a few hundred hyperthermal systems, probably fewer, but even at this rate of occurrence, geothermal energy would be significant.

Geological and geophysical exploration methods are being developed to prospect for geothermal systems, but to date, the number of successes and failures has been too few to permit selective refinement of techniques. The best geologic indicators appear to be thermal springs and detection of faults associated with intrusions. However, surface leakage from underground thermal reservoirs may be displaced laterally to considerable distances, and it is necessary to develop geophysical methods of locating buried geothermal systems. Methods that appear to offer some chance of success are resistivity mapping, micorseismicity studies, self potential surveying and gravity surveys. Surface temperature surveys or temperatures measured in shallow wells may also be helpful, particularly if the thermal system discharges directly to the surface.

In a water-laden geothermal reservoir, the effect of the elevated temperature is to decrease the resistivity markedly, at tempertures up to about 350°C. At higher tempertures, the resistivity of the ground water increases again, but rather slowly.

For geothermal reservoirs having temperatures in the range from
200°C to 550°C, resistivity is reduced 5 to 7 fold from the value
at normal temperatures, 20° to 40°C. Similar changes in the
resistivity of a rock may be evoked by increases in the salinity
of pore water, or by changes in water content. An area of low
resistivity is not necessarily a geothermal reservoir, but a geothermal reservoir almost certainly has a lower than normal electrical
resistivity.

It has been found in several geothermal areas that location
of the epicenters of small earthquakes can be useful in developing
a tectonic picture of a reservoir and in locating zones of fracture
permeability. The earthquakes used in such studies are so small
that they can be detected only by extremely sensitive instruments
located within a few kilometers distance from the earthquakes.
These small earthquakes possibly arise because the excess pressure
in the heated water rising along potentially active fault planes
permits the faults to move in many small steps rather than in
rarer, larger events that would be felt as normal earthquakes. The
epicenters of these small earthquakes can be used to define the
active fault planes in a geothermal reservoir. A microseismicity
study has an additional benefit in that the results can assist in
evaluating the seismic risk associated with structures such as a
power plant located near the geothermal reservoir.

Recent surveys in geothermal areas associated with active
volcanoes have revealed that large anomalies in spontaneous potential
are sometimes associated with active geothermal systems. These
voltages arise in part from the electrokinetic phenomenon, as
geothermal fluids move through rocks and in part from differences
in the chemical activity of the geothermal waters and the normal
waters which are in contact with them. The amplitude of spontaneous
potential anomalies observed in geothermal areas is one to two
volts.

Measurements of the gravitational field are often useful in
studying the regional structure around a geothermal area, as well
as for indicating directly the presence of a local intrusive which
may serve as the heat source.

Other geophysical methods which have not yet received much attention may also be useful in geothermal exploration. Because of their long usage in petroleum and minerals exploration programs, geophysical techniques are highly developed. At the moment, the technical challenge lies in the problem of integrating various combinations of exploration techniques to provide quantitative estimates of the temperature and permeability of a geothermal reservoir. With the low value per unit mass of geothermal fluids, the ability to estimate the temperature of the reservoir fluid to within 10° to 20°C would be extremely important.

Once a geothermal prospect has been delineated by exploration techniques, it must be proven by drilling and production tests run in these boreholes. This is not as simple a task as it may appear at first. First of all, the temperatures and corrosive fluids which may be found in a geothermal reservoir impose severe wear on drilling equipment. Because of large volumes of production needed from each bore, it is necessary to ensure that free flow can take place from any fractures that may be intersected by the bore. Unfortunately, the drilling muds often used in drilling normal holes bakes to a solid quickly at geothermal temperatures and may seal off the fractures needed for production. If no mud is used, or air is used, it may be difficult to control flow from a well if high pressure zones are encountered. A possible solution may lie in the use of starch base muds that can be broken down chemically.

Downhole measurements of physical quantities -- well logging -- is a highly developed science, particularly as applied to the detection of hydrocarbons in the rock around an oil well. The use of these same tools, or new ones developed specifically to meet the requirements of geothermal development, for logging of geothermal test bores and production wells poses a significant technological challenge. However, well logs may not provide an adequate evaluation of geothermal wells, and probably must be supplemented to a greater extent by flow tests than is the case with oil wells. A particularly important problem is met in

producing wells from which only steam flows. If the rock is steam
saturated, the life of the steam may be quite limited, but if the
steam is derived from boiling water, the lifetime of steam flow
will be longer.

     Once steam or hot water has been produced from a reservoir,
the problems of converting the heat to electrical energy is
straightforward. The efficiency of conversion allowed by thermodynamic is quite low for the temperatures likely to prevail in a
geothermal fluid, and only 8 to 15% of the enthalpy can usually
be converted to electrical energy. However, this lack of efficiency
is quite different than that encountered with thermal power systems,
where both the recoverable and non-recoverable heat are obtained
from combustion of a fossil fuel. The non-efficiency involved in
geothermal energy conversion is not in fact an inefficient use of
an energy source that would otherwise be available to us. The
choice of temperatures and pressures of the steam derived from a
geothermal fluid, or some secondary fluid used to extract heat
from a geothermal reservoir, is largely a matter of wise economic
design based on well-known design principles.

     Considering our present lack of knowledge of the nature of
geothermal systems in general, consideration of possible adverse
environmental effects is probably premature. However, a few rather
obvious problems can be forseen. Foremost among these is the
disposal of spent geothermal fluids, after the 8 to 15% of recoverable energy has been extracted. In addition to the heat, most
geothermal fluids will contain undesirable salts in solution, as
well as rare elements whose effect on ecology may not be well known.
At present, it appears that such fluids will have to be reinjected
in or near the geothermal reservoir from which they are taken.
However, before these fluids are disposed of so summarily, careful
consideration should be given to the use of the rejected heat in
these fluids for some type of manufacturing process or for a
greenhouse operation.

Another area of potential hazard exists with respect to induced seismicity which may accompany the production and reinjection of large volumes of geothermal fluids. The problem may be aggravated by the need for extensive rock fracturing, if it is found necessary to produce fluids from impermeable geothermal reservoirs.

In summary, however, the most pressing technological challenges appear to me to be those associated with our need to know and understand the nature of geothermal heat concentrations. Until we have this information at hand, problems related to production and environmental hazards will remain academic in nature.

SALT DOMES, PIT CRATERS, AND DRY STEAM FIELDS — HEAT PIPE APPLICATIONS

Jack Green *

Heat pipes, operating vertically, with the heat source at the lower end have, in theory, high efficiencies over pipe lengths exceeding 100 meters. Three geological applications are possible: (1) heat pipe — nuclear reactor couples in salt domes, (2) heat pipes penetrating the crust of lava lakes and (3) heat pipes tapping high enthalpy steam fields. Environmental aspects of these pollution-free energy sources are as follows: (1) Salt domes are seismically safe, can sustain positive or negative pressures, and exist at shallow depths in energy-need centers in the Gulf Coast and the American Arctic. Waste reactor heat in the Arctic could provide comfortable "astrodome" environments and ice-free ports for oil tankers. Heat pipes separate the radioactive source from the power generating source. Radioactive fluids cannot enter the hydrological or meteorological cycles should an accident occur within a salt dome. (2) Lava lakes provide extreme temperature gradients for high heat pipe efficiencies and are likewise in an energy-need center — Hawaii. (3) Regarding steam fields, high enthalpy (dry) steam fields are depleted by conventional borehole techniques and corrosive fluids and particulates are vented to the surface. Heat pipes would extend the life of the geothermal field while keeping the corrosive subsurface fluids apart from surface heat-exchange turbomachinery.

* Earth Information Services, McDonnell Douglas Corporation, Huntington Beach, California

INTRODUCTION

Many energy projections exist (Faltermayer, 1972; Freeman, 1973) as
do preferred power sequences. They range from nuclear to natural
gas to low-sulfur oil to low-sulfur coal or from conventional to hydrogen fuel (Hauz, et al, 1972). Concern is expressed here because it is
possible that our creativity might be geared to those projections. As
Lewis Mumford stated, "Trend is not destiny." In this paper a depar-
ture is made from trend in joining the relatively new technology of
heat pipes with that of geothermal energy. Geothermal energy is pur-
posely and loosely defined as that originating below ground. The scope
of the paper, therefore, is the application of heat pipes to high
enthalpy steam fields, to lava lakes, and to underground reactors
emplaced in salt domes. The reactor-heat pipe concept is developed
for a specific salt dome region, the American Arctic, where waste
heat could be most efficiently employed.

This paper emphasizes the use of heat pipe-turbine systems in geologically realistic situations, exclusive of coal gasification (Mills, 1972),
and permafrost (Waters, 1973) fields of interest. Specifications of
heat pipe, turbine, or reactor design are outside of the scope of this
paper.

HEAT PIPES

Heat-pipe technology has experienced a rapid development in the last
ten years. The concept, originally developed by G. M. Grover in
1964 at the Los Alamos Scientific Laboratories and reviewed by
Eastman (1968), was applied to early spacecraft heat transfer and
control. An excellent review of the literature, through April 1972,
is given by Marcus (1972, p. 226-233). Detailed analyses of problems ranging from hydrogen-gas generation to bubble occlusions to
tradeoffs of cross sectional area to pipe length have resulted in
sophisticated units of high efficiency that are possibly applicable to
the geothermal energy recovery.

Heat pipes operate on a completely enclosed boiling/condensing cycle
in which the recirculation of the condensate is accomplished by
capillary pumping within a wick. In geothermal applications the
recirculation is simplified by gravity reflux (Kaser, 1972). Because
heat is transported by way of the latent heat of evaporation instead of
the sensible heat, large amounts of energy can be transported without
an appreciable temperature drop. Heat pipes are not necessarily
pipes; they can be made into a variety of shapes and be flexible (Hughes
Aircraft Company, 1974).

Heat pipe design limitations depend upon a number of variables. The
power delivered is a function of heat-pipe size; conversely, power
recovered is a function of heat source area. The efficiency of a heat
pipe depends upon the useful temperature range of the working fluid.
If the pipe is subject to other than normal accelerations, the efficiencies vary. Heat-transfer limitations are imposed by sonic vapor flow,
the effect of friction on sonic velocity, liquid entrainment, wicking
ability, and boiling in the saturated wick. Sonic velocity of the vapor
generated from the working fluid is a critical parameter. In general,
heat pipes operate efficiently between 0. 1 and 0. 2 of the sonic velocity
(D. Knapp, personal communication, 1973). Regarding entrainment and
the geothermal application of heat pipes, i. e. with the heat source
always at the bottom of a vertical pipe, liquid entrainment problems
are minimal. Wick design affects all these properties (entrainment,
wicking ability, and boiling) except sonic velocity (Kemme, 1969).
Optimum tunnel wick design, for example, can produce very high heat-
pipe efficiencies of 100, 000 watt-inches (Edelstein, et al, 1972).

Heat pipes as applied to geothermal systems have many advantages.
Heat pipes have three orders of magnitude greater efficiency in heat
transfer capacity relative to the best metal conductor. Also, the temperature at the output end is uniform regardless of the nonuniformity of
natural heat sources. This temperature flattening characteristic is
valid because evaporation and condensation of the working fluid takes
place virtually at the same temperature. Not only is the temperature
uniform, but it can be controlled using variable-conductance heat pipes

in which thermal conductance is automatically varied by the blocking
action of the noncondensible gas in the condenser end of the pipe. As
a corollary to this temperature control, constant power can be achieved
at the heat-sink end regardless of source characteristics by excess
power radiators. The evaporation and condensation functions, their
area and pattern, are essentially independent. This heat-flux transformation means that the working fluid can be evaporated, or its vapor
can be condensed rapidly over a small area or slowly over a large
area.

Unidirectional heat pipes also have been designed by using multiple
wicks. Perhaps the most significant attribute of heat pipes to the
application of nuclear reactors in salt domes is the safe separation of
heat source and heat sink. Heat pipes emplaced in subsurface salt
melts produced by the primary loop could remotely energize secondary
loops at the surface (or in upper galleries) of salt domes.

CONVENTIONAL NATURAL THERMAL SOURCES

Many authors (Kiersch, 1964; Evrard, 1964; Lessing, 1969; Godwin,
et al, 1971) have reviewed the literature on conventional geothermal
power systems. Table 1 from Koenig (1971) lists the major geothermal fields of the world and their projected power resources.
Many new geothermal provinces are being explored (Nevin, 1973;
Keller, 1974). Geothermal fields have three major constraints similar to oil fields: a magmatic temperature source, a permeable
reservoir to store the steam or hot water, and an impervious cap or
seal to contain the fluid (Jaffe, 1971). Geothermal fields differ from
oil fields because the rate of oil replenishment is much slower than
water. The ideal geothermal field should have a high geothermal
gradient of more than 100°C per kilometer and a high enthalpy. White,
et al, (1971) indicate that a critical point of maximum enthalpy is
reached in the vapor-dominated portions of steam reservoirs where
pressures and temperatures are about 32 kg/cm^2 and 236°C, respectively. The field should have a diameter of greater than 1,500 meters

Table 1

MAJOR GEOTHERMAL FIELDS OF THE WORLD
(After Koenig, 1971, p.12)

Country, Field	1969 Capacity (kw)	Planned or Under Construction (kw)	Total Projected (kw)
Chile			
El Tatio	0	20,000	20,000
El Salvador			
Ahuachapán	0	20,000	20,000
French West Indies			
Guadeloupe	0	30,000	30,000
Iceland			
Namafjall	0	3,000	
Hveragerdi	17,000	0	20,000
Italy			
Larderello	365,000	Up to 50,000	~440,000
Monte Amiata	25,000		
Japan			
Matsukawa	20,000	7,000	
Onikobe	0	Tens of thousands	>50,000
Otake	13,000	0	
Hachimantai	0	10,000	
Mexico			
Pathé	3,500	0	
Cerro Prieto	0	75,000	78,500
New Zealand			
Wairakei	160,000	Not known	
Kawerau	10,000		>170,000
Broadlands	0	Tens of thousands	
Philippines			
Legaspi	0	10,000	10,000
Taiwan			
Tatun	0	10,000	10,000
Turkey			
Kizildere	0	30,000	30,000
United States			
The Geysers	83,000	769,000	852,000
Brady's Hot Springs	0	10,000	
USSR			
Pauzhetsk	3,000	20,000	29,000
Kunashiry	0	6,000	
TOTAL			>1,759,500

and a producing reservoir shallower than 3,000 meters; and for heat-pipe application, less than 600 meters. Although the magmatic heat source might add halogens to the meteoric water supply making up this reservoir, the contribution should ideally be minimal. The capping or sealing of these reservoir fluids usually results from (1) fracture filling by hydrothermal alteration of the wall rocks, (2) silica cementation in fractures by outwardly circulating hot fluids, (3) calcium carbonate and sulfate cementation by inwardly moving cooler waters, and (4) hydrologic characteristics of the overlying meteoric and condensate waters (White, <u>et al</u>, 1971; Garrison, 1972). According to Nevin (1972, p. 3), to produce 500 Mw or more, the ideal geothermal field should be able to produce about 4.5×10^6 kg of slightly superheated steam per hour from some 100 wells. From Table 2 it can be seen that 500 Mw would constitute a major field almost equal to the 1974 output of the Geysers field in California.

Relative to heat-pipe applications, and the matrix of geothermal field classification, we shall consider steam for power versus steam for chemicals, subsurface insulated fields versus surface hot springs, and high enthalpy (dry) steam fields versus low enthalpy (wet) steam fields. Table 3 presents a simplified version of geothermal fields where "dry" and "wet" are contrasted. Also note that even with a reduced efficiency of energy yield using heat pipes, the technique may still be competitive with other power systems.

In general, the "dry" steam fields are found in late Pliocene-Quaternary orogenic belts characterized by uplift up to 1,000 meters as typified by the Larderello, Geysers, and Mt. Amiata fields. The "wet" steam fields are found in tectonic zones covering a wider time span from late Tertiary to the Quaternary, and are characterized by subsidence of between 3,000 to 6,000 meters; Wairakei, Niland, and the Casa Diablo fields are examples. In all cases, an intrusion supplies the thermal energy. In the Geysers area (McNitt, 1963; Scott and Wood, 1972), superheated steam exists in faults in permeable graywacke similar to Larderello where the faults cut permeable

Table 2

ESTIMATED GEOTHERMAL- POWER GROWTH IN CALIFORNIA
(Modified from Wilson, 1973)

| | Megawatts | | | | |
	The Geysers	Imperial Valley	Other Areas	Total/ Year	Cumulative Total
Capacity at 12/31/71	184			184	184
Additional Capacity:					
1972	106			106	290
1973	400	10		410	700
1974	510			510	1, 210
1975	620			620	1, 830
1976	850	200		1, 050	2, 880
1977	>850	100	100	>1, 050	>3, 930
1978	>850	100		>1, 050	>4, 880
1079	>850	100	100	> 950	>5, 930
1980	>850	100		>1, 050	>6, 880
1981	>850	100	100	>950	>7, 930

Source: California Department of Conservation, Division of Oil and
Gas, August 1972 and Environmental Science, August 1973.

Note: This estimate reflects only present development trend, not
total state potential.

Triassic carbonates capped by impermeable shale. In the "wet" steam
fields as at Wairakei (Smith, 1958), saturated steam and hot water
occur in pumice breccias some 450 to 900-meters thick and capped by
impermeable mudstones. The pumice breccia reservoirs which are
recharged by water entering along peripheral faults are underlain by
a dense ignimbrite.

In the application of heat pipes to these thermal systems, one would
logically seek out the high-enthalpy fields for higher efficiency.

Table 3

COMPARISON OF "DRY" AND "WET" STEAM FIELDS

Field	Max. Reservoir Temperature (°C)	Enthalpy (kcal/kg)	Heat Discharge (kcal/sec)	Mass Discharge (kg/sec)
"DRY"				
Larderello	245	680	542,000	770
Geysers	208	667	>67,000	>100
"WET"				
Wairakei	266	250	240,000	960
Niland	300	315	-	-
Hveragerdi	233	250	123,000	460

System	Mills/kwh	
Geothermal	3 to 5.3[*†]	*Geysers Field
Thermoelectric	5.5 to 8.2[*†]	†W. W. Lyons, Deputy Under-secretary, Dept. of Interior, 9/14/73
Nuclear	5.5 to 11.6	
Hydroelectric	5 to 11.4	

Another consideration is the depth of the reservoir. Although long
heat pipes are being considered, the shorter the pipe the greater is the
efficiency of heat transfer. Supplementing the isobutane technique
in the Geysers area, heat pipes less than 600-meters long could
reach high-enthalpy horizons. The use of heat pipes is attractive
because only heat is transferred; the depletion of steam from the
reservoir is eliminated and the lifetime of the field prolonged. No
movement of high-temperature or corrosive fluids occurs outside
of the pipe.

In conventional systems, steam, water, and particulates impinge on
turbine blades. Even with the "bladeless" turbine approach (Possell,
1973) there is some corrosion of turbomachinery. Heat pipes do not
have this problem.

SPECIAL NATURAL HEAT SOURCES

Heat pipes to retrieve geothermal energy would work best where heat
gradients are maximized over short distances, as in volcanic
provinces.

Recently, drilling for geothermal energy was started in Hawaii.
Dr. G. Keller (1974) of the University of Colorado, drilling within two
kilometers of the Kilauea caldera rim, found optimum conditions for
steam recovery 200 meters below sea level at a depth of more than
1,200 meters where temperatures reached about 150°C. The earliest
work in Hawaii in probing volcanic fractures for thermal waters was by
Dr. A. Abbott in 1961 when the Hawaii Thermal Power Company drilled
power wells on the Puna Rift. In this case, wells shallower than
210 meters were drilled and bottom hole temperatures were less than
102°C.

A special case for heat-pipe application in Hawaii may exist where pit
craters are periodically filled with molten lava. Pit craters exist in
caldera areas where rapid subsidence of a magma pocket often produces cylindrical subsidence pits, some deeper than 200 meters. When
lava is extruded from nearby fractures or from the pit crater floor, a
"puddle" of lava is formed that could constitute a heat source for heat-
pipe exploitation over many decades. Figure 1 shows the chain of pit
craters, some of which are now filled, that existed on the flank of the
Kilauea caldera in Hawaii in 1967. Other pit craters occur near the
crest of Mauna Loa, some with clear evidence of subsequent pit floor
foundering (Fig. 2). Since the volcanic metabolism of Hawaii is very
high, it is not surprising that in the last 30 years, at least five of these
pit craters (Kilauea Iki, Halemaumau, Alae, Aloi, and Makaopuhi)
have been partially or totally filled with molten lava. Since the temperature of basaltic lava is on the order of 1,160°C and the conductivity of 500°C basalt is about 3.4×10^{-3} cal/cm°C sec, one can have
molten lava pools under thin, but strong, crusts on lava lakes within
six months after pit filling.

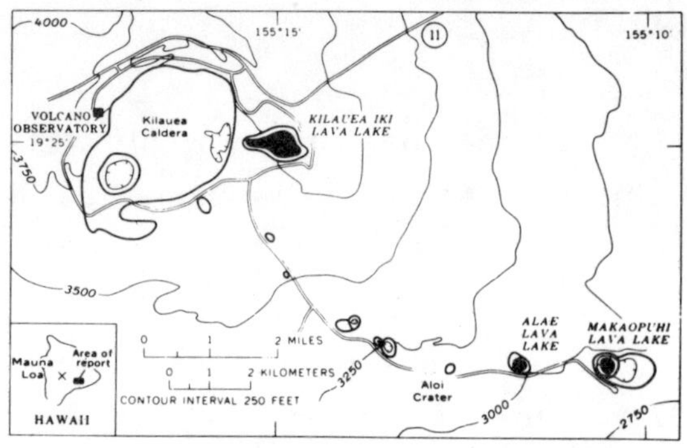

Fig. 1   Chain of Craters on Flank of Kilauea Caldera

Fig. 2   Lua Poholo Pit Crater, Northeast Rim of Mauna Loa Caldera. Pit is 310 m
in Diameter and About 150 m Deep. Floor has Foundered Producing the
Tilted Structure.

The August 1963 eruption of Kilauea formed a small pond of lava on the
floor of the Alae pit crater some 16 meters thick. Temperatures in
the cooling lava lake were studied by repeated thermocouple measurements in bore holes in the lava crust (Decker and Peck, 1967). The
lava lake cooled at a rate predicted by Jaegger (1961), with the last
interstitial melt solidifying in September, 1964. Figure 3 illustrates
the temperature profile with depth. The 900°C temperature at 10 meters
depth constitutes a 90°C per meter gradient in which heat pipes could
function efficiently.

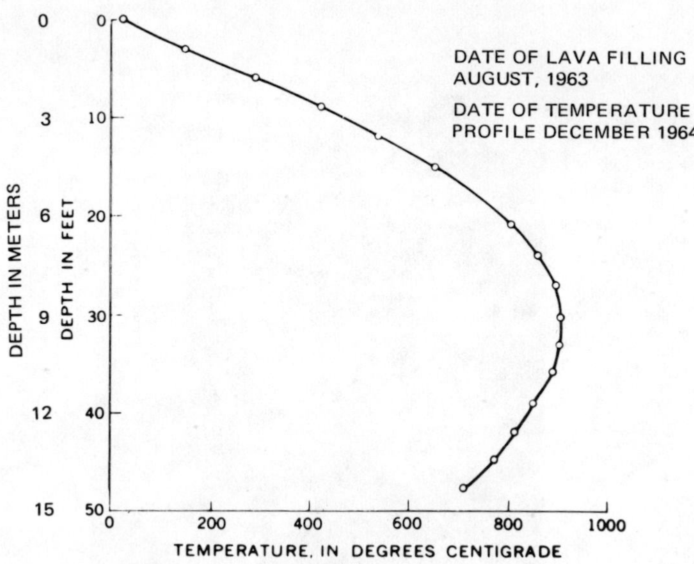

Fig. 3   Temperature Profile in Alae Pit Crater, Kilauea Caldera Flank
(After Decker and Peck, 1968)

In December of 1970 the entire Kilauea rift zone was flooded by one of the most voluminous flows in historic time — more than 10^8 km^3. Part of this flow entirely filled the Alae and Aloi pit craters. The source fissure has been called Mauna Ulu which is still intermittently active.

In March 1972, activity relating to the pit craters partially filled Makaopuhi crater to a depth of 20 meters. Figure 4 is a photograph of the spectacular lava falls entering Makaopuhi on March 21, 1972. The rate of infilling then was about 400,000 cubic meters per day. This was part of a larger flow which exceeded in cumulative total of approximately 250,000,000 cubic meters during the 1972 Kilauea eruption. Using order-of-magnitude solidification rates for rock melts exposed to the earth's surface (Daly, 1933), one can crudely calculate the lifetime of a heat reservoir if Makaopuhi were filled completely (Table 4).

From the western pit radius of about 300 meters and a depth of 200 meters, a specific heat of 0.1 and a useful temperature range (from a heat-pipe standpoint) from say 500° to 1,000°C, some 10^{15} calories would be available over a lifetime of many hundreds of years. (The solidification time would be about 1,000 years; core cooldown to 500°C

465

Fig. 4   Lava Falls Entering Makaopuhi Crater,
March 21, 1972 (Photo by J. G. Moore,
United States Geological Survey)

Table 4

RATE OF CRYSTALLIZATION WITH DEPTH OF BASALTIC MELT AT 1,100°C EXPOSED TO THE SURFACE (AFTER DALY, 1933)

Meters	Time
1	12 days
10	3 years
100	300 years
1,000	30,000 years
10,000	3,000,000 years

would certainly be over 1,000 years.) If a 0.1-percent efficiency of energy recovery could be realized by multiple heat pipes, 10^{12} calories or about one billion kilowatt hours would be locally available in the assumed decades of pipe operation.

The technique of drilling holes through the solid basaltic crust on lava lakes has been perfected. Heat pipes would be extremely efficient in this system because of the short pipe lengths (less than tens of meters) and high thermal gradient. On the other hand, the introduction of heat pipes into molten lava is not endorsed because of corrosion and freezing problems. Even introducing the pipe into recently solidified lava would probably be ineffective (Truesdell, A. H., U.S.G.S, personal communication, 2/28/74). The recently solidified lava at depth must first be fractured either by hydraulic or chemical explosion methods. Then the heat pipe could be emplaced and finally the working fluid—either water or one with a higher vapor temperature could be injected. Since large lava flows occur almost every year on Hawaii, lava damming and diversion (Finch, 1943; Finch and MacDonald, 1949; Mason and Foster, 1953) and chemical explosion techniques might provide stable melt ponds for limited heat-pipe research.

SPECIAL DRY GEOTHERMAL SYSTEMS

Several workers are examining dry geothermal systems (Hammond, 1973) wherein water injected under pressure, hydraulically fractures geothermally heated (but initially dry) rocks (D. Steward of Battelle; D. W. Brown, M. C. Smith, and R. M. Potter of the Los Alamos Scientific Laboratories; and B. Rayleigh of the U.S.G.S.). In these systems, the water heated by geothermal heat is not corrosive. If two adjacent wells are connected at depth (Brown, et al, 1972), a convecting system results. Heat pipes could also be used in this approach, but instead of any convecting system, the injected fluids to hydraulically fracture the rocks could be higher temperature working fluids instead of water. Storage of oil under geologically favorable conditions could be used as a heat sink fluid which could be tapped by heat pipes in a hydraulically fractured, high-temperature reservoir.

One of the main problems in reactor siting is the potential hazard of the release of radioactive and thermal waste to the hydrological, meteorological, and biological cycles (Dye, 1973).   The reality of the guillotine break in the primary loop in some reactor systems and the host of fail-safe techniques proposed in the event of such a break (Chave, 1973) are clear evidence of the seriousness of this problem.

People like to live near the sea and perversely nature commonly puts her faults there.   The inland location of reactors, remote from coastal faults, is not a solution to the problem because of the need for large quantities of coolant water which is at a premium in arid areas.   Does the emplacement of nuclear reactor-heat pipe systems in coastal zone salt domes tend to solve the environmental and safety problems of siting?   What are the characteristics of salt domes?

As with geothermal power, extensive literature exists on salt domes and their potential use for radioactive waste storage. A salt dome is a mushroom-shaped structure produced by slow upward salt growth due to density differences. A salt dome is also nature's shock absorber to faulting and is, in itself, a source of unlimited quantities of molten salt. Salt melt could be used as a coolant in reactors and incidentally as a material suitable for destroying organic waste (J. Greenberg cited in Lessing, 1969). By adding pipes to a reactor-in-a-salt-dome concept, the primary and secondary loops remove only heat and provide for safe reactor operation even if a guillotine break should occur in the primary loop. In the event of a run-away reactor, the radioactive waste would simply sink (unless artificially retained in crucibles) without contaminating the hydrological or meteorological cycles. Salt domes are also gas tight and completely dry. Thus, positive or negative pressures[*] could be maintained in reactor galleries. Many salt domes have large galleries already excavated. For example, reactors and their radioactive waste products could be accommodated in the salt workings of the Jefferson Island dome, Louisiana (Figure 5).

---

[*]The San Onofre reactor containment sphere is designed for an internal 3.26 kg/cm^2;

Fig. 5 Internal Structure of Jefferson Island Salt Dome, Louisiana Showing
Flow Structures in Rooms 30 Meters High (After Balk (1953))

Also, reactors emplaced in a salt dome are relatively invulnerable to hostile civil or military forces. Such reactors would be less affected by explosive shock because they would be buried in salt, which is the most common elasticoviscous material on earth.

Figure 6, taken from Trusheim (1960), illustrates a three-dimensional schematic of salt domes which are commonly aligned along ancient tectonic zones of weakness. The alignment reflects the disturbance of an original deep-seated, thick salt bed such as the Zechstein formation in Germany or Paradox formation in the United States. Because of the elasticoviscous nature of rock salt and its specific gravity, the salt rises thousands of meters over many millions of years to produce the typical dome or stock shape.

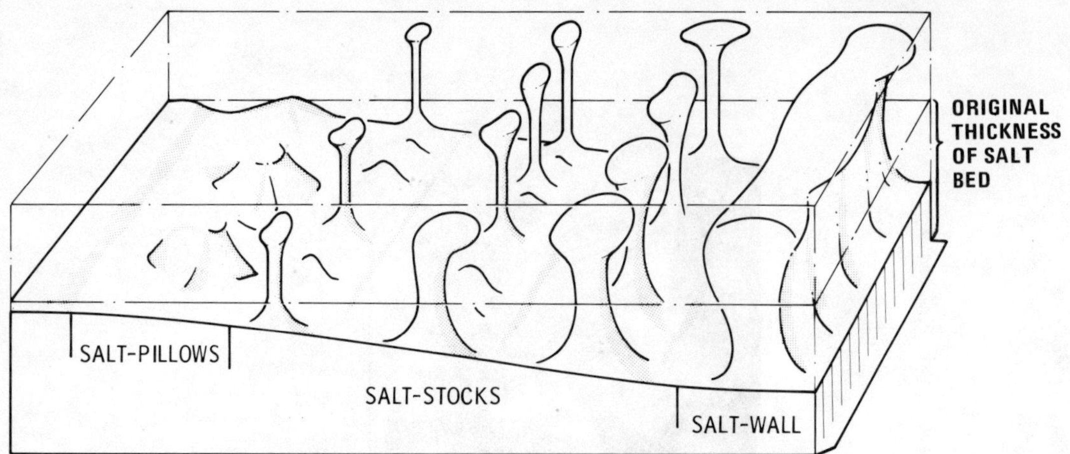

Fig. 6   Schematic of Salt Domes (modified after Trusheim, 1960)

Average growth rates are somewhat meaningless for salt domes because of the intermittent pulsed nature of dome growth. An average of from 0.002 to 0.005 cm/yr can be calculated; the maximum rate is unknown.

The plasticity of salt is greatly increased with increasing temperature. Experimental data show that by increasing the temperature of rock salt from 27 to 410°C the rate of steady-state creep is increased 75 times (Serata and Gloyna, 1959). High anomalous temperatures are recorded in the Pescadido salt dome and are thought by Heroy (1968, p. 625) to be caused when the mass of salt broke out of its normal stratigraphic position and rose upward through overlying strata. Because of the high specific heat of salt, it will retain its connate heat as it rises through the overlying strata.

The relatively low viscosity produces the mushrooming of the dome top which often serves to trap oil and gas on the dome flanks. The low viscosity of salt also serves to cause salt to self-seal when subjected to sudden stress similar to the behavior of silicone putty. A considerable amount of scale modeling research has been done to evaluate these parameters (Adams, 1951; Cloos, 1968; Currie, 1956; Dobrin, 1941; Escher and Kuenen, 1929; Hubbert, 1937; Link, 1930; Nettleton, 1934 and 1943; Nettleton and Elkins, 1947; Parker and McDowell, 1951 and 1955; Ramberg, 1963 and 1967).

As far as the effect of elevated temperatures in the reactor chamber
on the salt dome, creep will be accelerated but the effects on structural stability have not been accurately predicted (Bradshaw _et al_,
1968).

Figure 7 defines the general and specific locations of salt domes in the
United States. Figure 8 emphasizes that salt domes, which are invar-
iably associated with petroleum deposits, occur in the coastal areas
of the Arctic (Pitcher, 1973), both in Canada and Siberia (Lefond,
1969, p. 5). Figure 9 illustrates one of these domes in the Parry
Islands, the Isachsen dome (Heywood, 1955). Waste heat from
reactors emplaced in Canadian salt domes (Figure 10) could provide
habitable "astrodome" environments,* ice-free ports, and corridors
for oil tankers (Figure 11). Obviously, these corridors maintained by
buoyed hot water pipes would not have great (>300 km) length but would
permit interprovince oil transport to mainland pipeline heads. See
Breslau, _et al_, 1970, Edwards 1970.

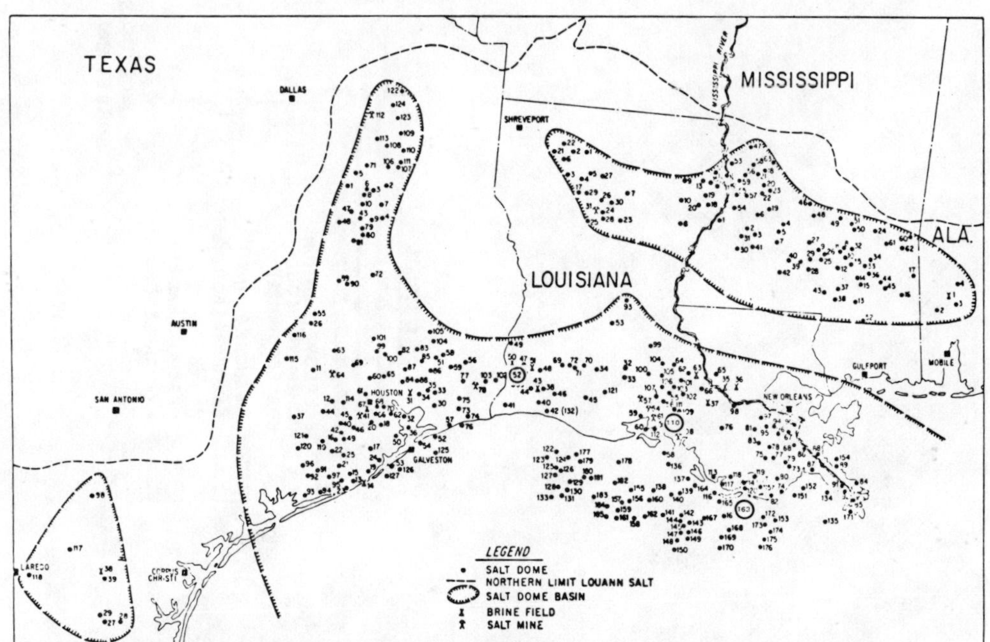

Fig. 7   Salt Domes in Gulf Coast Region of the U.S. (Lefond, 1969, p.5)
Encircled numbers are cited in text.

*The availability of molten salt in the Arctic would help solve waste disposal problems in new towns. Coincidentally, the molten
salt technique has been demonstrated for use in Eskimo villages in Alaska's North Slope by Anti-Pollution Systems, Inc (Lessing,
1973, p. 144, 146) showing a waste volume reduction of more than 95%.

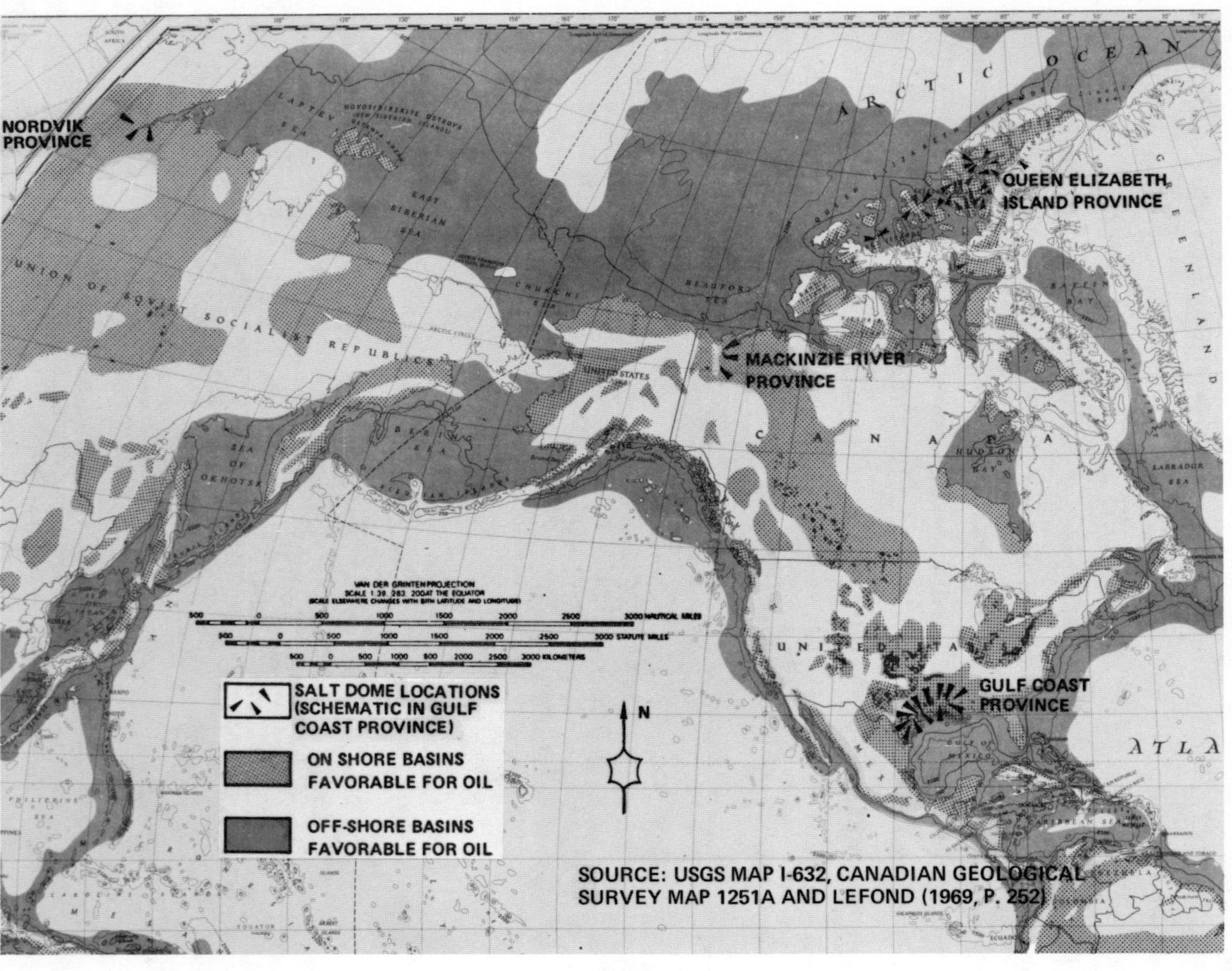

Fig. 8   Arctic Salt Dome Provinces

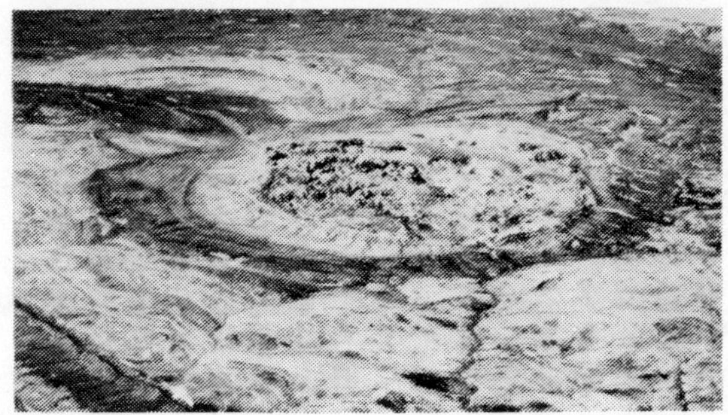

Fig. 9 Oblique View of the Isachsen Dome from the East on Ellef Ringnes Island of the Queen Elizabeth Islands. The Dome is About 7.5 km Wide by 8 km Long with the Dome Core being Composed Dominantly of Gypsum and Anhydrite — Typical Cap Rocks of Salt Domes. Photo Credit: Royal Canadian Air Force. (From Heywood, 1955, p. 28)

Initially a conventional reactor could be placed within a salt dome and heat pipes disposed to carry off the heat either to the surface or to an upper gallery within the salt dome (Figure 12).   What length heat pipes would be required? How deep would the reactors have to be?

The Vinton salt dome Calcasieu Parish, Louisiana (No. 52 on Figure 7) is quite shallow.   The cap rock is 117 meters below the ground surface and the salt 282 meters, with the maximum depth penetrated of salt down to 2,710 meters.   The Cote Blanche Island salt dome, St. Mary Parish, Louisiana (No. 110 on Figure 7) has cap rock only 78 meters below ground surface, salt at 91 meters, with maximum penetration of salt to 3,758 meters.   The offshore South Pelto, Block 20 salt dome, Louisiana (No. 163 in Figure 7) is 171 meters to salt, with deepest penetration of salt at 3,742 meters.

A more sophisticated approach would be to install a modified nuclear reactor in a salt dome (Figure 13).   It could conceivably generate a salt pool in a "swimming pool" type reactor or as a pool developed to one side of the reactor.   High temperature cladding materials, possibly SiN_4, could be employed to achieve the temperatures required to melt the salt.   The thermal properties of salt (Glassner, 1957) are shown in Figure 14.

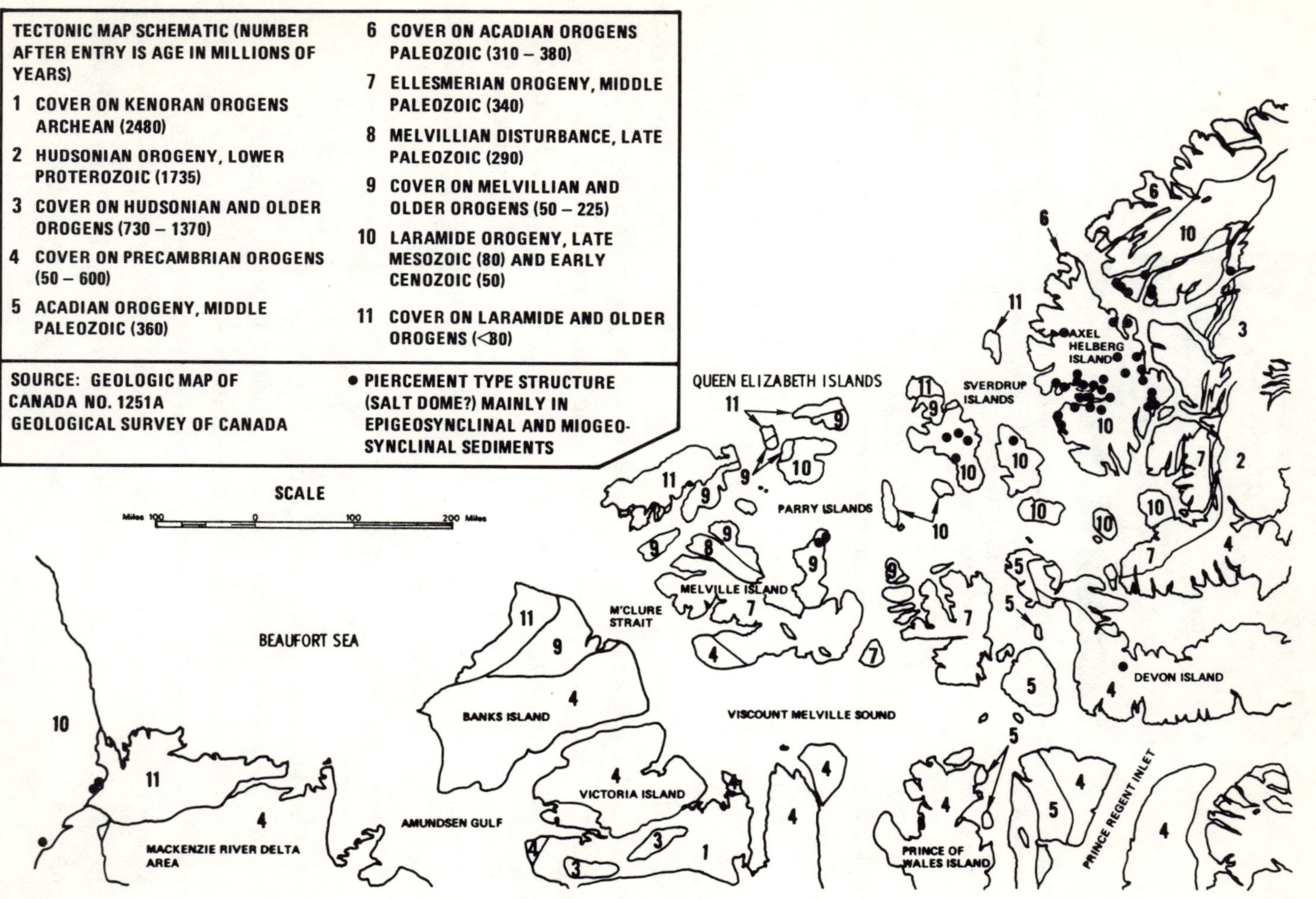

Fig. 10   Canadian Piercement Structures

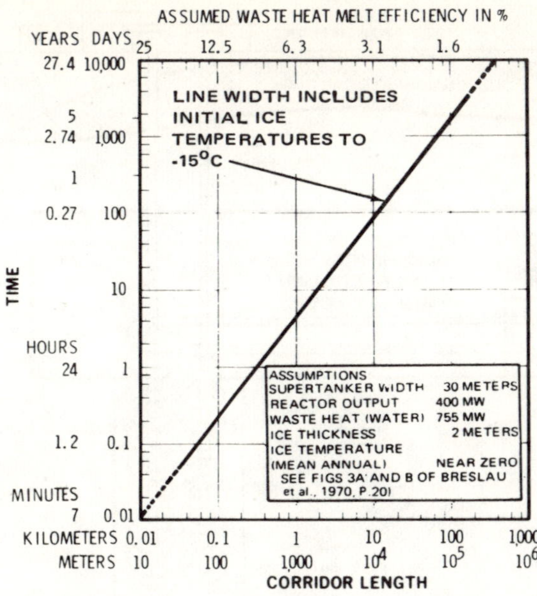

Fig. 11   Supertanker Corridor Lengths in Arctic Ice Melted by Reactor Waste Heat

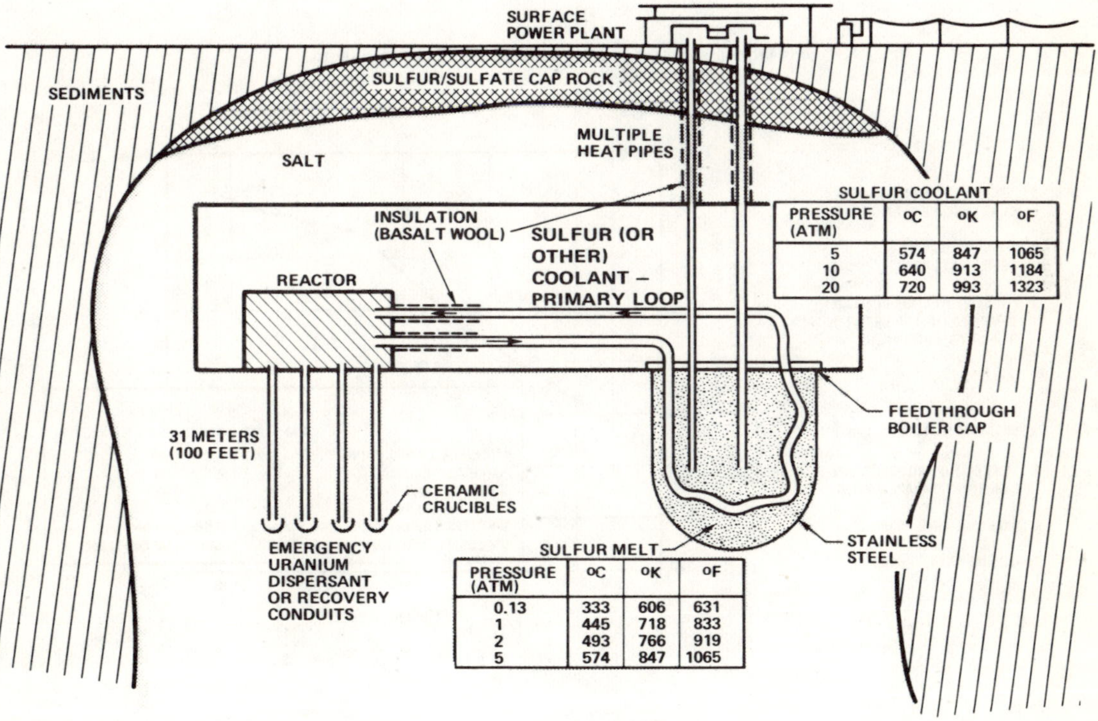

Fig. 12   Salt Dome Reactor Siting — Off-The-Shelf Configuration

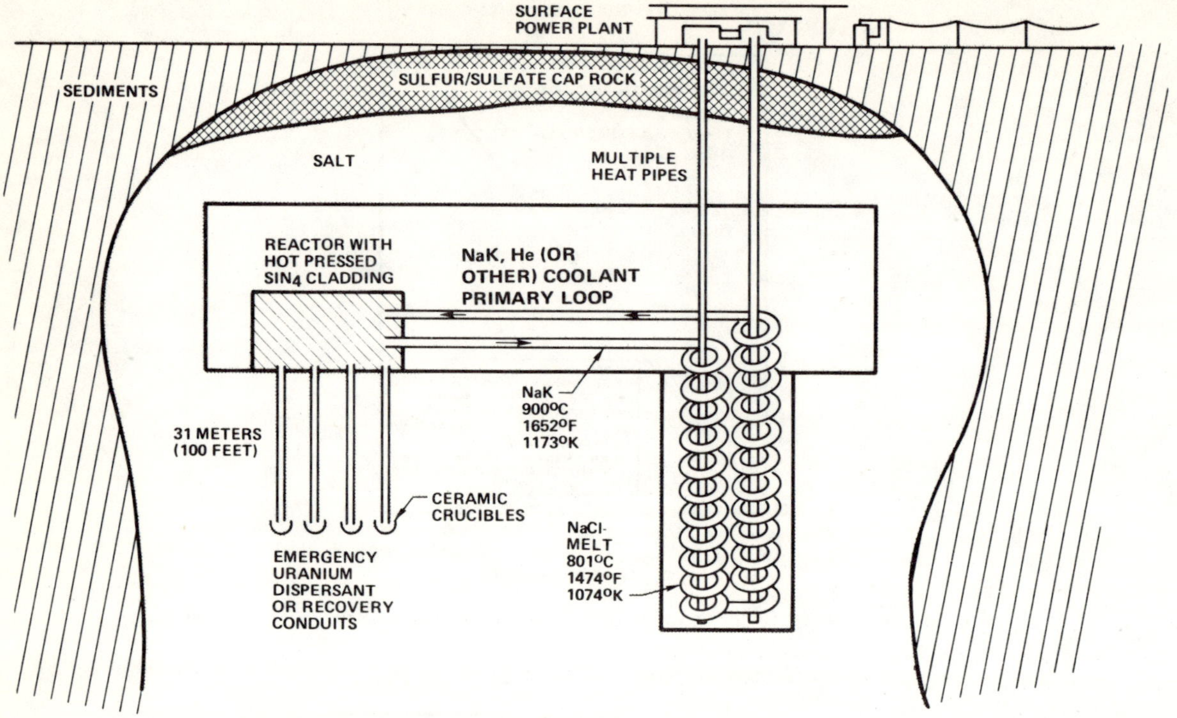

Fig. 13 Salt Dome Reactor Siting — Modified Configuration

1 PRESENTLY MAINTAINED CLADDING TEMPERATURE

2 PRESENT NaK OUTLET TEMPERATURE

3 PRESENT HELIUM COOLANT TEMPERATURE, FORT ST. VRAIN REACTOR, COLORADO.

4 PRESENT CLADDING LIMIT

5 "SAFE" HOT PRESSED SIN_4* CLADDING TEMPERATURE

6 SiN_4 MELTING POINT

7 UO_2 MELTING POINT

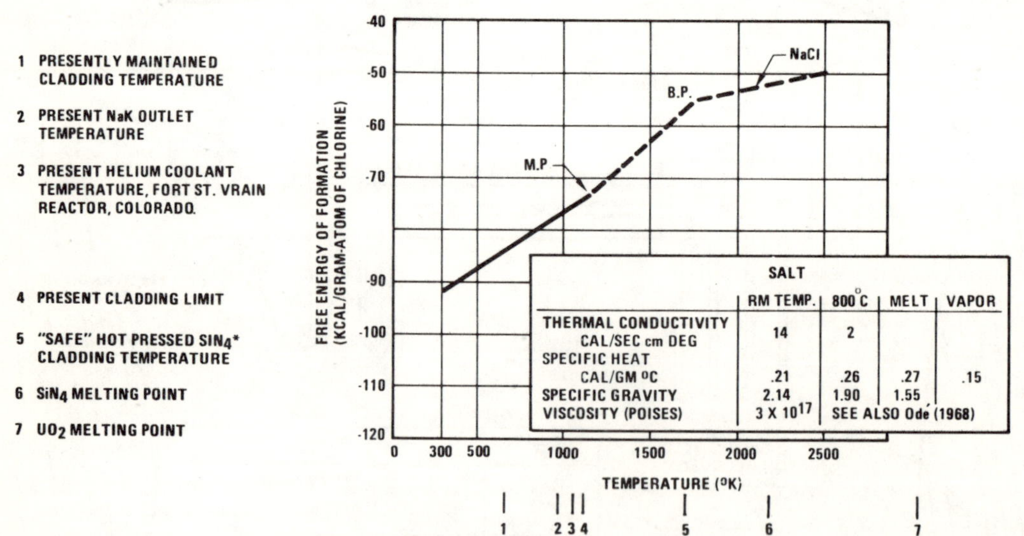

SALT	RM TEMP.	800°C	MELT	VAPOR
THERMAL CONDUCTIVITY CAL/SEC cm DEG	14	2		
SPECIFIC HEAT CAL/GM °C	.21	.26	.27	.15
SPECIFIC GRAVITY	2.14	1.90	1.55	
VISCOSITY (POISES)	3×10^{17}	SEE ALSO Odé (1968)		

*AS AN EXAMPLE ONLY; HIGH TEMPERATURE RESISTANT PYROLITIC CARBON AND SILICON CARBON ARE ALREADY IN USE FOR Th-U DICARBIDE FUEL ELEMENTS.

Fig. 14 Thermal Properties of Salt (NaCl)

To recapitulate, there are two conceptual design aspects. The first
is simply to put a conventional reactor system in a salt dome and
extract primary loop energy by heat pipes. The most desirable system would perhaps be similar to the Fort St. Vrain High Temperature
Gas-Cooled Reactor (HTGR) of Gulf General Atomic. Two power
generation systems can be adapted from HTGR technology: the gas-
cooled fast reactor of the Fort St. Vrain type and the helium-driven
gas turbine plant. The use of He as a coolant that can reach the melting temperature of NaCl is already a reality. The prestressed concrete reactor vessel, the fuel particles of thorium-uranium dicarbide,
and pyrolitic carbon and silicon carbide cladding permit operation at
higher pressures and temperatures in HTGR.

The second design concept is to modify a reactor, possibly of the
HTGR type, to specifically employ NaCl melts in heat-pipe and
organic-waste disposal applications. The secondary loop system, as
with the conventional system, could be within an upper gallery in the
salt dome or on the surface of the ground. For offshore salt domes,
the secondary loop would be seawater heated to steam by heat pipes.
Either type of reactor could be used to develop steam in a salt water
"secondary" loop above an offshore salt dome. In all systems there
is no contact of the primary with the secondary loops other than by heat
pipes. In all systems there is no contact of radioactive wastes with the
hydrosphere or atmosphere even if the most severe accident should
occur.

POTENTIAL HAZARDS

Three potential hazards involving reactors in salt domes are seismic,
thermal, and radiation.

Seismic Risk Analysis

It is necessary to know if a specific salt dome is, say, 100 times safer
from earthquake hazard than the San Onofre site.   To come up with
quantitative forecasts of this type, design earthquakes and site dynamic
response spectra must be calculated for the salt dome site in question.
These spectra are determined by the anticipated structural life of the
reactor site (L), and an acceptable nonoccurrence probability (p), of
the selected design earthquake intensity.

Earthquake response analyses of a reactor site require an evaluation
of the seismicity of the region, the regional geology, tectonic patterns,
and salt dome site dynamic characteristics (Hansen, 1970). Soil constants above the cap rock must also be evaluated on a salt dome by
salt dome basis.

Thermal and Radiation Hazards

Because of reactor heat in a closed room within a salt dome, the wall
temperature may well exceed 30°C, even with internal cooling systems.
It is obvious by the very design of the molten salt system that some
portions of the salt dome will be much hotter.   To assess the local
effects of the high-temperature reactor primary loops, molten salt
baths, heat pipe conduits, secondary loops, and surface turbines, a
detailed analysis of heat flow and dispersion will be required.   Also to
be considered, are elasticoviscous creep and related viscosity data
for rock salt of varying purities under mild hydrostatic loads for the
range of temperatures expected.

Other subjects outside the scope of this report are the radiation shielding effects.   Bradshaw, et al (1968, p. 655) show that salt behaves
essentially as light concrete both in weight and in gamma shielding
properties.   Przewlocki, et al (1957) discuss the monitoring of radioactivity in horizontal boreholes in salt domes in Poland.

Remote sensing and geoscientific field work are integral parts of site selection and monitoring. Specific salt domes should be evaluated relative to rock deformation, growth rates, cap rock mineralogy, salt permeability, and rock chemistry. Fracture patterns on and within specific salt domes (Tectonic Map Committee of GCAGS, 1972) need analysis as well as their seismic history.

Although 28 nations have salt dome provinces (Table 5), the Arctic and Gulf Coasts appear to be key regions for possible nuclear siting. Selected domes within these provinces, both on and offshore, should be considered, relative to their proximity to nearby cities, oil-producing centers, and petroleum pipelines.

Figure 15 illustrates an environmental geological map (Fisher, _et al_ 1973) that has been compiled for the Houston area. Here, salt domes are shown that already are used for oil and gas storage. Offshore salt domes are also defined in this figure. Salt domes with existing sulfur caps should be noted. As indicated on Figure 12, sulfur would make an acceptable cooling fluid for reactors as discussed by Saule (1960). The pressures required to achieve temperatures necessary to melt salt are very low compared to pressures currently used to pressurize steam lines in many conventional reactors.

In any project involving reactor siting and heat dispersion, a complete study must be made of the effect of such systems on the environment (Dette et al, 1973; Hubbert, 1971). The study must be thorough, with total analysis of the physical, biological, and socio-economic environments. For land-based operations, the thermal flux to the subsurface and surface environments must be assessed. Does the proposed program solve the siting problem of reactors where salt domes are present with respect to protecting the water and air resources? Are there problems of deformation affecting the reactor or its components and in turn, does the reactor provide cause for local surface

Table 5

NATIONS WITH MAJOR SALT DOME PROVINCES

Aden	Gabon	Tanzania
Afghanistan	Germany	Tunisia
Angola	India	United States
Australia	Iran	Alabama
Austria	Mexico	Colorado
Canada	Morocco	Louisiana
Chile	Peru	Mississippi
Colombia	Poland	Texas
Cuba	Portugal	Utah
Cyprus	Rumania	U. S. S. R.
Denmark	Saudi Arabia	Yemen
France	Senegal	

subsidence or disturbance by accelerating natural deformation of the salt? For offshore emplacement of reactors in salt domes, are the seals into the salt dome absolutely foolproof? Does the system provide the best separation of primary and "secondary" loops with minimum impact to the environment? How seriously is the local thermal regimen of the ocean affected? Would reactor-created ice-free Arctic ports affect the regional climate? How could permafrost be maintained in Artic reactor sites? What is the sociological and demographic impact of giving salt dome provinces in the United States (and elsewhere) the potential for becoming the energy centers for a nation or province? What is the time frame of salt dome - reactor program development relative to the availability of reactor fuel, solar energy sources, pipe line expansion, and other factors now under consideration?

Before any serious consideration is given to the use of heat pipes in any geothermal application, the environmental impact must be considered first.

480

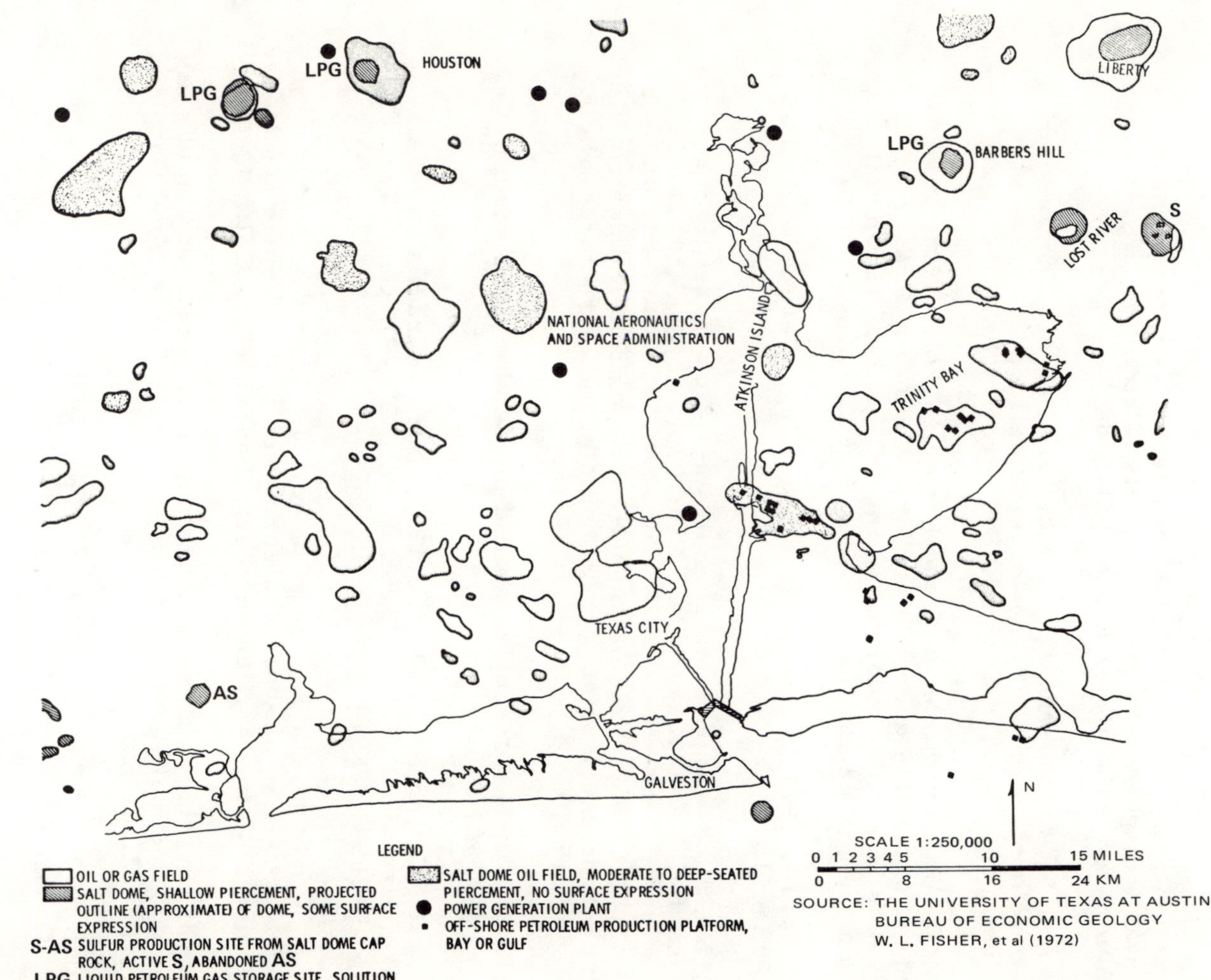

Fig. 15 Gulf Coast Salt Domes - Houston Area

BIBLIOGRAPHY

Adams, G. F., 1951, A working scale model of a salt dome: Jour.
Geol. Ed., vol. 1, p. 22-28

Balk, R., 1953, Salt structure of Jefferson Island, Iberia and
Vermillion parishes: Bull. Amer. Assoc. Pet. Geol., 37, p. 2455,
4474

Bradshaw, R. L. et al., 1968, Properties of salt important in radioactive waste disposal: In Saline Deposits (Mattox, R. B. ed.) Geol.
Soc. America Special Paper 88, p. 643-658

Breslau, L. R., Johnson, J. D., McIntosh, J. A., and Farmer, L. D.,
1970, Development of Arctic Sea transportation; Environmental
Research: Jour. Marine Technol. Soc., vol. 4, p. 19-43

Brown, D. W., Smith, M. C., and Potter, R. M., 1972, A new
method for extracting energy from "dry" geothermal reservoirs:
Los Alamos Scientific Laboratory, LA-DC-72-1157 (Preprint),
23 mimeographed pages, Los Alamos, New Mexico

Chave, C. T., 1973, Emergency core cooling: (letter): Science,
vol. 179, p. 1281

Cloos, E., 1968. Experimental analysis of Gulf Coast fracture
patterns: Bull. Am. Assoc. Pet. Geol. Vol. 52, p. 420-444

Currie, J. B., 1956, Role of concurrent deposition and deformation
of sediments in development of salt dome graben structures: Bull.
Am. Assoc. Pet. Geol. vol. 40, p. 1-16

Daly, R. A., 1933, Igneous rocks and the depths of the earth:
McGraw Hill, N. Y.

Decker, R. W. and Peck, D. L., 1967, Infrared radiation of Alae
lava lake, Hawaii: U.S.G.S. Prof. Paper 575 - D, Geol. Surv. Res.
1967, p. D-169 - D-175

Dette, J. T., Stahl, L. E., and Fischer, J. A., 1973, Site
investigations for an offshore nuclear power plant: Dames and Moore
Engineering Bulletin, 42, (September), Los Angeles, Calif.

Dobrin, M. B., 1941, Some quantitative experiments on a fluid salt
dome model and their geological implications: Trans. Am. Geophys.
Union, vol. 22, p. 528-542

Dye, L., 1973 (July 5), Nuclear waste perils thousands: Los Angeles
Times, Orange County Ed., vol. xcii, pt. 1, p. 1,3,32,33

Eargle, D. H., 1968, Stratigraphy and structure of the Tatum salt
dome area, southeastern Mississippi and northwestern Washington
Parish, Louisiana: In Saline Deposits (Mattox, R. B. ed.), Geol.
Soc. America Special Paper 88, p. 381-405

Eastman, G. Y., 1968, The heat pipe: Scientific American, vol. 218, p. 38 - 57

Edelstein, F., Kosson, R., Hembach, R., and Tawil, M., 1972, A tunnel wick 100,000 w-in. heat pipe. AIAA 7th Thermophysics Conference: San Antonio, April 10-12, 1972

Edwards, R. Y. Jr., 1970, The development of Arctic Sea transportation; Engineering Research: Jour. Marine Technol. Soc., vol. 4, p. 44 - 59

Escher, B. G. and Kuenen, Ph. H., 1929, Experiments in connection with salt domes: Leidsche Geologische Mededeelingen, vol. 3, p. 151 - 182

Evrard, P., 1964, La récherche et le development de l'énergie géothermique: Centre Internationale de Récherches Géothermiques, Publ. No. 1, 8 p.

Faltermayer, E., 1972, The energy joyride is over: Fortune, vol. lxxxvi, No. 3, p. 99 -101. 178, 180. 182, 184, 186, 188, 191

Finch, R. H., 1943, Lava rivers and their channels: The Volcano Letter, No. 480, p. 1 - 2

Finch, R. H. and MacDonald, G. A., 1949, Bombing to divert lava flows: The Volcano Letter, No. 506, p. 1 - 3

Fisher, W. L. et al, 1972, Environmental geologic atlas of the Texas Coastal Region: Bureau Econ. Geol., University of Texas, Austin, Texas

Freeman, S. D., 1973, Energy, drain on America first? (editorial): Science, vol. 179, p. 779

Garrison, Lowell E., "Geothermal steam in the Geysers - Clear Lake Region, California: Geol. Soc. America Bull., Vol. 83, No. 5, pp. 1449-1468, 1972

Godwin, L. H., Haigler, L. B., Rioux, R. I., White D. E., Muffler, L. J. P. and Wayland, R. G., 1971, Classification of public lands valuable for geothermal steam and associated geothermal resources: Geol. Surv. Circ. 647, 18 pages

Glassner, A., 1957, The thermochemical properties of the oxides, fluorides, and chlorides to 2500°K: Argonne National Laboratory, ANL - 5750, 45 pages, Argonne, Illinois'

Hammond, A. L., 1973, Dry geothermal wells: promising experimental results: Science, vol. 182, p. 43-44

Hansen, R. J., Ed., 1970, Seismic design for nuclear power plants, MIT Press, Cambridge, Mass.

Harris, E. S., 1973 (April 29) An expert looks at the energy crisis: Los Angeles Times, Part V, p. 10

Hauz, W., Leeth, G., and Meyer, C., 1972, The hydrogen economy: Proc. 7th Intersoc. Energy Conversion Conf., San Diego, Publ. Am. Chem. Soc., Washington, D. C.

Heroy, W. B., 1968, Thermicity of salt as a geologic function: In Saline Deposits (Mattox, R. B., ed.) Geol. Soc. America Special Paper 88, p. 619-629

Heywood, W. W. 1955, Arctic piercement domes: Can. Inst. Min. Met. Bull., vol. 48. p. 59-64

Hughes Aircraft Company, 1974, Heat pipe technology for industry: (Brochure), Thermal Products, Electron Dynamics Division, 3100 W. Lomita Blvd. Torrance, California 90509

Hubbert, M. K., 1937, Theory of scale models as applied to the study of geological structures: Bull. Geol. Soc. America, vol 48, p. 1459-1520

Hubbert, M. K., 1971, Energy resources for power production: Environmental Aspects of Nuclear Power Stations, International Atomic Energy Agency, Vienna, I IAEA - SM'- 146/1 (Preprint), p. 13-43

Jaffe, C., 1971 Geothermal energy, A review: Bull. Ver. Schweiz, Petrol. Geol. u. Ing., vol. 38, Nr. 93, pp. 17-40

Jaegger, J. C., 1961, The cooling of irregularly shaped igneous bodies: Am. Jour. Sci., vol. 259, p. 721-734

Kaser, R. V., 1972 (July), Heat pipe operation in a gravity field with liquid pool pumping: McDonnell Douglas Astronautics Company, Donald W. Douglas Laboratories, Richland, Washington, MDAC Paper WD 1912, 13 pages

Keller, G. V., 1974, New technology challenges in exploration, exploitation, and environmental impact of geothermal systems: Presented at AAAS Symposium, San Francisco Hilton Hotel, 25 to 27 February 1974

Kemme, J. E., 1969, Ultimate heat pipe performance: IEEE Transactions on Electron Devices, vol. ED - 16, No. 8, p. 717-723

Kent, P. E. et al, 1960, Evaporite piercement structures in northern Richardson Mountains, Northwest Territories (Abstract); Oil in Canada, vol. 12, p. 35

Kiersch, G. A., 1964, Geothermal steam: Cornell Univ., Ithaca, N. Y., Contract AF - 19 (628) for AFCRL, Bedford, Mass., 106 pages

Knapp, D., 1973 (March 7), Personal communication, Donald W. Douglas Laboratories, Richland, Washington

Koenig, J. B., 1971, Geothermal development: Geotimes, vol. 16, p. 10 - 12

Lefond, S. J., 1969, Handbook of world salt resources: Plenum Press, N.Y. 384 pages

Lessing, L., 1969, Power from the earth's own heat: Fortune, vol. LXXIX, No. 7, p. 138 - 141 192, 1986, 198

Link, T. A., 1930, Experiments relating to salt dome structure: Am. Assoc. Pet. Geol., Bull., vol. 14, p. 483 - 508

Marcus, B. D., 1972, Theory and design of variable conductance heat pipes: NASA Contract Report NASA CR - 2018, TRW Systems Group, Redondo Beach, Calif. 90278 (for Ames Research Center), 238 pages.

Mason, A. C. and Foster, H. C., 1953, Diversion of lava flows at Oshima, Japan: Am. Jour. Sci., vol. 251, p. 249 - 258

Mattox, R. B., ed., 1968, Saline deposits: Geol. Soc. America Special Papers 88, 701 pages

McGhee, E. and Johnson, J., 1972, Oil and Gas Journal map of North American Arctic areas, Petroleum Publishing Co., Tulsa, Okla

McNitt, J. R., 1963, Exploration and development of geothermal energy in California: California Division of Mines and Geology, Special Report 75, 45 pages

Mills, G. A., 1972 (Oct. 30 - 31), Coal gasification research: U.S. Bur. Mines, Fourth Pipeline Gas Symposium, Chicago, Ill.

Nettleton, L. L., 1934, Fluid mechanics of salt domes: Bull. Am. Assoc. Pet. Geol., vol. 18, p. 1175 - 1204

Nettleton, L. L., 1943, Recent experimental and geophysical evidence of mechanics of salt dome formation: Bull. Am. Assoc. Pet. Geol., vol. 27, p. 51 - 63

Nettleton, L. L. and Elkins, T. A., 1947, Geologic models made from granular materials: Trans. Am. Geophys. Union, vol. 28, p. 451 - 466

Nevin, A. E. and Sadlier-Brown, T. L., 1974 (April 18), Exploration and economic potential of geothermal steam in western Canada: 75th Annual General meeting, Canadian Inst. Mining and Metallurgy, Vancouver, British Columbia, preprint, 12 pages

Odé, H., 1968, Review of mechanical properties of salt relating to salt dome genesis: In Saline Deposits (Mattox, R. B.) ed.) Geol. Soc. America Special Paper 88, p. 543 - 595

Parker, T. J. and McDowell, A. N., 1951, Scale models as guide to the interpretation of salt dome faulting: Bull. Am. Assoc. Pet. Geol., vol. 35, p. 2076 -2086

Parker, T. J. and McDowell, A. N., 1955, Model studies of salt dome mechanics: Bull. Am. Assoc. Pet. Geol., vol. 39, p. 2384 - 2470

Pitcher, M. G., ed., 1974, Arctic geology: Proc. 2nd Internat. Symp. on Arctic Geology (Feb. 1 - 4, 1971), San Francisco, Calif., Am. Assoc. Pet. Geol. Memoir 19, AAPG, Tulsa

Possell, C. R., 1973, Bladeless turbines as a geothermal prime mover and potential reinjection pump: U. S. Federal Engineering and Manufacturing Co., San Diego, Calif., seminar USC, May 9

Przewlocki, K., Krzuk, J., Jurkiewics, I. and Owsiak, T., 1957, Apparatus for radioactive logging in horizontal boreholes for prospecting for potassium salts (in Polish); Acta Geophysica Polonica, vol. V, p. 283 - 307

Ramberg, H., 1963, Experimental study of gravity tectonics by means of centrifuged models: Bull. Geol. Institute of Univ. Uppsala: vol. 42, p. 1 - 97

Ramberg, H., 1967, Gravity, deformation and the earth's crust as studied by centrifuged models: Academic Press, London, 214 pages

Sawle, D. R., 1960, Sulfur-cooled power reactor: U. S. Atomic Energy Comm., AGN - 8015

Scott, S. and Wood, S. E., 1972, California's bright geothermal future: Cry California, vol. 7, p. 10 - 23

Serata, S., and Gloyna, E. F., 1959, Reactor fuel waste disposal project, development of design principle for disposal of reactor fuel waste into underground salt cavities: The Univ. of Texas, Dept. Civil Eng., Sanitary Eng. Research Lab., Tech. Rept. to the Atomic Energy Comm., 173 p.

Smith, J. H., 1958. Production and utilization of geothermal steam: New Zealand Engineering, (October 15), p. 354 - 375

Tectonic Map Committee of G. C. A. G. S., 1972, Tectonic map of the Gulf Coast region: Am. Assoc. Pet. Geol., P. O. Box 979, Tulsa, Okla. 74101, Scale 1:10^6

Tectonic Map of Canada Committee, 1969, Tectonic map of Canada: Geol. Surv. Canada, Ottawa, Ontario, Scale 1:1, 500, 000, No. 1251A

Trusheim, F., 1960, Mechanism of salt migration in northern Germany: Bull. Am. Assoc. Pet. Geol., vol. 44, p. 1519 - 1540

U. S. Atomic Energy Commission, 1973 (May), Nuclear fuel supply: AEC 1242, UC 51, Washington, D. C. 20545

Waters, E. D., 1973, Stabilization of soils and structures by passive heat transfer devices: Donald W. Douglas Laboratories, MDAC Paper WD 2043 to A. I. C. E. Mtng, New Orleans, 11-15 March 1973

White, D. E., L. J. P. Muffler, and A. H. Truesdell, 1971, Vapor-dominated hydrothermal systems compared with hot water systems. Econ. Geol., vol. 66, No. 1, pp. 75-97

Wilson, H. M., 1973 (January 29), California geothermal hunt heats up: Oil and Gas Jour., p. 3

SESSION X

INCREASING THE SUPPLY: ADVANCED CONCEPTUAL TECHNIQUES

Chairman: **Robert L. Gervais**
Project Manager, Douglas Missile & Space
Systems Division, McDonnell Douglas Astro-
nautics Co., Huntington Beach, Ca.

OCEAN THERMAL POWER AND WINDPOWER SYSTEMS -
NATURAL SOLAR ENERGY CONVERSION FOR NEAR-
TERM IMPACT ON WORLD ENERGY MARKETS

William E. Heronemus*

Jon G. McGowan**

The potential of two energy conversion systems
which use the natural solar collection of the
earth and its atmosphere over land and sea as
their power input is discussed. The first concept, for large scale power generation, is based
on a Rankine cycle heat engine driven by the
thermal difference (15 to 22 degrees Celsius)
which exists between the warm tropical surface
waters of the ocean and the great mass of cold
water below. Windpower, the second concept, is
discussed in the context of small to large scale
systems comprising a number of methods for extracting a portion of the kinetic energy of the
earth's atmosphere. A brief history and a current
status report on recent research developments as
well as the basic technical descriptions of these
concepts are presented. Suggested configurations
for energy distribution systems utilizing these
natural energy resources complete with energy
storage and transmission systems are given. In
addition to the potential impact of these systems,
the recognized technological and institutional
problems standing in the way of their potential
implementation are enumerated.

INTRODUCTION

In the light of our current and future energy demands and problems,

the time has come to utilize two methods of solar energy collection

which have been provided by nature for centuries. This paper will

examine the potential of the two most important for large scale

energy production: (1) The collection of solar energy by the earth's

atmosphere causing the winds, and (2) The natural collection of in-

coming solar radiation by the sea creating large temperature dif-

---

*Professor, Civil Engineering, **Associate Professor, Mechanical
Engineering, University of Massachusetts, Amherst, Mass.

ferences in the ocean. As will be discussed, technological systems
that can utilize these renewable energy resources are currently
available and with necessary development one or both of these
systems can make a significant contribution to the United States
energy supply in the near future.

DESCRIPTION AND HISTORY OF PROCESSES

Ocean Thermal Difference Power System

The fundamental law of thermodynamics which states that it is always
possible to place a heat engine between different temperature sources
to generate useful power is the basis of the ocean thermal difference
power plant concept. In this case the ocean surface (about 25 degrees
Celsius in the tropics) acts as a natural solar collector to provide
the high temperature source, while the cold water (about 4 degrees
Celsius) at depths at least 1000 or more feet acts as the cold sink.
More precisely, the surface temperature of the ocean stays at the high
temperature (about 25°C) because thermal equilibrium is established
between the solar radiation from the sun and such losses as evaporation
and convection. The cold water at the depths of the ocean is, in turn,
there because of melting ice at the polar caps, subsequent sinking of
the heavier cold water, and its movement toward the tropical parts of
the world. Thus, both the source and sink are renewed continually by
solar energy processes. Close to the eastern part of the United States,
the Gulf Stream provides an excellent natural pumping system for
replenishment of both hot and cold water.

The first idea of exploiting this natural energy source originated in
the 1880's with the French physicist D'Arsonval. The most significant
initial experiment and analytical work was carried out by George Claude
in the 1920's. Working off the coast of Cuba and maintaining only a
10°C temperature difference across his system, Claude was able to deliver 22 kw of power from his turbine using water as a working fluid.
Further work on similar (3500 kw) systems was carried out by the French
in the 1940's and 50's for systems intended for use off the Ivory
coast, but their work, like Claude's was plagued by technical difficulties. In more recent times, the Andersons of York, Pa. have revised this concept by proposing a closed cycle power system using the

ocean temperature difference of the Gulf Stream (1). Many of their
ideas have been incorporated in on-going feasibility studies sponsored by the RANN directorate of the National Science Foundation (2)
at the University of Massachusetts and at Carnegie-Mellon University.

The power cycle best suited to utilize the ocean thermal difference
is the simple boiling and condensing Rankine cycle. In this system,
the working fluid (ocean water for the open system; propane, ammonia,
or another typical refrigeration fluid for a closed system) is vapor-
ized by heat energy taken from the warm water. The vapor is then
expanded through a turbine to generate power. The exhaust vapor is
then condensed at a lower temperature by the cold water source, and
the condensed liquid is pumped back to the ocean (open system), or to
the boiler (closed system). Schematics of a closed cycle system are
shown in Figure 1 and it should be noted that key parts of this system
in addition to the four Rankine cycle components are the cold and hot
water delivery systems. That is, due to pressure drops on the hot and
cold seawater sides, large amounts of parasitic pumping work are needed
to keep the system operating.

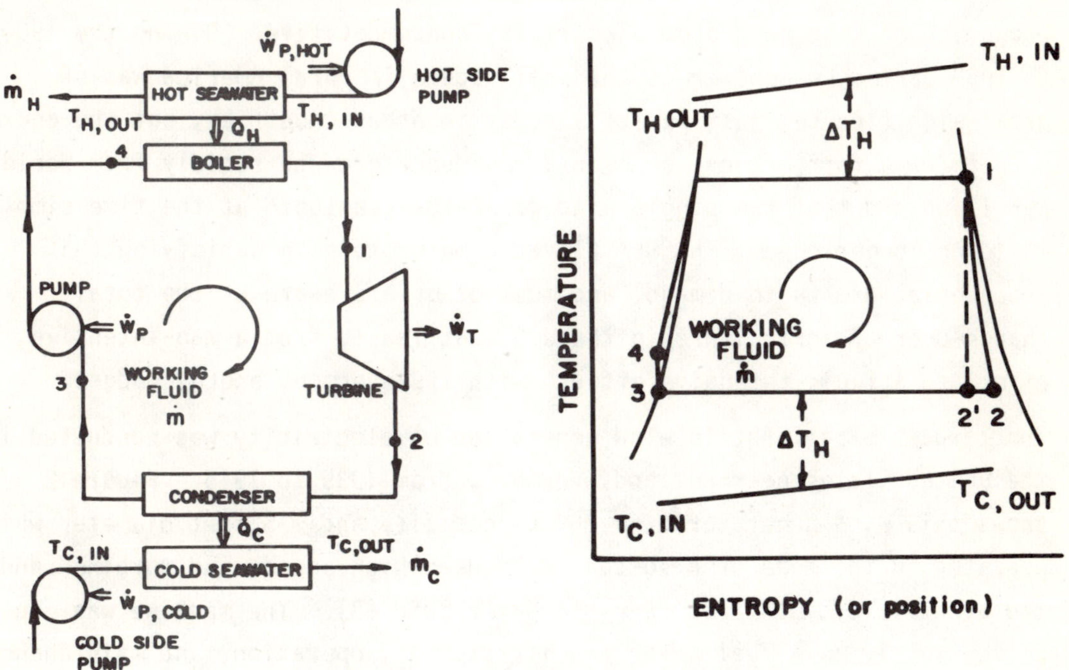

Fig. 1 Ocean Thermal Difference Power Cycle

Although the basic system contains but a few components, variations
in component design and placement can create an almost infinite
number of possible combinations. The containment and mooring of
ocean based plants necessitate additional groups of design variations.
Also, the energy transmission and storage systems linking these power
plants with the main power grid can have a distinct effect on overall
design.

<u>Windpower Systems</u>

Machines contrived to extract momentum from the winds have been amongst
the artifacts of the oldest of our civilizations. The significance of
windpower to national wealth and status probably maximized in the mid
sixteenth century when the Low Countries used windpowered manufactories
at home and wind driven merchant fleets to carry themselves to the peak
of affluence and power. Windpower was used to generate electricity very
soon after the wire-wound dynamo was invented, and many nations established
and maintained national programs aimed toward large-scale windpower systems.
Those systems did not materialize, however: first low-cost coal, then very
low-cost petroleum fueled power plants, and, finally, the optimistic
promises of the nuclear power advocates of the 1950-1970 era made the
economics of wind generated electricity noncompetitive. During the 1900
to 1950 period, windpower at the small scale in rural America was of
great significance, just as it is today in other countries; but the energy
budget for a farm, ranch, or even a residence grew so rapidly from World
War II onward that the single wind generator available at the time simply
could no longer cope. The REA played a major role in satisfying that
exponential growth in demand, and most of us are aware of the total
change-over of agriculture in the U.S., at least, from a man-intensive
effort to a fuels-intensive effort, with its enormous energy budget.

The largest experiment in wind generation of electricity was conducted in
the United States near Rutland, Vermont, from 1939 to 1945. Figure 2
shows this wind generator of 1,250 kw capacity and 175 feet diameter which
operated in the mode of a so-called "modern high-speed wind turbine" and
fed electricity into a sixty cycle power grid (3). The machine was conceived solely as a fuel saver and its capital, operation, and maintenance
costs had to compete against the differential cost of coal set down in

Fig. 2 Smith-Putnam Windpower Generator

Vermont. It did not meet that competition, and it suffered a fatigue failure in one blade resulting in abandonment of the program.

During the past four years, interest in windpower has been revived, mainly due to three different concepts evolving from work at the University of Massachusetts.

First, it has been shown that windpower electricity systems could be self-contained with their own storage subsystem and can deliver electricity upon demand at an average revenue per kwh that is competitive in today's market (4). Second, as discussed in Reference 5, it has been proposed that very large windpower electricity systems be installed offshore the Atlantic Coast (see Figure 3) and in the Great Lakes where open-water fetch is known to intensify winds considerably, thus increasing their productivity in momentum exchangers. The third concept introduced is that of the wind dam or wind barrage, which provides a method for wind generation by reaching up into the air with as dense a population of momentum exchangers as technology will permit.

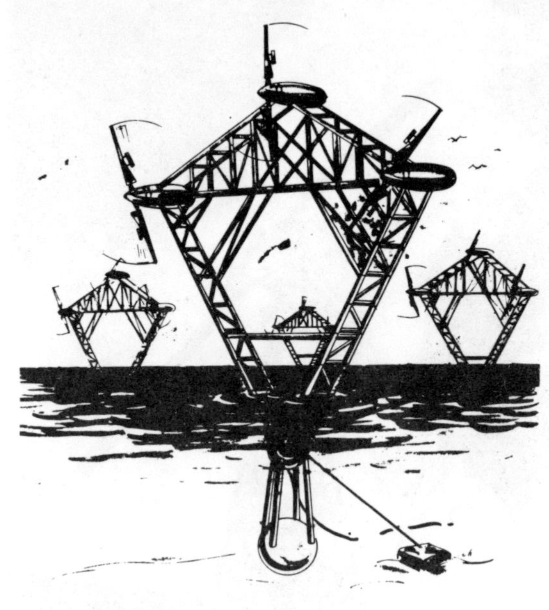

Fig. 3 Offshore Windpower Systems

Resource Availability and Site Selection

For a United States ocean thermal difference power system, a conservative
estimate of the maximum available power has been made based on the
utilization of the Gulf Stream alone with a flow rate of 33 x 10^6 m^3/sec.
If this stream were reduced in temperature by one degree Fahrenheit, a
heat energy input of 690 trillion kwh per year would be available.
Assuming a 2% overall conversion efficiency, which includes Carnot and
real machinery limitations, plus a 75% load factor, that available power
is reduced down to about 10 trillion kwh per year of delivered electricity.
Siting limitations (one line of 400 mw plants placed one mile apart along
550 miles of the Gulf Stream) reduce that available power to 1.4 trillion
kwh delivered per year. Studies indicate that the energy available for
the Gulf Stream is reasonably constant throughout the year, thus a base
load demand could be satisfied: ocean based compressed air or hydrogen
storage systems would readily add a peaking capability to that base load
capability. There are other locations near the United States (the Gulf
of Mexico, for example) where this concept could be used. However,
stagnation effects with the hot water supply would mean that moving plants

would have to be built. The fully mobile ocean thermal difference
power plant deployed into international tropical seas, connected to
the market via cryogenic hydrogen tankers, is a concept in which
all of the industrialized world should be keenly interested.

The Gulf Stream site selected for this first string of these power plants,
which we believe to be the best system with which the U.S. should start,
does present energy transmission problems. For example, the first
proposed site for a 400 mw plant is the Gulf Stream 25 kilometers east
of the Collier Building at the University of Miami Institute of Marine
Sciences. That location means that either ac or dc transmission by
cable could be used, or, the power plant could manufacture hydrogen
on site. In that case, the hydrogen could be transferred to shore via
pipeline or ship. Although these initial studies have tagged this
process to a regional resource, modern transmission technology should
make the user region much larger than the producer region.

The windpower resource is also regionalized and since power extracted
varies as the cube of wind speed, economics is served well by looking
for the strong and persistent wind fields. The worldwide resource is
huge: good estimates say that the instantaneous power available in
the world is of the order of 10^{11} mw. The annual energy potential
using practical systems in windy regions of continental U.S., the Aleutian
arc, and the Eastern seaboard is about 2 trillion kwh per year (6).

The United States' "wind belt" which includes a strip about 200 miles wide
along the northern boundary of the country (Canada has an even more
productive belt adjoining this to the North) is a particularly productive
region for conversion of potential energy and enthalpy into kinetic energy
in the winds. One reason for this is the very large generally low
pressure cell in the atmosphere which lies over the Atlantic below Iceland.
That cell acts as a huge suction pump drawing the winds across the land
and out to sea. This natural phenomenon in a sense magnifies the value
of the windpower resource over the land and continental shelf of North
America; if we are wise we can take advantage of a huge oceanic process
without venturing out into the midst of that ocean.

In comparison to these potentials, it has been estimated (7) that
the U.S. electrical generation load will be 3 trillion kwh per year
by 1980. Thus, the combined ocean thermal differences system plus
practical windpower systems are capable of making a very significant
impact on that demand.

CURRENT DEVELOPMENTS IN SYSTEM DESIGN

Ocean Thermal Power

Although there are many potentially feasible component choices and
arrangements for ocean thermal difference power systems (for example,
see Reference 1), we shall limit this discussion to the present results
of our ongoing research at the University of Massachusetts. This work,
which has as its goal preliminary system and component design, has
yielded information regarding the technical and economic problems to be
faced in the design of the thermal cycle, external seawater delivery, and
system containment components (8). The mockup of the first design is
shown in Figure 4 and is a 400 mw power plant designed for operation in
the Gulf Stream. Our studies show a 400 mw sized plant to be near the
maximum practical size, primarily based on the size of the heat exchangers
and the cold water supply tube.

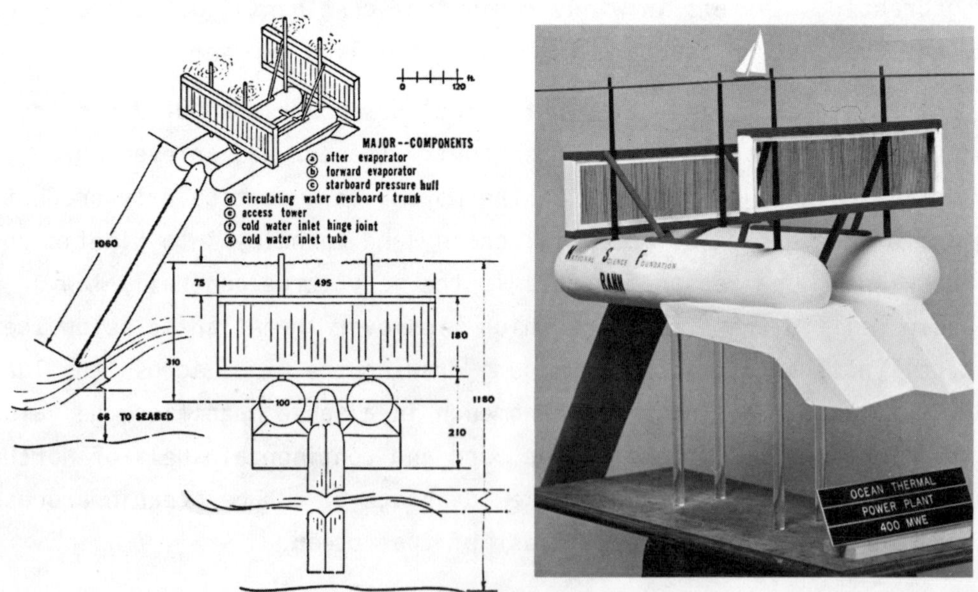

Fig. 4 Mark 1 Ocean Power System

This present design is based on a catamaran submarine hull (each cylinder approximately 100 ft. in diameter and 500 ft. long) with a pair of evaporators (approximately 400 ft. wide by 150 ft. high) exposed to the warm flowing water of the Gulf Stream. The single large diameter cold water inlet pipe (actually of streamlined cross section with an 80 ft. minor diameter) approximately 1100 ft. long serves as part of the mooring system. Four vertical access trunks reach from the hull top up through the ocean surface, and the hulls are fitted with the equivalent of a variable ballast system that will accommodate the expected fluctuations in Gulf Stream current velocities. A major part of each hull interior is a cold water distribution channel supplying 12 condensers each located beneath a turbine-generator. We thus have subdivided the power plant into a number of separate power modules each of whose turbines is of optimum size.

The key to a viable ocean thermal power system is a closed Rankine cycle using a conventional refrigerant as the working fluid. This type of cycle eliminates the major problems that plagued early experimental open-cycle systems and opens the way for more compact and economical component design. For example, in the closed cycle systems, the turbine-generator design becomes routine with small highly efficient turbines as compared against the fantastically large (greater than 200 ft. diameter) turbines that would be required for an open cycle using seawater as the working fluid. Specifically, based on the work of the Andersons and recent studies by United Aircraft Research Laboratories, turbine-generators for an ocean thermal power plant using propane or ammonia as a working fluid can be built economically using present technology.

The two large heat exchanger components, evaporator and condenser, are the most important and costly parts of the system. Economic consideration and size restrictions dictate designs based on plate-fin type heat exchangers (similar in construction to conventional automotive radiators) for both the evaporator and condenser components. The proposals of the Andersons are based on very thin, pressure-balanced (working fluid internal pressure and external pressure due to exchanger submergence) plates and fins: however, our designs are based on thicker, pressure proof components. Regardless of the type of exchanger construction, due to the small temperature drops available and low pressure drops allowable, studies have

shown the need for highly detailed analytical models of these components
and of the entire thermal cycle. Current designs are based on the use
of inherently corrosion and fouling resistant materials with reasonably
high water velocities to prevent fouling; experimental studies are
obviously needed to verify that fouling can be controlled.

The structural configuration of the cold water delivery system and the
hull are major design problems of an ocean thermal power system. This
first configuration is based on a cold water inlet pipe which is
essentially a long, deep "open-ended" ship's hull, designed and fabricated to be neutrally buoyant, not to vibrate due to the passage of
the Gulf Stream across it, and possessing enough axial strength to serve
as the major portion of the mooring system. The hulls, having the characteristics of pressure proof submarine hulls, are presently conceived
as very large diameter horizontal axis cylinders made from reinforced
concrete. The internal compartmentation bulkheads will also be cast in
concrete, and the Main Ballast tanks may be either reinforced concrete
or steel. Analogies for all of this work is taken from the rapidly expanding semi-submerged hull technology of the offshore oil industry.

Although our work and the ongoing work of the Andersons (more recently
via Sea Solar Power, Inc.) have shown the initial feasibility of a sea-
based power system, the scale of such a system requires a carefully
sequenced major development program to move from drawing board through
subsystems thence to test vehicles and finally to a demonstration prototype. Also, the potential environmental impact of such systems must be
considered, particularly if grand scale deployment is contemplated.

<u>Windpower Systems</u>
The block diagram of a proposed windpower system capable of providing
electricity on demand is given in Figure 5. The system would include a
relatively large number of rather small wind generators, say from 20 kw
to 40 kw each, made deliberately small to minimize cut-in speed and to
insure that blades can be turned out in matched mold processes of such
productivity that blade sales price will be only a few percent greater
than blade material cost. This large number of generators is to be suspended high in the air in wind dam fashion. Each wind turbine responds
individually to the momentum it feels in the oncoming air. The "generator"
will be an electric machine, an air compressor or a hydraulic pump which
will allow the turbine to maximize its momentum exchange with the oncoming

wind and which can feed, efficiently, into a common potential collecting
network.

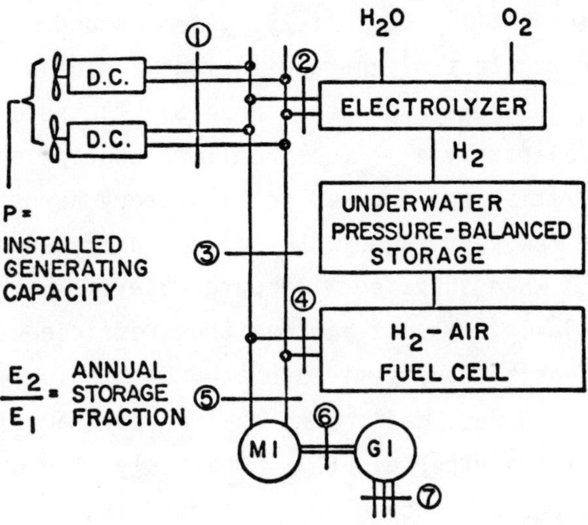

Fig. 5 Total Windpower System

The most economic system will probably be that one which delivers the
largest possible portion of its production as electricity fed into
a central power grid. That grid will undoubtedly be a 60 cycle grid
at one of the standard transmission voltages (137 kV to 745 kV). The
nature of the device(s) which sum up the productivity of the multitude
of generators and convert it into 60 cycle high voltage power, matching
the load demand curve, could be as old-fashioned as a Ward-Leonard drive
or as ultra-modern as a very large solid-state power integrator-inverter
device. If pneumatics or hydraulics are used to collect energy from
individual wind turbines, the summing device becomes a large cold air
turbine or a large hydraulic motor. The post WWII decades have brought
remarkable perfection and reasonable unit cost to all of those devices.
The potential for very low cost with large volume semi-automated manufacture
is great. The "Detroit" approach to wind generator manufacturing is
expected to deliver high-quality rugged hardware at very low cost. The
rather small wind generators (22 kw size, for example) described here for
large centralized systems will also possibly do an excellent job at the
individual residence level for heating. There will also be a need for
generators as large as 2000 kw: indeed families of generators from 20 kw
to 200 kw for several different wind regimes are required.

The storage subsystem shown in the diagram is based on preparation of
hydrogen by electrolysis, with storage during periods when wind productivity exceeds system demand, and recall and reconversion during periods
when wind productivity will not satisfy system demand. The block diagram shows a separate electrolyzer and fuel cell plus underwater pressure-
balanced storage of gaseous hydrogen. Probably the ideal hardware will
be the reversible electrolyzer - fuell cell, a concept suggested many
years ago by the British and reduced to laboratory hardware by Oklahoma
State University a few years ago. OSU has demonstrated electrolyzer
efficiencies of 95% when supplied with pure water.   Hardware available today commercially does not achieve those efficiencies. But again,
they are clearly feasible and semi-automated high-volume manufacture of
both electrolyzers and fuel cells make their future availability as very
low-cost items reasonably certain (9). Obviously, there are many variations in the storage subsystem loop that need study, many that might
well prove superior to the electrolyzer-gaseous storage-fuel cell combination shown. For example, perhaps methanol or liquid hydrazine could
be prepared and stored, or hydrogen and oxygen could be used in a coal
gasification or liquefaction process.

Decentralized generation of electricity is being practiced today and is
certainly feasible, too, even for residences with a 6 kw demand, but
economics are still on the side of the centralized windpower electricity
system. There appears to be good reason, however, to study windpower
systems (using smaller generators) for both electricity and heat at the
ten to one hundred home sub-division or village level: central electric
systems may decrease in size in future rather than increasing rapidly as
has been our recent experience. About one-half of the U.S. is usually
labelled as a region where the low temperature photo-thermal heating
process may be borderline on economics. Much of that region does experience reasonable winds, and they are strongest during the 8 or 9 month
heating season. Effort is being expended to identify a wind-furnace
system that could be economic, a system that might replace from 50 to 75
percent of the heating fuel required in a region, and that could be combined with a thermal solar heating system. This backing away from centralized utility systems is similar to that which terrestrial application of
photovoltaics will probably bring in a few decades.

502

The windpower systems discussed here suffer from a cascade of efficiencies between energy resource and energy consumer, and that cascade
represents considerable capital investment. Studies show that capital
to be well below that required for hydroelectric facilities - and we
have for years recognized long-term benefits from capitalizing hydroelectric facilities with low-cost public money. So, perhaps we will
come to accept the desirability of low-interest public investment in
extensive windpower systems. Studies also show that even if the capital required must be obtained at commercial interest rates the cost
of product from extensive windpower systems will be able to compete
with cost of product from nuclear plants whose capital costs are expanding at a rapid rate.

POTENTIAL IMPLEMENTATION

As discussed, the size of these natural energy conversion systems can
range from kilowatts for individual household windpower systems to
large scale 400 MW ocean thermal difference power plants. Such a size
range presents obvious differences in future implementation of such
systems. Separating the two types of systems yields the following
strategies for acquisition of these systems.

<u>Windpower</u>
At the present time the United States through the National Science
Foundation, with the assistance of NASA (as yet an informal arrangement,
not blessed by OMB) has embarked upon a national windpower research,
development and demonstration program. It was thought that bringing
demonstration plant into being at government expense would then lead to
adoption of the process by private industry. Progress during the first
two years of this program has been almost nil. This process, windpower,
should prima facie be the most able of all alternative energy processes
to make an early and sizeable impact because (a) there is a reasonably
good working technology today, and (b) manufacturing time for good
hardware components should be less than that for any other kind of power
plant. As a minimum this program should be accelerated. A better
strategy might be the deliberate creation of a public-held power authority,
federal, regional or state, which would set out to acquire a significant

windpower system whose product would be used to decrease need for proliferation of nuclear or fossil-fueled generation facilities in that area.
Confrontation between public and private utility interests seems inevitable if windpower generated electricity is to be made available to the
consumers. New England would be an excellent region in which to try
this strategy, and many other areas can be identified. Several of the
Prairie States, particularly those who may have some reservations about
nuclear proliferation and/or strip mining, might also adopt this strategy.
If the appropriate mechanisms were created, windpower systems could be
introduced into the energy vacuum in a very economic and salutary way
in the very near term.

<u>Ocean Thermal Power</u>

A national commitment properly organized, funded and managed, could see
a prototype ocean thermal power plant in operation within six years.
This task will probably have to be executed by some federal agency, but
there is some increasing possibility that a private group might assume
the responsibility if shielded as well from loss hazard as were the so-
called private nuclear pioneers. This program would have to start with
existing technology (including the results of the very recent research)
and move boldly through component, subsystem, system evaluations. A
number of candidates would have to be entered into competition at each
level, exactly as was done in the U.S. Navy's POLARIS program, for
example. Indeed we would not go wrong in assigning this ocean-flavored
task to the U.S. Navy whose vast resources, including shipyard, heavy
power plant fabrication and assembly capabilities plus ocean related
science and engineering capabilities, could be applied to almost all
facets of the program. The development, test, evaluation and trials
program envisioned above should be paralleled by research whose results
would improve the second, third or "n^{th}" model of the working system.
We can thus have a near-term good product, plus a continuing production
of improved products.

The combined potential for near-term meaningful impact on our energy
market by windpower and the ocean thermal differences process is felt
to be so significant that they should be elevated to the role of bonafide competitor with the breeder reactor.

SUMMARY

The following important parts concerning both systems should be
mentioned or re-emphasized.

1. Both systems make use of a renewable resource, energy from the
sun, and therefore only their construction and maintenance draw upon
the world's non-renewable resources.
2. Both systems are based on current technology and proven engineering
feasibility. Major efforts in product and manufacturing development must
be expended before either wind or ocean thermal power systems can be
placed into large scale operation.
3. Either of these systems provides an alternative or competition to
nuclear fission or fusion processes. The need for a well-balanced energy
resource program is apparent.
4. The need for improved methods of energy storage is apparent for both
systems. Work on hydrogen or other energy storage and delivery systems
should be placed on a high national priority.
5. The environmental impact of both systems should be determined before
any large-scale implementation. Such problems as potential weather
modification, changes in ocean environment, and visual pollution should
be considered in detail.
6. At the present time, capital cost estimates for each of these
systems, especially the ocean thermal power system are not of the level
of reliability needed before commitments are made. However, those estimates that have been made clearly suggest an excellent competitive
position for these processes, even without consideration of externalities.

Our discussion has probably only scratched the surface of the possibilities
for energy production from these two natural solar processes. There are
numerous possibilities where the combination of these solar processes with
each other or with other energy conversion processes might prove practical
(for example, see the proposed combined system of Esher (10)). There is
only one long-term pollution-free energy system available to mankind on
earth, a solar energy system. These two processes should be developed
as the first real step in the direction which we must take. The sooner
we get started, the better.

REFERENCES

1. Anderson, J.H. and Anderson, J.H., Jr., "Large-Scale Sea Thermal Power", American Society of Mechanical Engineers, Paper No. 65-WA/SOL-6, December 1965.

2. Proceeding of Solar Sea Power Plant Conference and Workshop, Carnegie-Mellon University, June 1973.

3. Putnam, P.C., Power from the Wind, D. Van Nostrand, New York, 1948.

4. Dambolena, Ismael G., "A Planning Methodology for the Analysis and Design of Windpower Systems", January 1974, Ph.D. Dissertation, Dept. of Industrial Engineering and Operations Research, University of Massachusetts (Amherst).

5. Heronemus, W.E., "Power from the Offshore Winds", Proceedings of the 8th Annual Marine Technology Society Conference, Washington, D.C., 1972.

6. Dupree, W.G. and West, J.A., "United States Energy Needs Through the Year 2000", U.S. Department of the Interior, U.S. Superintendent of Documents, December 1972.

7. An Assessment of Solar Energy as a National Energy Resource, NSF/NASA Solar Energy Panel, University of Maryland, Mechanical Engineering Department, December 1972.

8. McGowan, J.G., Heronemus, W.E., Connell, J.W. and Cloutier, P.D., "Ocean Thermal Difference Power Plant Design", American Society of Mechanical Engineers, Paper No. 73-WA/OCT-5, November 1973.

9. Hydrogen and Other Synthetic Fuels, A Summary of the Work of the Synthetic Fuels Panel, Superintendent of Documents, U.S. Government Printing Office, Washington, D.C., September 1972.

10. Escher, W.J.D. and Hanson, J.A., "Ocean Based Solar-to-Hydrogen Energy Conversion Macro System", ETA PT-33, Escher Technology Associates, St. Johns, Michigan, November 1973.

HYDROGEN—A CARRIER OF ENERGY

Derek P. Gregory[*]

As the United States and the rest of the world begin
to deplete their fossil-fuel resources, attention must
be paid to the best way to use the more abundant
energy sources such as nuclear and solar energy. At
the present time, technology is directed to converting these energy sources only to electricity. Electricity currently supplies only about 10% of our
energy needs, is relatively inflexible in use since
it cannot be stored, and is expensive to transmit
(even moreso if underground transmission becomes
required). It is hard to see how electric power can
be universally used for such things as air transportation, automobiles, and many other applications.

Nuclear or solar energy can be used to split water
into hydrogen and oxygen. Hydrogen is a fuel gas
which is easily transported in underground pipelines, can be stored by relatively inexpensive techniques, and can be used to meet most of the applications now met by oil and natural gas. By drawing
an analogy with the natural gas system, no insur-
mountable obstacles have been found to prevent the
universal use of hydrogen. As an aircraft fuel,
it is unexcelled on an energy-to-weight basis, and
as an automobile fuel it is almost entirely nonpolluting. The use of hydrogen is not without problems,
however, both real and imagined. The real problems include reducing its cost of manufacture and
finding a way to store it on board vehicles. The
imagined problems are mainly to do with exaggerated
concerns about its safety. When looking at the various
alternatives for a future scenario, the "Hydrogen
Economy" emerges as a very favorable choice.

INTRODUCTION

As this country's energy supply moves toward a nonfossil-fuel

system, we must consider how these new energy sources can best

be delivered to the consumer. Nuclear technology is currently

used to provide electric power, and most research and development

---

[*] Director, Energy Systems Research.
Dr. Gregory is affiliated with the Institute of Gas Technology,
3424 South State St., Chicago, Illinois

applied to the bulk generation of energy from solar, geothermal,
wind, tide, and biological sources is also associated with electric
power. When compared with the mixed system that we have today —
a mixture of chemical fuels and electric power — an all-electric
economy has some attractive features, but also has some disadvantages.

In the future, the nonfossil energy sources must be used in vastly
increased amounts, for this is the only way to provide the U.S. with
an energy supply independent of imports and an energy supply that
can meet the demands likely to be made during the 21st century. It
is reasonable to assume that a) for economic reasons these energy
sources must be harnessed or converted to usable forms in large
central conversion stations located remotely from the cities or load
centers and b) these sources will produce useful energy either at a
constant rate or at a regularly cycling rate, neither of which matches
the fluctuating demands for energy, either daily, weekly, or annually.
We conclude that a high-capacity storage and transmission system
is required to link the consumer with the generating station[1]. Moreover, the huge quantities of energy required in the future will have to be
utilized in a nonpolluting manner, and a large portion of this energy
will be needed in a "portable" form for both ground and air transportation applications.

A further consideration is the desirability of making as few basic
changes as possible in the way energy is actually used by the customer, so that his manufacturing techniques do not have to be rede-
veloped, and his energy-consuming appliances or equipment do not
have to be completely replaced.

Although electric energy can be produced from a wide variety of
energy sources and is extremely clean in end-use, it is not ideally
suited as a universal energy form. A complete transition to an all-
electric system seems undesirable from the following viewpoints:

a.   Unless power transmission technology is vastly improved,
     bulk transmission of electric power will require considerable
     land area, will be expensive, and will present serious insta-
     bility and realiability problems as the system grows.

b. Unless unprecedented breakthroughs occur, electric power cannot be used as a fuel for aircraft or for heavy over-the-road freight vehicles.

c. Complete electrification of residential and industrial energy needs would require enormous capital expenditures to replace current oil-, gas-, and coal-fired equipment.

d. Electrification of many manufacturing processes would require completely new process technology to be developed.

For these reasons it is wise to consider an alternative synthetic fuel that could be used in the much same way as our present fossil fuels, but could be made from the new energy sources and an unlimited supply of materials. Chemical fuels made from substances found in air and water are the only ones that can be considered from a pollution, raw materials availability, and materials balance point of view.

Water can be split into hydrogen and oxygen by an input of energy. The hydrogen can be used as a fuel directly, or it might be used as a raw material to produce ammonia, hydrazine, methanol, or hydrocarbons using either nitrogen or carbon dioxide from the atmosphere. If hydrogen itself can be used as a fuel, it is the simplest to make, requiring less wasted energy than the others, and is the cleanest to use [2,3].

Hydrogen is not a primary fuel. It does not occur naturally in the earth's crust in the uncombined state, and thus is not a potential energy "source" which could be used to relieve the energy crisis or to achieve U.S. energy independence. Hydrogen is merely a link between the new energy sources and the multiple users of energy, and it can thus ease and hasten the introduction of the new energy sources. Let us consider hydrogen as a tool with which to close the "Energy Delta."

PRODUCTION OF HYDROGEN

Hydrogen can be made today by the electrolysis of water[4]. This process is used industrially where specific conditions make it most attractive; for large-scale production this requires cheap electric power. Large-scale plants are in operation in Canada, Norway, and Egypt. As a result of fuel cell technology, advances in the cost and efficiency of electrolyzers have been demonstrated on a small scale[5,6]. Much remains to be done to fully exploit the potential for electrochemical improvements in electrolyzers. With power costs in the 4-7 mills/kWhr range, hydrogen costs of $2.00-$3.00/ million Btu are predicted. With reasonable assumptions on the results of long-term research improvements to the electrolyzer, costs of $1.50-$2.50 can be proposed[7].

These costs take no credit for the value of the oxygen produced in very large quantities as a coproduct. Oxygen is valuable in water-treatment and waste-disposal processes, as well as being highly important to the metallurgical and other industries.

An interesting possibility exists for splitting water into hydrogen and oxygen using the heat from a nuclear reactor or a solar furnace, without going through the electricity generation stage. The temperatures required to split water directly are too high for today's materials, but stepwise chemical processes, which achieve the same objective, have been proposed[8]. Although research is still in an early stage, it is now possible to make tentative estimates of costs or efficiencies.

Several so-called "thermochemical" water-splitting cycles have been published, their efficiencies calculated, and their chemistry investigated in the laboratory. Overall efficiencies, defined by the ratio of the fuel value of the hydrogen produced to the thermal energy consumed in the reactions, as high as 55% seem achievable[9]. Note that this is considerably in excess of conventional electric power generating efficiencies. However, to attain these values, heat sources in the 900^0-1100^0C range are required, and these are

within the reach of only the more advanced types of high-temperature
nuclear reactors. There is no reason why solar energy sources
could not be used to operate this or a similar process.

Today, industrial hydrogen is made from natural gas and other
fossil fuels by reaction with steam. The conversion of coal, shale,
and low-grade fuels to hydrogen is technically possible, and could
certainly significantly add to the sources of hydrogen fuel in the
near term. In this way, hydrogen could act as a common fuel to
bridge the gap between the fossil-fuel age and the nuclear or solar
age.

TRANSMISSION OF HYDROGEN

The technology for moving energy in an underground gas pipeline
has been well established by the natural gas industry. Also, the
cost of pipelining gas is considerably less, on an equal energy
basis, than typical electric power transmission costs. For example,
a typical 36-in. natural gas pipeline is capable of moving energy at
a rate of about 1 billion Btu/hr, or 12,000 MW, at a cost of about
1¢-2¢/million Btu-100 miles ($0.03-$0.07/MWh-100 miles).
Compare this to a high-voltage overhead transmission line that can
carry up to 2000 MW at a cost of 9¢-20¢/million Btu-100 miles ($0.3-
$0.8/MWh-100 miles). Underground power cables are likely to
cost from 10 to 40 times as much as overhead lines[10].

Industrial hydrogen is already being carried for short distances
in pipelines made of the same materials as natural gas lines. For
transmission over longer distances, pipelines would have to be
equipped with similar, but larger, compressor stations than those
now being used on natural gas lines. Because of hydrogen's low
density, the volume throughput of a given pipe is greater than with
natural gas. This almost compensates for the lower volumetric
heating value of hydrogen. Estimates of the cost of hydrogen
transmission in bulk pipelines that have been made[11] range from
3¢ to 4¢/million Btu-100 miles, or from $0.10 to $0.14/MWh-
100 miles. This cost is only about one-hundredth the cost of

underground electric power transmission.

There is cause for some concern about the likelihood of hydrogen
embrittlement, especially if high transmission pressures and high-
strength steels are considered. Embrittlement can occur both in
the pipes and in the compressor components. Further experimental
investigations are needed to define safe operating areas.

HYDROGEN STORAGE

A capability to store "electric" energy would be a great benefit to
today's electric utility industry and almost mandatory for the extensive use of nuclear, solar, or wind power. Where the geography
permits, pumped hydrostorage systems are being installed today.
There is also active research into the very difficult problems involved
with the development of inexpensive, high-efficiency, bulk storage
batteries. Hydrogen presents a possibility for storage on a scale
never before contemplated within the electricity industry.

The gas industry stores large quantities of gas for peakshaving in
underground porous rock formations such as in depleted gas fields
and aquifers where possible, and by cryogenic storage elsewhere.
In 1970, there was in use a total underground natural gas storage
capacity of 5.2 trillion SCF, representing 22% of the total annual
production of gas and almost exactly equivalent (1523 billion kWhr)
to the total annual production of electricity in 1970 (1638 billion
kWhr). In addition, the fast-growing use of liquefied natural gas
storage had reached a capacity of nearly 15 million SCF. In some
instances, more than 75% of winter peak-day sendouts came directly from storage.

With such a huge capacity, energy is stored on a seasonal rather
than on a daily basis. The saving to the gas industry in transmission
capacity and production facilities is considered. In contrast,
the world's largest pumped hydrosystem, at Ludington, Michigan,
is designed for a daily storage cycle. It has a relatively small
energy capacity (15 million kWhr) compared with a typical LNG
tank (300 million kWhr).

There seems to be no geological reason why hydrogen cannot be
stored underground in the same way that natural gas is stored today.
Moreover, liquid hydrogen technology is well developed for the
aerospace industry, and tanks up to 1 million gallons in capacity are
already in use. Two important precedents for underground hydrogen
storage already exist. A 13 billion SCF aquifer storage system has
been used for over 10 years at Beynes, near Paris, for the storage
of manufactured gas containing about 50% hydrogen. The other is
the 30 billion SCF of helium now being stored without problems in
Cliffside Field, a partially depleted natural gas field at Amarillo,
Texas.

Liquid hydrogen (LH_2) technology has advanced very considerably
because of the space program. The possibility of using LH_2 storage
in a way analogous to LNG storage seems perfectly feasible, although
sufficiently large storage tanks for the seasonal storage of LH_2 have
not been constructed yet. The largest single LH_2 tank is at Cape Ca-
naveral and holds 900,000 gallons. By comparison, the energy
content of this is only 4% of a typical LNG tank (Philadelphia Gas
and Electric Co.), but is 75% of the capacity of the Ludington
pumped hydroelectric storage plant. The costs for liquid hydrogen
storage, both in tankage and liquefaction plants, will be greater than
for LNG because of the lower boiling temperature of hydrogen
(-423^0F).

The efficiency of an electricity storage system using hydrogen is
probably the prime deterrent to its use. An optimistic estimate of
54% is considerably less than typical figures of 60-70% achieved
from pumped water. The outstanding advantage of hydrogen energy
storage, however, is its availability on an enormous scale at almost any geographical location.

UTILIZATION OF HYDROGEN

The use of hydrogen as a fuel presents some problems together
with some distinct technical advantages. There seems to be no
reason why pure hydrogen could not be used for all the purposes
served by natural gas today. It burns smoothly and easily, when
mixed with air, on burners closely resembling today's gas burners.
Natural gas burners would have to be modified to account for the
different combustion properties of hydrogen, but the data required
to do this are available. The combustion products would be water
vapor and traces of nitrogen oxides. The latter can be eliminated
if combustion temperatures are kept low enough. A promising
way to do this is to carry out the combustion on a catalyst bed in
which flameless oxidation is obtained. Simple domestic heating
devices operating on this principle have been demonstrated by
the Institute of Gas Technology[12]. Catalytic combustion of hydrogen
is an easier technological task than catalytic combustion of natural
gas.

Engines operate well on hydrogen. Already, several teams have converted automobile engines to run on hydrogen [13]. The main modifications required are carburetion and ignition. It has been shown[14]
that a laboratory engine operating on hydrogen has far less nitrogen
oxides emissions than a gasoline engine. No other pollutants are
possible. The main problem concerning hydrogen as a vehicle fuel
is tankage. Compressed hydrogen tanks are too heavy and bulky
to be seriously considered. Liquid-hydrogen tankage at present is
expensive and rather bulky, although technologically quite feasible. A promising area is the use of chemically bound hydrogen in
metal hydrides, which can be decomposed by the heat of the exhaust
to provide pure hydrogen. This work is in its early stages at
Brookhaven National Laboratory[15] and elsewhere.

Gas turbines can be converted to hydrogen fuel with little trouble.
This was done quite successfully in the 1950's. Indeed, a B-57
aircraft flew with liquid-hydrogen fuel several years ago. Hydrogen is a mandatory fuel for hypersonic transport aircraft of Mach 6

and over, where its cooling characteristics can be used. The bulk
of liquid hydrogen is a disadvantage for low-speed aircraft, but
hydrogen's energy-to-weight ratio is far higher than that of any
known fuel, so that for such large-volume aircraft as the wide-
bodied subsonic transport, it should retain considerable possibilities.

A number of aircraft manufacturers have carried out preliminary
design studies for DC-10 or L-1011 type wide-bodied passenger
aircraft and have concluded that a liquid-hydrogen-fueled version
would have about 60% of the gross take-off weight of the conventional
version[16]. This would result in considerable advantages in operating
economy, noise suppression, runway lengths, and endurance or
load-carrying capacity.

Hydrogen is a raw material for the chemical industry of vast and
growing importance. It is used in the manufacture of fertilizers and
foodstuffs, for upgrading petroleum fuels, and for many other purposes. Its use as a direct metallurgical reductant, for example,
in producing iron from ore, has been technically proved, but it
has only been used in special cases, where hydrogen happens to
be cheaper than coke. As fossil–fuel prices rise, hydrogen will be
used more, with its attendant decrease in pollution levels.

Hydrogen is an ideal fuel for fuel cells, eliminating the need for the
fuel-conditioning stage required for hydrocarbon fuels and increasing the efficiency somewhat. Local or dispersed generation of
electric power by fuel cells is under active study by both the gas
and electricity industries. Hydrogen is also an attractive fuel for
more conventional electricity generation techniques, using advanced
gas or steam turbines.

SAFETY

Perhaps the most controversial issue concerned with the use of
hydrogen is that of safety. Hydrogen is a hazardous and dangerous
material, but it has been used so extensively in industry and in
aerospace that very clearly defined codes of practice have been
developed. Compared with natural gas, its lower flammability

limit in air is about the same (4% and 5%), but its flammability
range is far higher (up to 75% for hydrogen and up to 15% for
natural gas). Hydrogen has a very low ignition energy, so that a
static spark will ignite it, and a very high flame speed. Compared
with propane and gasoline, its lower flammability limit when mixed
with air is higher,and it is far lighter than air so that it diffuses
away from a leak or spill. With odorization to make leaks easily
detected and with proper handling techniques, pure hydrogen should
be no more hazardous than the old town gas or manufactured gas,
which was composed of 50% hydrogen and which had the added hazard of toxicity because of its carbon monoxide content.

CONCLUSIONS

Hydrogen appears to be a very promising universal fuel. Its price
today is high compared with that of fossil fuels, but hydrogen from
nonfossil sources should become relatively cheaper as the fossil
fuels become more expensive. It has very considerable advantages
in pollution and in energy transmission. An energy system centered
around a hydrogen transmission and distribution system is attractive
because hydrogen could be introduced to the system from a wide
variety of sources — nuclear, solar, geothermal, wind, and tide
power — as well as from low-grade fossil fuels, while the consumer
would experience no changes in his fuel quality. Such flexibility will
be very important to the energy industry during the next 50 years
or so when very considerable changes in energy sources must be
made. Hydrogen is a clean chemical fuel which can be produced
from wholly domestic energy sources without importation. While
interest in this concept is rapidly growing within the energy industries, the problems in implementing such a system are immense
and would have to be thoroughly planned well in advance.

REFERENCES

1. D. P. Gregory, "Hydrogen-Transportable Energy Medium",
    <u>Astr. and Aeron.</u>, vol. 11, Aug. 1973, p. 38

2. D. P. Gregory, "The Hydrogen Economy," <u>Sci. Am.</u>, vol. 228,
    Jan. 1973, pp. 13-21

3. D. P. Gregory <u>et al.</u>, <u>A Hydrogen Energy System</u>, A. G. A.
    Cat. No. L.21173, American Gas Association, Arlington,
    Virginia, Aug. 1972

4. A. K. Stuart, "Modern Electrolyser Technology in Industry."
    Paper presented at the American Chemical Society Annual
    National Meeting, Symposium on Non-Fossil Chemical Fuels,
    Boston, April 1972

5. J. H. Russell, L. J. Nuttall, and A. P. Fickett, "Hydrogen
    Generation by Solid Polymer Electrolyte Water Electrolysis."
    Paper presented at the American Chemical Society Meeting,
    Division of Fuel Chemistry, Chicago, August 1973

6. W. C. Kincaide and C. F. Williams, "Storage of Electrical
    Energy Through Electrolysis." Paper presented at the 8th
    Intersociety Energy Conversion Engineering Conference,
    Philadelphia, Aug. 13-18, 1973

7. J. E. Mrochek, "Economics of Hydrogen and Oxygen Pro-
    duction by Water Electrolysis and Competitive Processes,"
    in Grigorieff, W. W., Ed., <u>Abundant Nuclear Energy</u>,
    U. S. Atomic Energy Commission, Washington, D. C.,
    1969, pp. 107-122

8. C. Marchetti, "Hydrogen and Energy," <u>CEER (Japan)</u>, vol. 5,
    Jan. 1973, pp. 7-25

9.   J. B. Pangborn and J. C. Sharer, "Analysis of Thermochemical Water-Splitting Cycles." Paper to be presented at The Hydrogen Economy Miami Energy (THEME) Conference, Miami Beach, Florida, March 18-20, 1974

10.   The Transmission of Electric Power, A report to the Federal Power Commission by the Transmission Technical Advisory Committee, Feb. 1971, p. 123

11.   J. Wurm, "The Transmission of Gaseous Hydrogen." Paper presented at the 48th Annual Fall Meeting of Society of Petroleum Engineers of AIME, Las Vegas, Nevada, Sept. 30-Oct. 3, 1973

12.   J. C. Sharer and J. B. Pangborn, "Utilization of Hydrogen as an Appliance Fuel." Paper to be presented at The Hydrogen Economy Miami Energy (THEME) Conference, Miami Beach, Florida, March 18-20, 1974

13.   R. E. Billings and F. E. Lynch, "History of Hydrogen-Fueled Internal Combustion Engines," Publication No. 73001, Energy Research, Provo, Utah, 1973

14.   R. G. Murray and R. J. Schoeppel, "Emission and Performance Characteristics of an Air Breathing Hydrogen Fueled Internal Combustion Engine." Paper 719009 presented at the 1971 Intersociety Energy Conversion Engineering Conference, Boston, Aug. 1971

15.   K. C. Hoffmann, et al., "Metal Hydrides as a Source of Fuel for Vehicular Propulsion." Paper SAE 690232 presented at the International Automotive Engineering Congress, Detroit, Jan. 1969

16.   G. D. Brewer, "The Case for the Hydrogen-Fueled Transport Aircraft." AIAA Paper No. 73-132 presented at the AIAA/SAE 9th Propulsion Conference, Las Vegas, Nevada, Nov. 1973

PROSPECTS OF PHOTOSYNTHETIC
ENERGY PRODUCTION

Bessel Kok[*]

Not too long ago serious consideration of direct solar energy conversion was limited to two economic and technological extremes: underdeveloped nations, and highly developed spacecraft. The vast region in between -- the densely populated, industrialized world -- found it far more economical to use fossil fuels to meet virtually all their needs. Now, with the price of crude oil climbing to fantastic levels, solar energy conversion could become economically competitive, if not downright essential.

So far, the cheapest, largest, and most successful solar energy conversion system available is photosynthesis. It seems reasonable to take another look at this natural system that has provided man for eons with food, oxygen, and fuel.

One approach is to search for increased yield of combustibles (e. g., woody plants). Another perhaps more exotic, approach might be to interfere with photosynthetic processes so that combustible gases are produced (e. g., hydrogen or methane).

The truth is that photosynthesis is not nearly as efficient as we would like and that we still have a long way to go in our understanding of solar conversion in plants before this method of energy production can make a greater impact on a highly industrialized society. But we must begin now if the promise of clean, inexpensive, and replenishable fuel supplied

---

* Associate Director and Head of Bioscience Department, Martin Marietta Laboratories, Baltimore, Md.

by photosynthesis is to be realized at some reasonable time in the
future.

The purpose of this brief paper is to review the relevant aspects of
photosynthesis, to evaluate prospects for expanded exploitation of
natural solar conversion systems, and to suggest some avenues for
needed research.

MECHANISM AND EFFICIENCY OF PLANT PHOTOSYNTHESIS

In order to do photochemical work, light must be absorbed. Of several
pigments used to this end in the plant kingdom, chlorophyll _a_ is the most
important and general. Its absorption spectrum, reveals two bands, one
near 680 nm and one near 430 nm. Among the other so-called acces-
sory pigments, the yellow carotenoids are ubiquitous in plants. A leaf,
because of its high pigment concentration, readily absorbs _all_ wavelengths shorter than 700 nm.

The chlorophyll molecules are densely packed in thin lamellae so that
absorbed photons can "migrate" to trapping centers in the pigment
array. In these centers, a special chlorophyll is excited by light,
donating an electron to an associated acceptor molecule. In rapid,
consecutive reactions, the electron (a reduced chemical moiety) and
the hole (an oxidized chemical moiety) are processed into more stable
forms, and the trapping center is restored to its photosensitive state.
Photochemistry only occurs from the first excited singlet state of
chlorophyll _a_ (the red absorption band at 700 nm), which represents
about 1.8 volts. The chemical (midpoint) potentials of the oxidizer
and reduced products differ by 1 volt, so that the energy efficiency of
the primary photoacts is 50%. But, as will be discussed, the overall
efficiency of photosynthesis is much lower.

A chemical potential, well in excess of 1.2 volts, is required to (1)
split water into oxygen and hydrogen (or its physical equivalent) and
(2) generate additional energy in the form of ATP. This work is performed by _two_ photoacts, with the charge separations or "solar batteries"

operating at different potential levels and connected in series. One
photosystem produces a mold oxidant and a strong reductant on the
level of the hydrogen electrode (an iron-containing enzyme, ferredoxin).
The second photosystem produces a strong oxidant (a manganese-
containing enzyme capable of liberating oxygen from water) and a mild
reductant. The weak photooxidant of the first system reacts with the
weak photoreductant of the second system, which closes the electron
transport chain from water to ferredoxin. The ATP, generated by the
photosystems, aids reduced ferredoxin to reduce CO_2 to sugar and to
synthesize the multitude of products which constitute new plant material.

Rather independent of the nature of the ultimate product, be it plant
material (symbolized as $\overline{}CH_2 O$ in eq. (1), sugar, methane or hydrogen, the energy fixed in the overall process corresponds to 120 kcal
per mole of O_2 evolved.

$$H_2O + CO_2 \xrightarrow{\text{light}} \left[CH_2O\right] + O_2 \tag{1}$$

Since two photons are used per electron equivalent, it takes eight quanta
to produce an O_2 molecule. For average solar radiation ($\bar{\lambda}550$ nm), a
mole of quanta represents ~50 kcal, and the "theoretical" light conversion efficiency, therefore 8 x 50/120 = 30%. Because wavelengths
longer than <700 nm (about half of the solar emission) are not absorbed
by chlorophyll, the best possible conversion of natural sunlight would
be ~15%. Although the quantum yields of the photoacts approach unity
and other steps also are quite efficient, the actually measured optima
of plant growth are closer to 10%. (An efficiency of 10% of the average
solar flux in the USA corresponds to ~70 kw/acre).

For many reasons, this optimal value seldom is realized in the field.
Plants, like animals, must maintain themselves and do so by burning
organic materials in respiratory processes. Thus, in dark and in
very weak light their energy balance is negative. On the other hand,
the efficiency also drops in very strong light where the enzymic processing reactions cannot keep pace with the quantum influx. In most

plants, photosynthesis "saturates" at intensities below that of full sunlight at noon.

In addition to light, plant growth requires water, CO_2, and a variety of inorganic nutrients. The low concentration of CO_2 in the atmosphere (0.03%) generally limits the rate in strong light, and the availability of other ingredients, temperature levels, etc., often are suboptimal. Moreover, many land plants consume a significant fraction (up to one-half) of their "gross production" for their own maintenance, often in large, nonphotosynthetic structures such as stems and roots. As a result, the total production of organic matter in the field seldom exceeds 2% of the solar radiation received. Typical values actually are 0.5-5% (3-10 kw/acre). In many crops, only part of this total production (e.g., only the seeds) are harvested.

PHOTOSYNTHESIS AS A SOURCE OF ENERGY

Until the last century, contemporary photosynthesis, agriculture, and forestry filled nearly all of society's energy needs. Today it primarily fills our demand for food. Per capita, we use $\sim$ 3000 kcal of food per day, corresponding to 0.15 kw. We use a much greater amount of energy ($\sim$70x) for all other purposes -- $\sim$10 kw per person (of which $\sim$ 2.5 kw is converted to electricity). Practically all of this non-food energy is derived from fossil fuel. (The $2 \cdot 10^8$ people in the U.S. use 70×10^{15} BTU, corresponding to $\sim$ 2000 gw.)

At this moment, we wonder whether we have gone overboard, not only in thoughtlessly increasing our demands, but also neglecting renewable resources. Should we again consider using contemporary photosynthesis for broader energy purposes? A few figures reveal a considerable potential:

The sunlight falling on the U.S. is abundant, but dilute. It exceeds our total energy consumption 700-fold; e.g., if converted with 15% efficiency, only 1.5% of our land area would supply all our energy needs. On the other hand, this 1.5% amounts to 28 million acres -- a formidable

area, corresponding to the entire state of Pennsylvania or about twice
the area of all our highways.

Consider, however, that U.S. farmers and foresters already work a
much larger area (<10x). More than 300 million acres in the U.S. are
devoted to agriculture and forestry (1.5 acre per person). During a
growing season of 5 months, these areas receive 24×10^{17} BTU of
sunlight. A harvest, equivalent to $\leq 3\%$ of this sunlight, would equal
the total energy consumption in the U.S. Actual above-ground harvests
generally are below one-third of this amount.

We should consider, however, that conventional plant breeding aims for
high yields of special products or parts of the plant, such as seeds,
roots, sugar or stemwood. It is not unreasonable to expect considerable
gains if, instead, genetic selection and other approaches were aimed at
the most abundant production of combustible material -- in any form, as
long it can be harvested (and dried) conveniently. Proposals have been
made to grow special crops in "energy plantations" with the sole purpose
of combustion in electrical power plants. For 2% efficient conversion
of light to combustibles and $\leq 40\%$ efficient conversion to electricity, a
1000-Mw power plant would require ≥ 250 square miles (1600 acres).

However, in view of the pressures for increased supplies of food and
wood, this may not be too appealing. At least as a start, harvesting of
combustible side products of already economically worthwhile crops
seems to be closer at hand. It might not add a prohibitive expense to
harvest the stalks along with the kernels.

In this respect, we should consider that the food on our table represents
only a fraction of the total caloric harvest. It has been estimated that
the major crops (corn, wheat, soybeans, etc.) leave 5.10^{8} tons of
stalks, stubble, and leaves, representing 4.10^{15} BTU. In addition, we
are quite careless with our soil, and we insist on eating a great deal
of meat which requires up to 20 times more land area than the nutri-
tional equivalent of vegetable food, notably soybeans. Some of this loss

is readily retrievable as solid wastes from the farm animals, and urban, organic wastes already are being collected. These sources might relieve up to 10% of the total U.S. energy demand.

I should mention that burning of organic products to generate steam for electricity is not the only -- or necessarily the best -- approach. By pyrolysis or chemical reduction, organic materials can be converted into oil or gas. Methane (or ethanol) can be produced from wet materials, using biological fermentation. The above numbers and ideas may not be novel or dramatically promising in terms of getting large amounts of energy on short notice, yet we can ill afford to neglect them forever.

PHOTOLYSIS OF WATER AS A SOURCE OF ENERGY

Recently, attention has been drawn to the possibility of using only that part of the photosynthesis apparatus which carries out the photolysis of water (Eq. 2).

Water photolysis:
$$2\,H_2O \xrightarrow{\text{light}} O_2 + 4H^+ + 4\,[e^-] \tag{2}$$

Hydrogenase enzyme:
$$4\,H^+ + 4[e^-] \longrightarrow 2\,H_2 \tag{3}$$

Overall:
$$2\,H_2O \xrightarrow{\text{light}} O_2 + 2\,H_2 \tag{4}$$

Thermodynamically, this is feasible, since the potential of the primary reductant of phogosynthesis ($[e^-]$ in Eq. 2) is below that of the hydrogen electrode.

If plants or algae could be made to carry out reaction (3) instead of CO_2 reduction, hydrogen, a clean, readily applicable fuel would be generated. Actually, it has been long known that many algae, under proper conditions, evolve some hydrogen when illuminated. Also, it is standard laboratory practice to isolate chloroplasts from leaves which are no longer able to carry out CO_2 reduction (Eq. 1) but retain full capacity to split water, generate oxygen, and reduce low potential agents. These observations raise the tantalizing prospect of a 10% efficient solar generator of hydrogen at reasonable cost. In one conceivable scenario, algae are grown in the normal way, which can be done readily and cheaply; then their metabolism is tricked to produce H_2 rather than sugar.

Another approach is to move away from the living system. In this case, chloroplasts are isolated from green leaves and stabilized together with the hydrogenase, for instance, on a solid substrate, so that activity can be maintained for a long time outside the living cell.

Many gaps in our present knowledge of the system give these scenarios a somewhat futuristic flavor. For instance, bleeding off hydrogen would deprive the plant of its precious reducing power, and such a drastic derangement of the photosynthetic apparatus might imply a major feat of genetic manipulation. Hydrogenase, the enzyme which equilibrates hydrogen gas with metabolic redox agents (Eq. 3), although present in many algae, might not be coupled directly to the photo-synthetic machinery. Outside the living cell, hydrogenases are strongly inhibited by oxygen, and, conversely, the photosynthetic photoact responsible for O_2 evolution is "turned off" under reducing conditions.

In the living system, such problems are solved by structural arrange-ments. The photosynthetic system consists of a two-dimensional array of functional units (assemblies of pigments, trapping centers, and pro-cessing enzymes) arranged on a thin (60 Å) lamella, doubled as a flattened sac. The oxidized products of both photoacts are formed on the inside, and the reduced products are manufactured on the outside, apparently to prevent backreactions. Ideally, each photochemical reductant molecule in the array should be tightly associated with a compatible hydrogenase molecule. To arrange such a coupling artifi-cally seems beyond the present art of molecular engineering.

A defeatist might enumerate many other unknowns and difficulties. Still, it seems realistic to predict that some day we will be able to accomplish this feat. It certainly would not hurt to try. The time table will depend on the amount of effort and the direction of research, not only in photosynthesis, but also in genetic manipulation and

stabilization of biological catalysis _in vitro_, areas showing considerable progress and promising radically new technologies. Photolysis
of water may be one of these, but we should not count on it within the
coming decade.

SESSION XI

THE ECONOMICS OF ENERGY ISSUES

Chairman: James Blackman
            National Science Foundation,
            Washington, D.C.

DEMAND FOR ENERGY AS A FUNCTION OF PRICE

Hendrik S. Houthakker[*] & Michael Kennedy[**]

One unfortunate byproduct of our present energy problems is a veritable deluge of bad economics. We are overwhelmed with calculations as to how much energy we will "need" in the future and how much we are "wasting" now. Rarely do we see any recognition that energy, in common with virtually all other commodities, is used in larger volume when it is cheap than when it is expensive. Energy is only one of several inputs into the processes of production and consumption and is therefore subject to substitution and complementarity. Thus an individual who wants to go to his work needs not just energy (e.g., in the form of gasoline) but also time and equipment (such as a car). He may be able to save energy by using public transportation, but generally this will take more time. If gasoline is cheap in relation to the value of his time he will take his car, but if the price of gasoline goes up he will be more inclined to take the bus even if he loses some time. There is no sense in saying that his switching to the bus shows the gasoline was not "needed" and hence "wasted"; what is rational in one situation may not be rational in a different situation. We cannot consider the "need" for energy independent of its price.

[*] Prof. Economics, Harvard University, Cambridge, Massachusetts
[**] Graduate Student, Economics, Harvard University

We need to bear this in mind when evaluating dire
predictions now being made of the implications for economic
growth and the balance of payments of recent increases in the
price of crude oil. Indeed, a mechanical projection of past
import trends leads to staggering bills in a few years, on
the order of 100 billion dollars per year. Is it reasonable
to expect such drastic shifts in world income and payments
patterns to occur?

Many discussions of this issue ignore what has just
been said about the reaction of consumers of oil to changes in
its price. If high prices lead to reduced oil demand, the
implications of recent OPEC actions could be much less drastic
than we are usually led to believe. This paper attempts to
measure the demand functions of OECD nations for oil products,
and draw conclusions from the functions estimated about the
magnitude of the future oil problem.

After reviewing the treatment of demand in widely
circulated projections of the petroleum market, and finding
that it lacks an adequate analysis of the response of consumers
to price, we outline the demand model used here. It is a
dynamic economic demand system in which current consumption
depends on income, price and the past patterns of these variables.
Some of the statistical problems encountered in estimating the
model are also discussed.

Finally, the estimated functions are presented.
Included are equations for gasoline and residential electricity
in the U.S., and gasoline, kerosene, distillate fuel and

residual fuel in a subset of OECD countries. The basic
finding is that demand for these products shows a significant
response to price changes, i.e., substantial decreases in
consumption will result if prices are maintained at current
levels. Thus, existing forecasts of the future oil market are
seriously deficient in ignoring price as a determinant of demand.

Existing Forecasts of the World Oil Markets

There are currently several existing forecasts of
the world oil market, which project consumption, supply and
trade flows for various regions in different years. One such
study was done by the OECD (1966), and its findings have been
updated by Schurr et al.(1971), and Darmstadter (1971). These
investigations project demand for energy as a whole, both
globally and by demand sector (transportation, industrial,
residential and commercial, and electric utility), and then
confront those figures with assessments of available supplies.
Projections along similar lines have been made by oil companies
(e.g., Shell, 1973), and by industry-government groups (NPC, 1972).
All these studies have merit in that they carefully consider the
relation of energy use to the sectoral composition of GNP, and
make explicit the assumptions about technical change used to
modify historical energy-output relationships.

Unfortunately, such studies are predicated on an
unchanged environment for the world oil market, and the sudden
change in Persian Gulf pricing patterns is a drastic rupture of
that environment. There is no explicit mechanism in these
forecasts for adjusting the conclusions to a change in
underlying economic conditions, so they have limited usefulness

when trying to assess the likely outcome of the oil trade under

today's circumstances. In particular, one cannot apply the new

OPEC export tax on oil (seven dollars a barrel) to the volume of

imports forecast under the assumption of a much lower impost.

For example, Schurr predicted OECD imports of 29.2 million

bbl/day in 1980; Shell Oil Company predicted 48 million

bbl/day. At seven dollars a barrel, these imply import bills

of 75 billion and 123 billion dollars a year, respectively.

However, the three and a half fold increase in the Persian

Gulf price of oil is a violent change in world pricing patterns,

and the law of supply and demand suggests that as a result

large changes will occur in production, consumption and trade

flows.

Consequently these studies of the oil market do not

give us much guidance in determining the magnitude of the

changes that will occur. They do contain some discussion of

the prices needed to draw forth various amounts of energy

production, but almost completely ignore the role of price in

influencing consumption patterns. For example, when discussing

the energy-GNP ratio, there is little mention of the fact that

real energy prices (that is, prices in relation to prices of

other goods and services in the economy) had been falling for

the last fifteen years until very recently, and that this had

presumably stimulated energy consumption. A reversal of this

price trend would affect the growth of energy demand in the

opposite direction. One of the most important considerations

for policy makers today is the magnitude of the response of
consumption to recent changes in oil prices.

Studies of the Demand for Energy

There does exist a considerable literature concerned
with estimating the demand for certain energy sources as a
function of income and price. Those for the United States have
typically disaggregated the data by states, and covered a period
of several years. Important studies were done by Fisher and
Kaysen (1962) in electricity and by Balestra and Nerlove (1966)
in natural gas, but their results have not been incorporated
into energy demand projections. A review of subsequent work
on demand for electricity and natural gas, and some new findings,
are given in Anderson (1973), and all of the results show
significantly high price elasticities, usually in the neighborhood
of -1.$\underline{1/}$ Demand studies for energy products done at a more
aggregated level, namely for time series covering the U.S. as
a whole, have proven less successful. Houthakker and Taylor (1970)
generally did not find prices to be significant in determining
consumers' demand for energy products as classified in the U.S.
national income accounts; Adams and Griffin (1972) reported that
price was not significant in equations which explained the
consumption of five types of refined oil products in the U.S.
This failure of price to help in explaining aggregate demand is
usually attributed to population shifts and state-to-state price
variations, caused by differential state taxes and transport
charges. Therefore, an important step in explaining demand for
energy is to achieve regional disaggregation wherever possible.

There has been less work done in estimating demand
for energy in other OECD countries. Aside from the analysis
of British electricity demand by Houthakker (1951) and the
Robinson report (OEEC, 1960), both out of date, we have found
no studies which effectively include the the influence of price
of energy consumption. Adams and Miovic (1968), and Darmstadter
(1971) present econometric evidence on the size of the "energy
elasticity," i.e., the percentage change in energy consumption
resulting from a one percent increase in GNP. (This number is
more generally called an income elasticity.) These studies do
not, unfortunately, attempt to assess the influence of price on
energy demand. These studies do indicate an income elasticity
of less than or equal to one for developed countries, and
Adams and Miovic point out interesting problems in the
aggregation of different energy sources. Since they ignore
energy price, which was typically falling over the sample
period, these studies probably overstate the income elasticity
of demand.$\underline{2/}$

Finally, spurred on by the energy crisis, there is
considerable work in progress attempting to measure the response
of consumers to oil price changes. Berndt and Jorgenson (1973)
have found significant substitution by producers toward energy
inputs in the face of real energy price decreases in U.S.
historical data. Erickson, Spann and Ciliano (1973) report
that relative prices of electricity, natural gas and fuel oil
have significant effects on the consumer's home heating
decision measured in a time series of individual state data
for the U.S.

A review of the literature thus reveals that there are still serious gaps, particularly in the case of the demand for refined oil products, and the demand for energy outside the U.S. We do note, however, that most studies which incorporated price variables found significant negative influences of price on consumption.

The Model Used Here

All the results we report are based on a dynamic flow-adjustment model of demand, described in Houthakker and Taylor (1970). This model postulates that demand for an energy product depends on underlying economic factors (such as price and income), but that the effect of these factors is spread out over time.

We begin with a level of <u>desired demand</u> for a given energy product (q*), which depends on income (y) and price (p).

$$q* = f(y,p) \tag{1}$$

(Desired demand may depend on other factors as well, such as temperature, but these are momentarily ignored for clarity of presentation.)

We specialize this equation to the log-linear form.

$$q* = \alpha \, y^\beta \, p^\gamma \tag{2}$$

This specialization can be viewed as a first order Taylor series approximation to the function

$$\ln q* = \ln f, \tag{3}$$

and is particularly convenient in that β and γ are immediately interpreted as income and price elasticities.

The actual level of demand in one year, however, is

not necessarily equal to the desired level. In particular,
actual demand adjusts toward desired demand according to the
equation

$$q/q_{(-1)} = (q^*/q_{(-1)})^{(1-\lambda)} \tag{4}$$

where $q_{(-1)}$ is last period's consumption. This lagged adjustment
of actual demand to desired demand is the result of (i) an
existing stock of equipment which uses a specific form of energy
at a specific efficiency, and which cannot be replaced immediately[3]
and (ii) an unwillingness by consumers to view price and income
changes as permanent until they have continued for some time.

Substitution of (4) into (2) generates the estimating
equation.

$$\ln q = \ln a + (1-\lambda)\beta \ln y + (1-\lambda)\gamma \ln p + \lambda \ln q_{(-1)}. \tag{5}$$

In this equation, β and γ are long-run elasticities of demand, and
$(1-\lambda)\beta$ and $(1-\beta)\gamma$ are short-run elasticities.

Statistical Implementation

The dynamic flow-adjustment model has been applied to
six sets of data. Using a cross section of U.S. states over
different time intervals, Houthakker, Verleger and Sheehan (1974)
estimated the demand by consumers for gasoline and residential
electricity. Along with a summary of their work we shall use
cross-sections of OECD member countries over recent years, to
estimate demand functions for gasoline, kerosene, distillate
fuel oil and residual fuel oil.

Both the U.S. and OECD studies combine time series
from different areas. This is valuable in determining the

behavior of energy users because the variation in income and
price within states and countries is greater than the variation
in aggregates over time. It requires that the structure of demand
be the same in different locations, but this powerful assumption
allows the econometric techniques to use all available data to
determine those responses to income and price which are in best
agreement with the evidence. Previous international cross-
country studies, such as those by Houthakker (1965) and Goldberger
and Gamaletsos (1967), support the hypothesis that patterns of
demand in different countries can be usefully related to
variations in prices and incomes.

Several interesting statistical problems arise in
the estimation of these demand curves.[4/] One is the problem
of multicollinearity, or correlation among the exploratory
variables. In recent years, real prices of energy have been
falling, while income has been rising. In the cross section of
countries, low oil prices are associated with high per capita
income. Since high income and low price affect consumption
in the same way (positively), there may be some ambiguity in
interpretation of the data - did consumption rise because price
fell or because income rose? This ambiguity, if it exists,
will be reflected in relatively high standard errors of our
estimates of price and income elasticities. However, the use of
time series for different states and countries tends to reduce
the problem of multicollinearity.

Another problem we have is that of left out price
variables. In economic theory, demand for a product or factor

of production is a function of its own price, income, and the
prices of all other goods and services. Our model takes some
account of this in that prices are deflated by either the CPI
or WPI; technically, this means we assume all other prices
increase at the same rate, the general rate of inflation. One
price of particular importance in the fuel oil equations is the
price of coal. Consistent time series of market coal quotations
could not be found, although there is some evidence (see Gordon,
1970) that the overall price increase of coal was comparable
to that of the general level of inflation during the time period.
If this is so, our estimates are unbiased. If the price of coal
rose faster than the CPI, our price elasticities are biased
upward in absolute value, and vice versa.

There are two more problems associated with estimation
of the equations, each of a technical nature. They are the
presence of a lagged dependent variable on the right hand side
of equation (5), and the fact that there is another equation
determining the price of energy products, the supply equation.
Standard techniques exist for handling these problems, and our
approach to them can be found in the references in footnote 4.

Results of Demand Studies

The results of the U.S. studies have been reported by
Houthakker, Verleger and Sheehan (1974), so only a brief review
of them will be given here. The gasoline equation was fitted
on quarterly data for 48 states over the time period 1963-1972.
The results (with standard errors in parentheses) are:

U.S. Gasoline

$$\ln q = .593 + .303 \ln y - .075 \ln p + .696 \ln q_{-1}$$
$$\quad\quad\quad (.017) \quad\quad (.013) \quad\quad (.019)$$

$$R^2 = .92$$

This equation indicates a long-run price elasticity of -.24,
and a long-run income elasticity of .98. The value $\lambda = .696$
indicates a rather sluggish response of demand to changes in price
and income, which is consistent with the long life of automobiles,
the main gasoline using equipment. All estimates are statistically
significant by the usual standards. In particular, the price
elasticity of -.24 indicates that an increase in the pump price
of gasoline from 40¢ to 80¢ should decrease demand about 15% in
about two years. This is precisely the magnitude of the
gasoline shortage now quoted in Washington.

Houthakker, Verleger and Sheehan have also derived
an estimate of the demand for residential electricity. The data
used here were annual observations of states over the period
1961 to 1971. Due to the declining average cost of electricity
to the user as more is used, estimates of marginal prices had
to be derived, and these were used as the price variables in the
equation.

U.S. Residential Electricity

$$\ln q = .072 + .143 \ln y - .089 \ln p + .913 \ln q_{-1}$$
$$\quad\quad\quad (.026) \quad\quad (.020) \quad\quad (.015)$$

$$R^2 = .99$$

The long-run price elasticity implied here is -1.0, and the
long-run income elasticity 1.6. There is a very long adjustment

lag implicit in the estimate λ = .913. Since residential
electricity demand is mostly used in conjunction with long-lived
appliances for which there is no second-hand market, this long
lag is reasonable. Again, all results are statistically significant. The evidence of a high long-run price elasticity is
particularly important here, for most projections of electricity
use show it increasing at 6 to 7 percent until the year 2000.
This may be an accurate extrapolation of past trends, when the
real price was generally declining. But higher rates due to
primary energy price increases and a flatter rate structure
will put a large dent in that yearly demand growth, and we can
confidently predict that the growth of electricity consumption
will slow down markedly in the coming years.

The next set of equations were estimated from annual
time series for a number of OECD countries. These results have
not been presented before, so they will be described in more
detail. The income terms have all been deflated to 1963 constant
value home currency figures, then converted to U.S. dollars
at 1963 exchange rates. Prices have been similarly deflated
and converted. Left hand variables are either in per capita
terms or deflated by income level as an ad hoc adjustment for
heteroskedasticity. The estimation procedure, which follows
Dhrymes (1971), takes account of autocorrelation and the
presence of a lagged dependent variable. The data sources are
given in an appendix.

The equation for gasoline demand was run on twelve countries[5], and covered the years 1962-1972. Consumption is measured in per capita terms, and the income variable is real per capita consumption expenditures. The full estimated equation also included variables for population density and temperature.

<u>OECD Gasoline</u>

$$\ln q = .213 + .744 \ln y - .465 \ln p + .442 \ln q_{-1}$$
$$\quad\quad\quad (.101) \quad\quad (.105) \quad\quad (.068)$$

$$R^2 = .93$$

The implied long-run income elasticity is 1.3, and the long-run price elasticity is -.82. The figures are significantly higher in absolute value than those found for the U.S. alone, although this sample includes the U.S. One explanation is that the log-linear function is only a local approximation, and that the lower income levels and higher price levels typically prevailing in Europe lead to higher elasticities. Evidence for this interpretation is provided by a Chow (1960) test, which shows that the coefficients for the U.S. and for all other countries are not significantly different.[6] At any rate, this is strong evidence for the influence of price on gasoline consumption, other things being equal, and lays to rest the idea that Americans drive big cars because of advertising pressure or other psychological factors. Americans drove big cars because gas was cheap, and now that gasoline is becoming more expensive they are rapidly changing the pattern.

The remaining equations were fitted over a smaller
sample of nine countries,[7] for the years 1965-1970. The U.S.
was not included. In the kerosene and distillate fuel equations,
dependent variables were expressed in per capita terms, and per
capita consumption expenditures and prices, both deflated by
the consumer price index, were used as explanatory variables.
This specification assumes that the demand for these products
can be captured by variables relevant for residential consumers.
In further research, the demand for the fuels should be disaggregated by consuming sector . In any event, during the
time period under consideration, residential use did predominate
in middle distillate consumption.

OECD Kerosene

$$\ln q = 2.36 - \underset{(.082)}{.207} \ln y - \underset{(.186)}{.173} \ln p + \underset{(.039)}{.927} \ln q_{-1}$$

$$R^2 = .93$$

The long-run income elasticity here is -2.5, indicating that
kerosene is an inferior good, in agreement with common sense.
The long-run price elasticity is a rather high -2.0, indicating
the existence of good substitutes, no doubt #2 heating oil and
coal. These results are intuitively quite plausible when the
role of kerosene as heating oil is considered, but the increasing
use of kerosene as jet fuel makes them suspect for projection
purposes. As jet fuel, the income elasticity/would likely
                                        of kerosene demand
be over one, and the lack of good substitutes should make the
price elasticity low in absolute value (although clearly not

zero). This appears to be a case where changing underlying
structure vitiates the projective power of historial data.

In the distillate equation, a separate dummy variable
for each country was included.

OECD Distillate Fuel Oil

$$\ln q = -.231 + 1.43 \ln y - .388 \ln p + .471 \ln q_{-1}$$
$$(.436)(.193)(.150)$$
$$R^2 = .90$$

These results are quite plausible. The long-run income
elasticity is 2.8, indicating that distillate fuel is a
superior good. Since its substitutes, coal and kerosene, are
much less convenient, this is very likely. A lower elasticity
might be expected in the forecast period, however, as (i)
saturation of fuel oil heat begins to appear, and (ii) an even
more convenient fuel, natural gas, comes onstream in some
European countries. The long-run price elasticity is -.76.

In the equation for residual fuel oil, GNP was taken
as the income variable, and prices were deflated by the wholesale
price index. The dependent variable was deflated by GNP, so the
income elasticity is one plus the coefficient on ln y. Again,
all uses of resid are lumped in one equation.

OECD Residual Fuel Oil

$$\ln q = .884 + .396 \ln y - 1.05 \ln p + .335 \ln q_{-1}$$
$$(.146)(.426)(.207)$$
$$R^2 = .90$$

The long-run price elasticity is a very high -1.58, reflecting
the close competition with coal in this market. The income

coefficient of 1.6 in the long-run implies that either residual
fuel is superior to other fuels due to convenience factors, or
that the sectoral composition of GNP was shifting toward
residual-intensive uses over the sample period. Both factors
seem likely, as the share of manufacturing output was increasing
in OECD countries at this time. The adjustment coefficient,
λ = .34, is quite low. This reflects the rapid growth in
industrial capacity, and thus turnover of energy using equipment,
during this time. The use of the WPI as a deflator in this
equation may bias the price elasticity estimate downward in
absolute value, since the price of labor, an important variable
in mechanization decisions, rises much faster than the WPI.

All of the equations presented in this section tell
the same story. Price is an important factor in demand for
energy, in particular for refined oil products. Blind extrapolations of past consumption trends are nonsensical, because recent
history is a reflection of falling real prices, coupled with
price elasticities in the neighborhood of -1. Therefore, oil use
grew much faster in recent years than any measure of income.
The abrupt recent turnabout in oil prices will sharply curtail
demand growth, just as previous price weakness encouraged it.
The world's economy is no more "locked into" oil use today
than it was locked into coal use in 1950. Relative price changes,
as reflected in the historical evidence presented here, will
determine the use levels of various forms of energy.

<u>Implications of Demand Functions</u>

Having criticized current market forecasts for
ignoring the influence of price on consumption, we must
produce something better. Therefore, we have constructed an
economic model of the world oil market which, for any target
year, generates equilibrium values of consumption, supply,
trade and price of oil products in various regions of the
world, given the underlying economic variables. Among the
most important of these variables are the demand functions for
various products described above, the supply functions for crude
in different world regions, refining costs, shipping costs,
excise taxes, and, of course, the export duty in the Persian
Gulf and other producing regions.

The model embodies the standard economic supply and
demand scheme in which price effects are explicitly allowed for.
Its projections have the property that all consumption flows are
on the demand curve, which incorporates income and price, and
production in each region is also dependent on the local price
(or conversely). In addition, the refiner's margin must be
sufficient to produce an adequate return on capital, and any
import trade must be profitable enought to cover long-run
shipping costs plus import or export duties. Transportation,
treated in considerable detail, is an inherent part of the model.

The model has two virtues. Any projection it generates
is automatically forced to be internally consistent, which is
difficult to achieve informally when we are dealing with several

refined products and regions. Also, if there is uncertainty

about the future value of any underlying parameter, simulations

can be run over alternative parameter values to produce a range

of outcomes. The projection models we reviewed in the early

part of this paper seem to lack both the simultaneous

determination of price and quantity, and the on-line capability

to generate new forecasts in the face of changed circumstances.

Of course, our model in turn lacks some of their richness of

detail (such as projecting oil use by consuming sector) and is

presently only concerned with the oil market. Work is

continuing on strengthening the model.

To illustrate the effect of the price elasticities

presented above on the future of the world oil market, we present

the results of some model simulations in Table 1. These

simulations take 1980 as the target year, and are long-run in

nature; that is, they assume that seven years is enough for all

price effects to be worked out. The prices are given in 1973

U.S. dollars, so a general inflation factor must be allowed for.

Four results are presented. Cases A and C assume the Persian

Gulf export tax is $3.50/bbl., approximately the level prevailing

in November 1973, and B and D put it at $7, the current level.

There is much uncertainty about the response of

North American oil supplies to changes in price. Most observers

feel that the long-run response is quite high, with a long-run

price elasticity equal to or greater than one (see Erickson and

TABLE 1

Results of 1980 Simulations (in 1973 dollars)

Cases	A	B	C	D
Persian Gulf royalty ($/bbl)	3.50	7.00	3.50	7.00
Elasticity of supply in U.S. and Canada	.25	.25	.67	.67
Persian Gulf and North Africa exports (MM bbl/day)	17.7	8.4	16.4	6.9
Persian Gulf and North Africa revenues ($ bil/yr.)	22.8	21.7	20.7	17.7
European consumption (MM bbl/day)	15.1	12.5	15.1	13.6
Japanese consumption (MM bbl/day)	6.6	4.5	6.6	4.5
U.S. price of crude ($/bbl)	5.36	7.08	5.00	6.42
U.S. Production (MM bbl/day)	14.3	15.8	16.0	19.4
U.S. consumption (MM bbl/day)	16.9	14.7	17.7	16.0
U.S. imports (MM bbl/day)	2.6	-1.1	1.7	-3.4

Spann (1971), Mancke (1970) and U.S. Cabinet Task Force (1970)).
However, with our recent history of declining drilling rates
and the need to go offshore for most new supplies, this could
be considered optimistic. Therefore, we simulate the model
over North American supply elasticities of both .25 (Cases A
and B), and .67 (C and D). Other crucial supply assumptions
in this run are that non-Persian Gulf and North African OPEC
members will not increase production, and that Alaska will
supply 2-1/2, and the North Sea 4 million barrels a day in 1980.

The results of these runs are in sharp contrast to
currently circulating projections. The revenues of Persian
Gulf and North African suppliers never get above 23 billion
dollars a year, and are in fact higher at the lower royalty
level. Of course, Persian Gulf and North African production
is much lower when the royalty is 7 dollars, so it is conceivable
that there will be more aggregate revenue over the very long-run
because existing reserves last longer. But with the tight
control over production implied here, real problems of cartel
stability arise. The 1980 exports of 8.4 million barrels a day,
under a seven dollar royalty and low North American supply
elasticity, are less than the current production of North Africa
and Iran, to say nothing of Saudi Arabia and Kuwait. How the
OPEC members will manage to split up the dwindling market is
hard to predict, particularly after the member countries
experience the very high revenues they will receive in the next

two or three years, before supply and demand elasticities fully
take hold.

What causes these sharp decreases in export volume at
the high OPEC price? Table 1 reveals that it is precisely the
significant price elasticities of demand described above. A fair
generalization of our results is that demand is likely to stagnate
in future years, since any impetus toward increased consumption
due to income growth will be restrained by the high price. Along
with this pattern of roughly constant demand, we have the
increased supplies from Prudhoe Bay and the North Sea, and
whatever extra comes from the lower 48 and Canada as a result
of the higher price. Again, this does not assume any production
increase from other OPEC regions, such as Indonesia, Nigeria and
Venezuela.

A final remarkable implication of the model is the
re-emergence, after 30 years, of U.S. exports, given that OPEC
nations stay with the present high royalty. Of course, the
probable internal tensions brought about by erosion of Persian
Gulf markets may well bring this seven dollar price down before
any tankers bring Louisiana crude to Rotterdam. But the prospect
of U.S. exports by 1980 if the $7 export duty stays in force
should not be surprising, for this implies a U.S. price more
than double the price prevailing a few months ago. Such drastic
changes in underlying circumstances lead to equally drastic
changes in economic results.

<u>Conclusion</u>

This paper has argued that forecasts of the world oil market that ignore the impact of price on consumption are of little help in predicting the consequences of any sharp change in factors underlying the market, and may indeed be thoroughly misleading as a guide to policy. In particular, these forecasts can't be easily adapted to incorporate the recent increase in the price of OPEC crude oil, and simple modifications of the figures, such as multiplying previously predicted trade volumes by the new export tax, will lead to absurd results.

After reviewing existing studies of energy demand as a function of price, we presented some recently derived results of our own. These generally showed significant effects of price on consumption, with several elasticities in the neighborhood of -1. Finally, the implications of these elasticities for a model of the world oil market were shown. One result of interest is that the present OPEC crude price of $7 per barrel will cause a sharp curtailment of Persian Gulf and North African export volume by 1980, creating problems for the allocation of production among cartel members.

<u>APPENDIX</u>

Data Sources for OECD Regressions

<u>Consumption</u>

Gasoline and kerosene from U.N., <u>World Energy Supplies</u>. Distillate and residual fuel oil from OECD, <u>Oil Statistics</u>.

<u>Price</u>

Gasoline prices from <u>Petroleum and Petrochemical International</u>, formerly <u>Oil and Gas International</u> (December prices). Distillate and residual prices from ENI, <u>Energy Yearbook</u>, 1971. Kerosene prices computed as distillate price less tax plus kerosene tax as reported for 1964 and 1970 in U.S. Bureau of Mines, <u>International Petroleum Annual</u>.

<u>Income and Population</u>

GNP from Un.S., <u>Yearbook of National Income Statistics</u>. Consumption, population, deflators and exchange rates from IMF, <u>International Financial Statistics</u>.

<u>FOOTNOTES</u>

<u>1</u>/A price elasticity is the percentage change in consumption
caused by a one percent increase in price; thus, price
elasticities are normally negative. Income elasticities are
similarly defined.

<u>2</u>/See Griliches (1957) for the effect of left out variables on
the estimates of other parameters.

<u>3</u>/See Anderson (1973) for an example of demand analysis in
which stocks of different energy using equipment were explicitly
included in the analysis. This is a preferred line of attack,
but data deficiencies are severe. Anderson was limited to only
two years in his cross section of U.S. states.

<u>4</u>/For a more extended and more technical discussion of these
problems, see Houthakker, Verleger and Sheehan (1974), and
Kennedy (1974).

<u>5</u>/Portugal, Italy, Austria, Belgium, Denmark, France, West
Germany, Netherlands, Norway, Sweden, U.K., U.S.

<u>6</u>/The computed value of the test statistic is .17, while the
critical F value is F (6,108) = 2.19. Essentially nothing is
lost by disaggregation. It must be recalled, of course, that
these are data for the U.S. as a whole, and that we only cannot reject
the possibility that the two samples are from the same
population.

<u>7</u>/Japan, Italy, Belgium, Denmark, France, Germany, Netherlands,
Sweden, U.K.

REFERENCES

Adams, F.G., and J.M. Griffin, "An Econometric-Linear Programming
    Model of the U.S. Petroleum Refining Industry," Journal of
    the American Statistical Association, 67 (September 1972)
    542-51.

Adams, F.G., and P. Miovic, "On Relative Fuel Efficiency and
    the Output Elasticity of Energy Consumption in Western Europe,"
    Journal of Industrial Economics (November 1968) 41-56.

Anderson, K.P. Residential Energy Use: An Econometric Analysis,
    The Rand Corporation, R-1297-NSF, (October 1973).

Balestra, P., and M. Nerlove, "Pooling Cross Section and Time
    Series Data in the Estimation of a Dynamic Model: The Demand
    for Natural Gas," Econometrica, 34 (July 1966) 585-612.

Berndt, E.R., and D.W. Jorgenson, "Production Structure," Ch.3
    Energy Resources and Economic Growth by H.S. HOuthakker and
    D.W. Jorgenson, Data Resources, Inc. 1973.

Chow, G.C., "Tests of Equality between Sets of Coefficients in
    Two Linear Regressions," Econometrica, 28 (July 1960) 591-605.

Darmstadter, J., et al., Energy in the World Economy (Baltimore:
    Johns Hopkins Press) 1971.

Darmstader, J. "Appendix" in Schurr, S.H., ed., Energy Economic
    Growth and the Environment (Baltimore: Johns Hopkins Press) 1972.

Dhrymes, P.J., Distributed Lags: Problems of Estimation and
    Formulation (San Francisco: Holden Day) 1971.

Erickson, E.W., and R.M. Spann, "Supply Response in a Regulated
    Industry: The Case of Natural Gas," Bell Journal of Economics
    and Management Science, 2 (Spring 1971) 94-121.

Erickson, E.W., R.M. Spann and R. Ciliano, "Substitution and
    Usage in Energy Demand," in M.F. Searl, ed., Energy Modelling,
    Resources for the Future, Inc., EN-1, (March 1973).

Fisher, F.M., and C. Kaysen, The Demand for Electricity in the
    United States   (Amsterdam: North Hollan Publishing Company),
    1962.

Goldberger, A.S., and T. Gamaletsos, "A Cross Country Comparison
    of Consumer Expenditure Patterns," Social Systems Research
    Institute, University of Wisconsin, November 1967.

Gordon,R.L. The Evolution of Energy Policy in Western Europe:
   The Reluctant Retreat from Coal (London: Praeger), 1970.

Griliches, Z., "Specification Bias in Estimates of Production
   Functions," Journal of Farm Economics, 39 (February 1957) 8-20.

Houthakker, H.S., "Some Calculations on Electricity Consumption
   in Great Britain," Journal of the Royal Statistical Society,
   Series A (1951).

Houthakker, H.S., "New Evidence on Demand Elasticities,"
   Econometrica 33 (April 1965) 277-288.

Houthakker, H.S., and L.D. Taylor, Consumer Demand in the
   United States, Analyses and Projections (Cambridge: Harvard
   University Press) 1970.

Houthakker, H.S., P.K. Verleger, Jr., and D.P. Sheehan,
   "Dynamic Demand Analyses for Gasoline and Residential Electricity,"
   to be published in the American Journal of Agricultural
   Economics, 1974.

Kennedy, M. An Economic Model of the World Oil Market (unpublished
   Ph.D. dissertation, Harvard University, 1974.

Mancke, R.B., "The Long-run Supply Curve of Crude Oil Produced
   in the United States," Antitrust Bulletin 15 (Winter 1970)
   727-756.

National Petroleum Council, U.S. Energy Outlook, N.P.C.
   (Washington, D.C.) 1972.

Organization for Economic Cooperation and Development, Energy
   Policy: Problems and Objectives (Paris: OECD) 1966.

Organization for European Economic Cooperation, Towards a New
   Energy Pattern in Europe (Paris: OEEC) 1960.

Schurr, S.H., et al., Middle East Oil and the Western World
   (New York: American-Elsevier) 1971.

Shell Oil Company, The National Energy Outlook, 1973.

U.S. Cabinet, Task Force on Oil Import Control, The Oil Import
   Question (Washington: GPO) 1970.

ENERGY SUPPLY AS A FUNCTION OF PRICE

Milton F. Searl*

Increases in domestic energy output through
1985 must come in large measure from the conventional energy sources: crude oil, natural
gas, coal used in the form of coal, and present
nuclear reactor types. In the absence of a
government-sponsored crash program of plant
construction, the new sources and new technologies now under development are not likely
to contribute substantially to energy supply
by 1985. Beyond 1985, oil shale and the new
technologies can be expected to begin to contribute significantly to energy supply.
Domestic resources of each of the fuels are
adequate to support any production levels
which the nation may desire in 1985 without
undue depletion of the resource base. The
constant dollar prices, at which crude oil
and natural gas may be available in 1985 and
beyond--under an assumption of nonprohibitive
environmental policies--appear to be lower
than might be expected from current price
behavior. Government policies will play a
major role in determining future prices.
Both resource costs and market prices should
be considered in establishing government policy.

Curiously enough, in view of the importance of energy to the
long-term development of the economy, little attention was
paid to the relationship between price and energy production
during most of the post-World War II period, or for that
matter, to energy price and consumption. Even long-term
forecasts going to the year 2000 did not give a prominent
role to price. Price did not seem to matter.

---

*Mr. M. Searl is Manager, Program of Energy Supply Studies,
Electric Power Research Institute, P.O. Box 10412, Palo Alto,
California 94304

However, economic forces were at work. In the first decade
after the war, drilling for oil and gas doubled in response
to favorable price and profit expectations. The nation's
crude oil productive capacity outstripped markets for domestic
production. By 1958, the nation's productive capacity for
crude oil and natural gas liquids exceeded production by
nearly 3 million barrels/day--about 40 percent. Eventually,
the cost of maintaining excess capacity and federal regulation
of wellhead prices of natural gas caused profit expectations
to turn unfavorable and the drilling of oil and natural gas
wells began a long-term decline which may only now be ending.

The nation's postwar experience in energy has once again
demonstrated what economists have long known--price relative
to cost does matter. However, normal price-output relations
in the oil, natural gas and electrical industries have been
complicated by extensive intervention of governments--federal
and state--in industry economics. Government intervention
has generally had a price stabilizing effect, sometimes
deliberately, sometimes consequentially. This has meant that
economic adjustments which more normally might have taken the
form of price increases have instead taken other forms,
particularly declines in the ratio of reserves to production
(i.e., instead of price increases there has been a slowing
of the rate of capacity addition). This has been particularly
true for crude oil and natural gas. The second order nature
of economic adjustments instead of direct changes in price
with output contributed to the illusion that price did not
matter.

The postwar economic history of the energy industries also
provides evidence that it is not necessary to postulate any
major decline in the quality of the resources remaining to
be found and developed in order to account for the present

plateau in domestic production of oil and gas. Indeed,
while some older oil and gas provinces may have passed their
peak, the outlook from a resource standpoint is quite bright.
On the North Slope of Alaska, the most prolific oil field in
the nation's history was discovered a few years ago. It also
contains large amounts of natural gas. One or more major
fields have been found in the Santa Barbara Channel. Large
reserves are known to exist in the Elk Hills Naval Reserve.
Moreover, it is highly probable that other major fields exist
on Alaska's North Slope, particularly in Naval Petroleum
Reserve No. 4. Very large potentially oil- and gas-producing
structures are known to exist in the Gulf of Alaska, in the
Gulf of Mexico as far east as Mississippi and Florida, and
along the Atlantic Coast. Geologically, some of these areas
may rival prolific Middle East producing areas.

Recent history contains numerous examples of new areas
becoming major sources of production. A decade ago the
North Sea did not look very promising. Now it is a major
producing province with estimates of its ultimate capacity
being continually revised upward. And, a little earlier the
Groningen gas field, one of the world's largest, was discovered in the Netherlands in formations which were not
expected to be productive.

There are many reasons why potentially very productive areas
have not been developed to date. In some areas there are
environmental problems. Historically, the government has
not seen fit to lease many of the attractive offshore areas.
In many of the better offshore areas, water and weather
conditions are treacherous and exploration and development
is difficult and requires large investments, but development
is by no means beyond the capability of a nation determined
to have access to these resources. And one can only speculate

as to why the vast Naval Petroleum Reserves remain closed
to development.

Moreover, oil resources in the hundreds of billions of
barrels--30 to 40 times current annual consumption--exist in
known fields but have been uneconomic to recover at historic
prices. Large amounts of natural gas are also known to exist
in tight formations awaiting favorable economics and/or
improved technology for their recovery.

It is, I believe, abundantly clear that the nation will never
physically run out of oil, natural gas, or coal resources.
To talk in those terms is scare strategy. What will eventually happen is that costs of recovering these resources will
rise, as may or may not be presently occurring, faster than
the ability of technology to offset the cost increases and
other sources and technologies will take over the market--
but there will always be additional amounts of all of the
fossil fuels recoverable at a price. We will never run out
in the sense of physical depletion of our fossil fuel
resources.

If important options for meeting national energy requirements
are not to be foreclosed, it is important that the public and
government officials realize that the price mechanism does
work and has been working to the extent allowed by government
and other restraints, and that the nation is not resource
limited. The fact that production from the Alaskan North
Slope and the Santa Barbara Channel might now be flowing at
the rate of 3 million barrels/day, if they had been promptly
developed, illustrates the point. However justified the
delay in the production of these sources may have been, it
is economic nonsense to say that price has failed to bring
forth the necessary resources.

As our subsequent discussion will indicate, the results of a
Resources for the Future analysis in which I participated led
to the conclusion that the most rapid course to increase
domestic output lies in allowing price mechanisms to work.
The only alternative, if there is a viable alternative, is
massive, government-subsidized crash programs which in all
likelihood will cost the nation more than allowing prices to
perform their economic function. And, if through subsidy
such a program results in low prices to consumers, it will
probably once again stimulate a high rate of growth of
energy consumption.

Any projection of future prices and quantities must, of
course, be conditioned not only by considerations of the cost
of labor and capital and the quality and quantity of the
nation's energy resource base, but also by the prospects for
successful development of new energy technologies, as well
as evolution of existing technologies; by the costs and
constraints imposed by environmental and health and safety
measures; by government tax, leasing and royalty policies;
by the availability and cost of oil and natural gas imports,
and by government policy toward imports. Moreover, price
projections must distinguish very carefully between short-
and intermediate-term prices and long-term prices. Misunder-
standing of short-term price changes, such as have occurred
in the last few months, can rather easily result in estimates
of prohibitive prices in the future which could result in
drastic changes in our way of life and lead to counter-
productive government policies.

In the short and intermediate term, which in the current
case may be several years, prices are determined more by the
elasticity of demand than the elasticity of supply. By
definition of the short run, major additions to capacity are

not possible; price, unless established by government edict,
is set by the value of the commodity to purchasers--the
amount they are willing to pay and not by changes in the cost
of supply. It is, however, this difference between price
and cost which drives producers to expand output. The larger
the difference, the more rapidly output will be expanded
although the precise form of the relationship is unclear.
Only as the increases in supply brought about by price and
profit expectations begin to be realized do we begin to get
some indication of where long-term equilibrium prices lie.
However, due to both demand and supply elasticities, long-
term equilibrium prices will almost certainly lie below
short-term marginal prices.

The analysis by Resources for the Future of energy supply
prospects upon which this paper draws did not seek to
establish specific levels of production for the various
sources and forms of energy in the future, but rather sought
to develop supply curves--that is, the relationship between
price and quantity--primarily for the year 1985 so as to
answer questions as to the cost of various levels of output
in the future. As the various energy forms are discussed,
estimates of price for certain levels of production will be
given, but these consist merely of points on the supply
curve corresponding to possible supply strategies and have
no special significance as forecasts of price or quantity.
As far as quantity is concerned, there is no assurance that
the government will adopt supply strategies calling for
those quantities and as far as price is concerned, no
assurance that environmental, leasing, tax and regulatory
policies on which the prices are based will be in effect.

It is appropriate to start with liquid fuels since they are
currently of most concern. Historically, liquid fuels have

consisted of crude oil and natural gas liquids. Liquids
from coal and oil from shale will likely contribute to
liquid fuel supply in the future.

Technologies which might be used to produce synthetic crude
oil or petroleum products from coal are at an early stage
of development. Their complex nature and present status
makes it unlikely that liquids from coal will make a
significant contribution to domestic liquids supply for
several decades. Prices of synthetic crude oil from coal
have been projected in the $7 to $8 per barrel range in
1972 dollars, although costs could be much higher. Coal is
attractive as a source of synthetic crude oil because of the
nation's large resources of coal, a potential yield of 2 to
3 barrels of oil per ton of coal, and the wide geographic
distribution of coal resources.

A much more likely source of synthetic petroleum within the
next decade is shale oil. Shale oil resources are very
large, dwarfing even Middle East oil reserves. High-grade,
readily accessible shale oil resources in thick seams are
estimated at about 300 billion barrels--fifty times current
annual domestic oil consumption. Intermediate-grade
resources are generally estimated to add a trillion and a
half barrels, and low-grade resources are very much larger.

There are some serious environmental problems involved in
the production of oil shale, particularly the disposal of
the residue from retorting which has a greater volume than
the shale originally mined. If disposal of the spent oil
shale in canyons with watering and other measures to establish vegetation comparable to that existing in the area is
environmentally acceptable and if other environmental
concerns do not prove more costly to solve than now seems

likely, the RFF analysis indicated that shale oil prices
might be in the $4.50 to $5.00 per barrel range (again in
1972 dollars), once the technology has been demonstrated in
several plants. The $4.50 to $5.00 per barrel figure does
not include allowance for lease bonus payments to the federal
government, which owns most of the best shale oil resources,
nor more than minimal royalties. Lease bonuses and royalties
for shale oil at rates approaching those for offshore oil
could raise the cost of shale oil to the $6.50 to $7.00 per
barrel range. Even at such prices, shale oil would be
saleable on today's market where prices are climbing up the
short-run supply curve but would be only marginally competitive with the prices of crude oil in 1985 projected in the
study. Recent bidding on government shale oil leases indicate
that bonuses may indeed approach those for offshore oil,
although a wide difference of opinion appears to exist
between companies as to the near-term prospects for shale oil.

The Resources for the Future study agreed with the National
Petroleum Council projection that shale oil would provide
only a few percent of liquid fuels supply by 1985. However,
shale oil production could be perhaps 3 million barrels per
day in 1985 if such output were recognized as a national
necessity now.

There does not appear to be any alternative to depending upon
crude oil for the bulk of domestically-produced oil in the
next several decades. The amount of oil which will be
required by the nation in 1985 and beyond is, of course,
subject to considerable uncertainty. Not only is the effect
of higher prices on consumption levels uncertain, but there
is also the possibility that much of the previously expected
growth in petroleum consumption might be shifted to natural
gas and nuclear-produced electricity with growth in liquid

fuels consumption largely limited to transportation and
other uses where it has particular advantages. The possibility of substantially increased natural gas consumption
was an unexpected finding of the study about which more will
be said later.

Given that expansion of domestic liquid fuels output must
come mainly from crude oil, how may domestic crude oil output
respond to increased prices? What prices are necessary to
insure that crude oil production is sufficiently profitable
to induce the necessary capital flows into exploration and
development?

The most certain and quickest path to major increases in
domestic crude oil production lies in bringing into production such major known resources as the Elk Hills Naval
Reserve, the Santa Barbara Channel fields, and the Prudhoe
Bay field on Alaska's North Slope. It will take three to
four years to bring these resources into full production
once legal and environmental questions are settled. Beyond
this, additional large new fields need to be found and
developed, much of which will be offshore. Confidence has
already been expressed that the resources are there to be
developed. The drilling of all wells will need to expand
rapidly--10 percent per year for at least a decade would not
be too much. This would require a similar rapid expansion
in the nation's capacity to produce drilling rigs, offshore
drilling platforms and oil field steel, as well as a rapid
expansion in the supply of oil field workers. A review of
the nation's ability to expand output of crucial items during
World War II, and, under the stimulus of favorable prices, to
expand drilling from 27,000 wells in 1945 to 58,000 in 1956
indicates that rapid expansion is not physically out of the
question.

However, it is doubtful whether the nation will, in the near
future, establish the policies necessary to bring about a
sustained expansion in exploration and drilling. One key
element is access to offshore leases at an accelerated rate.
There is some reason for optimism in this area. However, an
adequate rate of leasing of promising acreage appears feasible
only if leasing can be shown to be environmentally sound and
if new methods of bidding are adopted.

Other questions dealing with what have been labelled excess
profits, with industry taxation, with the competitiveness of
the industry, and with the possibility of the establishment
of a subsidized national petroleum company all affect the
expected future profitability of investment in oil and
natural gas production and will constitute a drag on the rate
of expansion. Until there is a clear resolution of these
sorts of questions, oil companies and the outside sources of
capital on which they must draw are going to be reluctant to
make the required long-term capital investments, and suppliers
of drilling rigs, offshore platforms, and steel products are
likewise going to be reluctant to make the investments to
expand capacity, or at least higher risk premiums on investment will be required.

Assuming that tax policies were frozen, that leases were made
available, that environmental questions were settled, and the
producing industry judged to be workably competitive, what
sort of supply response to price might be required to elicit
various levels of domestic output in 1985? An independent
estimate of crude oil supply derived by Dr. Henry Steele of
the University of Houston for the RFF study concluded that
at less than $5.50 per barrel in 1972 dollars, crude oil
production in the lower 48 states and Southern Alaska would
be at about current levels of production. Domestic crude

oil prices are now approaching that level in current dollars.
The addition of North Slope crude (3½ million barrels/day),
natural gas liquids (1½ million barrels/day), and shale oil
(3/4 of a million barrels/day) brings domestic liquids
production to nearly 15 million barrels/day at this price
level. The analysis indicates that a further increase of
less than 10 percent in 1985 prices would increase domestic
output from nearly 15 million to somewhat over 17 million
barrels/day. Congressional testimony given by Dr. Vincent
McKelvey, Director of the U.S. Geological Survey in the fall
of 1972 and based on rapid development of largely known major
oil resources indicated such quantities might be available
at even lower prices.

There is, of course, substantial uncertainty about future
oil prices. National Petroleum Council studies for roughly
the same levels of output can be interpreted as requiring
prices about $2.00 per barrel or 5 cents per gallon higher.
No convincing basis exists for choosing between the estimate
of Dr. Steele for the RFF study and the estimates of the NPC
study.

At 17 million barrels/day, production in 1985 would equal
current levels of consumption. Imports of crude oil and
products of perhaps 3 million barrels/day, a relatively
modest level, could probably be had from secure sources and
would bring total oil availability to 20 million barrels/day
in 1985. This is far short of the 26 million barrels/day
called for by the standard forecasts. However, the results
of the RFF analysis indicate, as discussed subsequently, that
the difference can probably best be made up by other fuels
from both an economic and environmental standpoint.

Natural gas ranks second only to oil as a source of energy for the economy. Indeed natural gas, if we include by-product liquids, provided two-thirds of the increase in domestic output of energy in the postwar period.

It has been evident for some time that the nation faced shortages of natural gas. The ratio of proven reserves to production has been declining since the end of World War II and from an operational standpoint is at an effective minimum. Average prices of natural gas at the wellhead which even in 1972 were only one-third of the price of crude oil on an energy content basis were not sufficient to develop new reserves at the same rate production was expanding. Although limited price increases have been authorized by the FPC, according to the calculations of practically all economists who have studied the situation, presently authorized prices are still below market clearing conditions (i.e., conditions which will balance supply and demand).

A major finding of the RFF study is that large increases in natural gas production are one of the nation's best options, economically and environmentally, for meeting energy needs. This was a somewhat surprising conclusion given the general pessimissm about natural gas supply. As with crude oil, natural gas resources seem adequate to support a large expansion in output for at least several decades. The Potential Gas Committee, a committee representing the entire gas industry from distributors to producers, estimated that at the end of 1970,gas resources, including proven reserves "recoverable without fundamental changes in economics or technology," were nearly 2,150 trillion cubic feet,which is equal to 90 years supply at current rates of consumption. The U.S. Geological Survey, as of the same date, estimated

resources recoverable under current conditions, i.e., prices
and technology to be some 25 percent larger. The Potential
Gas Committee did not estimate total gas resources. The
Geological Survey estimated total remaining natural gas
resources at about two and one-half times those recoverable
under 1970 conditions. There does not appear to be any
responsible basis for pessimism about the ability of natural
gas resources to support large increases in output.

The five recent major studies of natural gas supply differ
greatly in methodology and conclusions. Yet all, as interpreted in the RFF study, indicate that as much as 30 trillion
cubic feet of natural gas could be produced in 1985 at prices
in 1972 dollars between 60 and 90 cents per MCF, roughly three
to four times current average prices from conventional sources
of gas in the lower 48 states and Southern Alaska.

The addition of gas from the Alaskan North Slope, modest
amounts from nuclear stimulation of tight natural gas
formations,and gas imports at 15 percent of total consumption
would give an estimated 40 trillion cubic feet of natural
gas in 1985, nearly twice present gas supply.

Since each 2 trillion cubic feet per year of natural gas is
roughly equal to 1 million barrels per day of oil, forty
trillion cubic feet of natural gas per year would go a long
way toward relieving the need for large imports of oil. Even
at the upper end of this price range, natural gas would still
be at no more than parity with crude oil at the wellhead.
And in view of its environmental superiority, natural gas
should probably command a premium.

Turning now to coal, which is frequently considered our
largest energy resource available without substantial

technology development, resources of coal are vast. A
commonly accepted figure is 3.2 trillion tons--some 500
times current annual production.

Coal is currently said to have two problems--it can't be
mined and it can't be burned. To these two, a third might
be added--it probably can't be transported either. There
are indications that the nation's railroads are not
prepared to move coal output should it be rapidly expanded.

Surface mining, the lowest cost means of coal mining from
which 50 percent of the nation's output now comes, is being
seriously challenged on environmental grounds--some have gone
so far as to suggest that it be banned entirely. This would
shut down about one-half of the nation's electricity output
and shut down some electric systems entirely. It appears
unlikely that such extreme measures will be adopted, but the
conditions under which surface mining is permitted to continue
will greatly affect the price of coal and the rate at which
output can be expanded. Underground mining suffers from
some environmental problems but miner health and safety
and low productivity are the major problems in most areas.
These result in high costs per ton relative to surface mining.
A rapid expansion of underground coal output would also
require recruitment and training of new workers--a possibly
difficult problem.

On the utilization side, there have been severe limitations
on the utilization of high-sulfur coal, and low-sulfur coal
is in short supply. Moreover, much of the low-sulfur coal
is located in the West far from the big markets of the
Midwest and East. The use of equipment to clean sulfur from
the stack gases of coal-fired plants is a possible solution.
The technology is claimed to be workable by the Environmental

Protection Agency and unreliable by many electric utilities.
Both agree it will be expensive. In any event, it appears to
be the most nearly developed technology permitting use of
high-sulfur coal in conformity with air quality standards as
set before the Arab oil embargo.

It appears unlikely that the production of low heating value
gas from coal or various approaches to the use of coal other
than with stack gas cleaning technology will be available
and commercially viable until well into the 1980s and their
costs are very uncertain at this time.

The main uncertainties beyond the environmental ones in
projecting coal costs in constant dollars are the level of
productivity in underground mining and the extent to which
wages will rise faster than general inflation since wage-
related items are a major element of coal cost. Resource
depletion is not expected to be a major factor in rising
coal costs.

Coal costs and prices differ widely between areas, with type
of mining, and with coal quality. It is therefore not
particularly meaningful to give dollar per ton figures. The
estimates in the RFF study were for price increases in
constant dollars of up to 17 to 26 percent,depending upon the
region and type of mining.

The future cost of uranium from the mill, which corresponds
to the price of other fuels at the mine or wellhead, is
subject to widely differing views. To some extent, the
differences in view stem from uncertainty as to the effect
on uranium supply of the rapid growth in nuclear power which
will deplete known reserves of low-cost uranium rather
rapidly. It will also commit much of the additional low-cost

resources which the Atomic Energy Commission estimates will be found. However, these estimates are incomplete from a geological and geographic standpoint and some geologists believe that large additional amounts of low-cost uranium will be found as prices firm up.

Actually, from a total cost viewpoint, the cost of uranium is such a small portion of the total cost of nuclear power that a doubling or tripling of uranium prices, at which large additional resources exist, will have little effect on total power costs. Late in the century, the breeder reactor may make costs of uranium for new power plants even less important.

To summarize, domestic resources of all fuels are adequate to support and sustain major expansions of output for at least several decades.

The cost and price changes necessary to bring forth large increases in domestic supply will be heavily influenced by government tax, leasing, and regulatory policy. With no change in tax policy, and with market clearing prices for crude oil and natural gas, crude oil prices may well stabilize for a long period of time at between 50 and 75 percent above 1972 levels, in constant dollars. Average natural gas prices at the wellhead may stabilize at three to four times average 1972 prices in constant dollars. This level of prices is already being reached by prices for new gas not controlled by the Federal Power Commission.

The prospects are that the nation can be adequately self-sufficient in energy in 1985 at constant dollar prices which do not appear economically prohibitive and without complete disregard of environmental values.

THE ENERGY GAME AND THE ROLE OF UNCERTAINTY[*]

Arnold H. Packer[**]

A fundamental "fact" about energy is the great uncertainty that characterizes projections of the future. This fact and the market power of the players in the energy game limit the applicability of pure market theory to energy policy. A more appropriate model is that of players engaged in games under uncertainty.

The important players are the energy companies, the oil-exporting countries, and the industrialized oil-importing countries. Uncertainty over market structure will reduce the investment of the energy companies in risky ventures. Oil countries will consider oil-in-the-ground as part of their investment portfolio and therefore plan production as a function of the uncertain, expected appreciation rate. Oil-importing countries should treat energy as an inventory problem. The cost of carrying spare capacity would be balanced against the cost of a stock outage in an uncertain world. An inventory policy by the importing countries would mean that exporting countries face a high elasticity of demand and, thus, it would stablized world energy markets.

THE HAZARDS OF FORECASTING

Fifty years ago some experts announced that all the oil within the United States had already been discovered. Thirty-five years ago others predicted exhaustion of reserves by 1950. However, at the end of 1972, proven U.S. reserves were still 28 billion barrels on-shore, production in the lower 48 states was running about three billion barrels annually,

---

* The views expressed are not necessarily those of the officers, trustees, or other members of the Committee for Economic Development. Appreciation is expressed to Robert Kilpatrick and Frank Schiff for helpful comments on earlier drafts.

** Committee for Economic Development, Washington, D. C.

and another 335 billion barrels were estimated to be ultimately recoverable[1]. The find-rate of recoverable new oil (exclusive of the North
Shore or higher recovery rates on old fields) has averaged about one and
one-half billion barrels annually over the last 20 years[2].

In 1970, a high-level government task force predicted that: "Without
import controls the domestic well head price /¯of oil_7 would fall from
$3.30 per barrel to about $2.00 which would correspond to the world
price...we do not predict a substantial /¯price_7 rise in world oil
markets over the coming decade."[3]

However, when import controls were removed in April of 1973 the world
price was already rising. By the end of 1973, as price controls were
being removed, the controlled price on oil from old domestic wells was
well below the world price, which by that time was over $10.00 landed in
New York.[*]

The point is <u>not</u> that events often make forecasts look foolish, but
rather that any projection of the energy future is highly uncertain.
Uncertainty is the single most important "fact" about the energy crisis,
and dealing with it is the crucial policy issue.

Economic theory deals primarily with what may be called "uncertainty in
the small". In such cases, the theorems of "certainty equivalence" can
be applied because there are so many firms, buyers, and transactions that
uncertainty in individual cases can be ignored. This traditional economic theory is useful in analyzing the economics of oil exploration
because enough wells are drilled so that the law of large numbers
applies. Thus, though it is impossible to foretell accurately whether
any particular exploratory well will be successful, it is possible to
estimate the average amount of oil that will be found on-shore in the
U.S.; e.g., 400 barrels of oil found per exploratory foot drilled.[2]

---

* At this writing the controlled price of domestic "old" oil is $5.25 a
  barrel while "new" oil sells for $10.00.

The exploration part of the energy industry can be expected to respond
as normal markets do. An increase in the price of oil increases the
profit per well which will lead to an increase in exploratory drilling
and, given time, to an increase in the domestic supply of crude oil.
Similarly, higher prices will lead to the application of secondary and
tertiary recovery methods to existing fields so that a larger proportion
of the reserve will be exploited. In this sense, the oil business is
hardly unlike any other. But, in other important ways the differences
are striking. We turn next to the unusual uncertainties that affect
the oil market.

SOURCES OF UNCERTAINTY

<u>Geological Uncertainties</u>

The law of large numbers means that proven reserves will increase as a
function of exploration. However, the parameters are not sufficiently
well-known so that a world-wide supply curve can be drawn with any
certainty. The vast size of the Mideastern oil reserves was a happy
surprise while recent find-rates in the U.S. have been disappointing.

Immediate uncertainties are the cost and quantity of oil to be found in
the North Sea, Alaska, Russia, the China Sea, and in the Outer Continental Shelf of North America. On a world-wide basis, the estimated
total remaining recoverable oil is four times as great as the quantity
that is proven and currently recoverable. Moreover, proven reserves are
30 times current consumption levels.[1] *

However, a reserve-to-production ratio of 30:1 is not as much of a
cushion as it sounds. For one thing, almost two-thirds of the Free
World's proven reserves are in the Mideast. For another, reserve-to-
production ratios of one to ten or so must be maintained if economic oil
flow rates are to be achieved. Finally, consumption has been increasing
at a rapid rate, more than five percent annually over the last decade.

---

* These estimates were made prior to the price escalation that took
  place subsequent to October 1973.

<u>New Technology Uncertainties</u>

The possibilities for additional resources beyond those noted above are enormous. New exploration techniques may greatly improve the find-rate. Beyond that, even the "ultimately recoverable" crude oil figures are based on recovery rates of less than 40 percent. Only one-third of the expected addition to domestic lower 48-state oil reserves is likely to come from new oil; the rest is expected from improved recovery methods. Uncertainty about recovery rates translates into substantial uncertainty about how much oil will be available. And no one knows what the new oil prices will do to recovery-rate technology.

In summary, there is considerable uncertainty about the supply curve for conventional crude oil. The supply of oil from non-conventional technologies is also highly speculative. There are shales and tar sands and heavy oil which can supply oil in enormous quantities. Estimated recoverable Western Hemisphere oil from these resources exceed two trillion barrels.[1]

In addition to these vast potential sources of oil, there are even greater energy resources in other forms -- coal, natural gas, uranium, solar, and geothermal. Yet, because energy investments are lumpy, technological problems are difficult, and institutional obstacles (e.g., the provision of ample water) important, the supply situation is highly uncertain.

<u>Market Uncertainties</u>

The technological uncertainties have the strongest influence on the long-range energy picture, and the geological uncertainties color the mid-range (5 to 15 years) picture most heavily. But, over the shorter run, uncertainties over the market structure and behavior of the oil-exporting countries are more significant.

The cost to extract oil from Mideastern fields has been estimated at less than 50 cents a barrel, which could translate into a price, landed in New York, of $1.50. Five years ago the landed price was close to

$2.00 and this price is the minimum that could be expected. Though few
consider it likely now, some eminent economists think a price of $2.00
is possible and even suggest that imported oil would now be $2.00 a
barrel if appropriate policies had been followed.

The argument is not that $2.00 a barrel oil is likely but that the
possibility of such an outcome cannot be ignored. This outcome might
require a substantial political change in the Mideast and major discoveries elsewhere. However, neither of these is unthinkable.

On the other hand, because recent find-rates are diminishing and the
cohesiveness of the Oil Producing and Exporting Countries (OPEC) is
increasing, there are many who think the world price of oil will continue to increase and that, for 1980, a price of $20.00 a barrel is much
more likely than $2.00.

Thus, decision makers might have a Bayesian probability distribution for
the 1980 price of oil in which $2.00 and $15.00 would be two standard
deviations away from a mean of, say, $8.50 per barrel. That is a great
deal of uncertainty relative to the price expectations for other
important commodities. Moreover, oil will be sold for a price that will
depend on expectations as well as realized events. What is believed
will be as important as what is true. Investors face the threat that
exporting countries will reduce prices as far as is necessary (down to
the $2.00 floor) to undercut the price of competitive fuels from shale,
coal, or other sources. Therefore, they will be reluctant to invest
without guarantees.

PLAYERS IN THE ENERGY GAME

The degree of uncertainty and the importance of individual actions in
energy decisions suggest that a model of game-playing describes the
situation better than do models of competitive behavior. The energy
game has the following four groups of actors, each with its own set of
instruments and objectives.

Group A is the oil companies and the remainder of the private energy
industry. Their objective is to stay in business and maximize profits.
Their primary instrument is the capacity to invest in exploration,
development, production, and distribution of energy materials.

Group B is the exporting countries or OPEC. Their objectives are the
stability of their regimes and economic development, and their (not quite
separate) instruments are the price and quantity of the oil they will sell.

Group C is the government of the industrialized oil-importing countries,
the OECD countries. Their objectives are to obtain dependable energy
supplies at a minimum financial, environmental, and geopolitical cost.
Their instruments are research and development of domestic sources of
energy, influence -- via taxes, tariffs, and controls -- on total and
imported energy consumption, and their ability to affect the stability
and development objectives of the exporting countries.

Group D is the governments of the less-industrialized oil importing
countries. Their objective is to avoid a balance-of-payments squeeze
arising from higher energy costs combined with a reduction in trade and
aid. The reductions are likely as the industrialized world seeks to
concerve foreign earnings so as to finance its own energy purchases. At
present, these less fortunate countries have no powerful instruments.

The three powerful groups -- the energy companies, the governments of
the oil exporting countries, and the governments of the industrialized
oil consuming countries -- are engaged in the energy game. The outcome
will affect the game's observers, the peoples of the world who are the
most numerous and least powerful of all those involved.

THE COMPANIES: UNCERTAINTY AND INVESTMENT DECISIONS

Risk, in economic theory based on uncertainty in the small, only means
that a risk premium must be added to the pure interest rate to determine
the rate of return required on uncertain investments. The assumption is
that the loss on a few bad investments will not make the investor insol-
vent before the "law of averages" comes into play. However, when

investments are very large and lumpy, the law of large numbers may not apply to a single firm. In this instance, even investments with a high expected value will not proceed.

The existence of large companies with great financial resources and consortiums of large companies for special ventures has traditionally made it possible to spread the risk sufficiently so that investment would go forward. However, uncertainty has been more pervasive in recent years. In addition to the externally caused uncertainties mentioned previously, there has been a series of regulatory uncertainties. These include regulating the price of natural gas (at the well-head and the city gate) for many years and controlling the price of petroleum products during the last few years. In addition, air, water, and safety regulations covering the operation of energy facilities and the use of fuels have recently been in a state of flux. Finally, uncertainty surrounding the siting of energy facilities -- including extraction, ports, refineries, power plants, and the Alaskan pipeline -- has made planning difficult.

The high degree of uncertainty has led to a short-fall in energy facilities. For example, exploration on the Alaskan slope was delayed pending resolution of the pipeline issue and refinery construction has been postponed in light of uncertainties about availability of crude oil and environmental standards.

The effect of uncertainty (for a given expected value) is thus to impede the energy industry's investment in exploration and production facilities. In some cases uncertainty may completely prevent development. For example, the knowledge that Mideastern oil can be profitably sold for substantially less than $5.00 a barrel can prevent the construction of synthetic fuel plants that could produce oil at $5.00 a barrel. Moreover, if the threat is successful, the imported oil need never actually be sold for that low a price.

THE HAND OF THE EXPORTING COUNTIRES: UNCERTAINTY, PRICE, AND QUANTITY

The governments of the exporting countries seek to use their oil reserves to develop their country's economy and to maintain their own position in

power. Thus, the finance ministers or portfolio managers of these countries will want to know the production rate that will maximize development
for a petroleum exporter. Assume that a schedule of domestic projects is
available, that those projects can be listed in order of their rate of
return, and that the exporting country can invest in foreign assets.
Then a marginal efficiency of capital (MEC) curve can be drawn. (The
construction of this curve has its own uncertainties, but those familiar
to economic theory). The curve will slope down to the right and, for
most countries, will become horizontal when the rate of return of the
next internal development project is so low that it becomes equal to the
return from foreign investments. That is, once there are no projects
yielding, say a 10 percent return, any further petroleum earnings should
be invested in foreign assets returning that rate.

For Saudi Arabia, however, the funds after internal projects providing
more than a 10 percent return may be so large that the external foreign
investment line may slope downward also. In other words, there may be
enough funds to depress the world-wide return on capital. Thus, the MEC
curve will have a kink at the intersection between internal and external
uses for the capital.

The MEC curve represents the countries' demand-for-capital schedule. For
most oil exporting countries, the supply of capital is their net earnings
from oil; i.e., quantity, time, price less the extraction cost. The cost
of extraction follows the typical U-shaped marginal cost curve. If there
were also no positive real interest rate, if oil in the ground maintained
a constant value, and if demand were completely inelastic, each producer
would set his production at the lowest-cost point of the marginal cost
curve. Production would be determined by engineers and not finance
ministers, and countries would produce at the minimum-unit-cost quantity
each year until the oil ran out.

What happens when each of these three assumptions is relaxed? If there
is a positive real interest rate, each producer is encouraged to produce
more than the minimum-unit-cost production quantity. Production will be
increased until the marginal cost of producing an additional barrel this

year instead of next is equal to the present value of the income stream
generated by the next dollar of investment. Thus, a country whose development plans establish a high demand-for-capital curve will be encouraged
to produce more and higher-cost oil now at the expense of future production.

Next let us relax the assumption that oil prices do not change. If oil
in the ground appreciates, then the marginal cost of producing now
instead of later must be evaluated in terms of the difference between
the rate of return on the next investment and the rate that oil in the
ground appreciates. A prospective change in oil prices means that this
asset must also be considered in drawing the MEC curve. The OPEC finance
minister who is trying to achieve an optimum portfolio must now determine
how much of the national wealth should be invested in internal development, how much in foreign assets, and how much should be kept in oil.

The portfolio question need not be answered by every oil producing
nation. The appreciation rate for oil will have to be very high to
interest nations who face a very high capital demand curve. It will be
obvious, in these cases, that the optimum portfolio is the one that
finances the maximum number of development projects. Thus, the Iranians
tend to push production to the point where their production cost curve
rises sharply.

However, relative rates of return are critical to the production plans
of Saudi Arabia and Kuwait. If the expected rates of return on capital
and on oil in the ground are equal, then there is no incentive to produce
this year instead of next. The optimum production level in this instance
is the minimum point of the cost curve -- just as it was when there was
no real interest rate. (Note, however, that if the nominal interest
rate were ten percent, the price of oil would have to double every seven
years to provide an equivalent yield.)

The next-to-last assumption to be relaxed is the perfect inelasticity of
demand. The interesting case is those countries that face the portfolio
question; e.g., Saudi Arabia. Elasticities are crucial to the Saudi

production and pricing decisions because this country supplies such a
large portion of the world's demand that it is likely to be the swing
producer. That is, Saudi Arabia may absorb a substantial portion of
the marginal changes in world demand.

Because the demand elasticity of oil is considerably less than one, a
production cutback will mean a revenue increase to all producers combined.
For example, adding $4.00 to the price of crude oil adds less than 10
cents to the price of gasoline. Thus, recent doubling of crude prices
may mean only a 25 percent increase in U.S. gasoline prices and 10 percent increases in Europe and Japan. The impact on world-wide demand for
crude might, thus, only be 10 percent. However, if Saudi Arabia and a
few of their neighbors are the marginal producers and they produce 20
percent of the market, then <u>their</u> demand might be cut in half by a
doubling of price. (Thus, the fragility of cartels.)

Moreover, there are positive supply elasticities for both conventional
fuels and synthetics. A substantial increase in oil produced outside the
Middle East would erode the position of the swing producers. Therefore,
successful revenue-increasing production curtailment is dependent on the
cohesiveness of a cartel with many actors.

The final assumption to be discarded is the absence of uncertainty. The
producing countries must plan their strategies in face of uncertainty
about demand and supply elasticities, including the uncertainty about
find-rates for conventional fuels and for technical fixes. More generally, there are uncertainties about the effect of their production cur-
tailments or price increases on the expected appreciation rates for oil
in the ground. In other words, the mapping from their policy instruments to their objectives for future revenues is unclear. However, some
general characteristics of the mapping function can be predicted.

If the future price of oil is determined by technological progress then
price appreciation taken today means that less will be available tomorrow.
In these circumstances, current price increases reduce the expected
appreciation rate. Moreover, if an increase in the price charged by the

Saudis induces the development of oil reserves and oil substitutes elsewhere, that development will also reduce the expected appreciation rate.
Note that if future alternatives to Mideastern oil depend on current
prices, then the producer countries' optimum price will be less than the
short-term market clearing price.[*] Because price is not constrained by
costs and the elasticities of supply and demand are unkown, price and
quantity need not uniquely determine each other.

It is easy to imagine the following scenario of instability. The Mideast
begins a process of raising their price thereby creating an anticipated
high rate of appreciation which, initially, discourages current production. However, as the price rises, other producers and substitute
fuels enter the market until the expected future price is less than the
current price. Appreciation becomes negative and oil-in-the-ground
becomes an unattractive asset. As a result, production is encouraged,
and prices fall further. In this situation, there is rush to unload and
a glut develops. High priced sources would then be eliminated and, as
supply diminishes, the cycle begins again.

Thus, uncertainty itself influences the price charged and quantity produced by the exporting countries. Greater certainty about future prices
means more stable production plans.

ENERGY CONSUMING NATIONS

The governments of the energy-consuming nations represent the third
potent player in the energy game. It would seem that as long as the
international oil companies were substantially stronger than the oil
exporting countries, the consuming country governments could play a
passive role. However, now that the power relationships have changed,
their role must change also.

An examination of the reasons for the increase in the relative strength
of the oil-exporting nations may indicate what the new policy should be.

---

[*] For example, the producer countries may currently have set prices
well above the optimum from their own best interest.

The dramatic change of the last few years is the result of the substantial reduction in the industrial world's short-run elasticity of demand for Mideastern oil. The end of the surplus in domestic U.S. oil production capacity meant that the industrial world became dependent on Mideastern oil. Therefore, the Arab nations could extract a high economic and political price for their product.

A successful U.S. and OECD strategy would have maintained a high elasticity by treating energy as an inventory problem. Then safety stocks would have been maintained that balanced the cost of the inventory against the risk and penalty of a short-fall. Some of the stock would have been in final form (e.g., oil in tanks and salt domes), some in intermediate form (e.g., shut-in wells and reserve fields), and some as spare capacity (e.g., synthetic fuel industries). The quantities at each stage would have been a function of the time required to make the oil in the next inventory stage available for use.

This policy is quite different from establishing rigid import quotas. In fact, in past years when Arab oil was cheap we could have developed the inventories at low cost. An inventory policy designed to create independence via a high short-run elasticity would have avoided the use of import quotas and depletion allowances. Environmental protection laws that increased dependence on imported oil could not have proceeded until further stockpiles and standby capacity were available. Thus, a delay in the Alaskan pipeline or the leasing of off-shore oil properties would have meant a mandatory delay in implementing the Clean Air Act <u>or</u> a reduction in consumption.

These actions would have increased the elasticity of U.S. demand for Mideastern oil. At the same time we could have decreased the elasticity of the Mideast's demand for oil earnings by putting the following arrangement in place: We could have eliminated the quotas and placed a tariff of, say, $2.00 a barrel on Arab oil. The proceeds of that tariff could have been used to finance development projects throughout the Arab world (in the way that the proceeds of Iran's oil sales finance Iranian projects.) The projects would have naturally made the Arab world more

dependent on Western technology as well as on the continuation of the
oil trade. This fund would have increased the Arab demand schedule for
oil revenues. More importantly, the internal position of the rulers
would have been enhanced, rather than threatened, by cooperation with
the western nations.

The objective of the U.S. and other oil importing nations should be to
minimize the production, environmental, and geopolitical costs of obtaining a dependable supply of energy. Moreover, the consumption of energy
should be such that the social benefits of using the marginal BTU would
be equal to the social cost. Uncertainity will inevitably add to these
costs. For example, undependable imports may lead to import restrictions.
This, in turn, may lead to extracting oil from shale which will have
higher economic and environmental costs than available imports. The
optimum policy will minimize the uncertainty cost by developing appropriate inventories. This will stabilize the world price for oil and
reduce uncertainity for all consuming and exporting countries.

REFERENCES

1. H. E. Linden and I. D. Parent, "Analysis of World Energy Supplies."
    Presented before Conference on Energy: Demand, Conservation, and
    Institutional Problems, Massachusetts Institute of Technology,
    Cambridge, Mass., February 12-14, 1973.

2. National Petroleum Council, U.S. Energy Outlook: An Initial
    Appraisal 1971-1985, Washington, November 1971.

3. The Cabinet Task Force on Oil Import Control, The Oil Import
    Question, Washington, February 1970.

SESSION XII

PANEL DISCUSSION: RECOMMENDED STRATEGIES
FOR MINIMIZING THE ENERGY "CRUNCH":
1975 - 1990 AND BEYOND

Chairman: George W. Morgenthaler
            Corporate Director of Research &
            Development, Martin Marietta Corporation,
            Denver, Colorado

PANEL DISCUSSION[*]

RECOMMENDED STRATEGIES FOR CONTROLLING THE

ENERGY "DELTA": 1975 - 1990, AND BEYOND

PANELISTS:

Prof. Samuel Z. Klausner Sociologist University of Pennsylvania	Dr. Roland Schmidt Cryogenic Physicist General Electric Research and Development Laboratories

Mr. Walter S. Sullivan
Science Editor
New York Times
	Dr. John Andelin, Jr.
	Physicist, Administrative
	    Assistant to Congressman
	    Mike McCormack

Dr. Chauncey Starr Electrical Engineer President of Electric Power Research Institute	Dr. Edward Teller Nuclear Physicist University Professor University of California System

MODERATOR

Dr. George W. Morgenthaler
Corporate Director of Research
and Development
Martin Marietta Corporation

---

* Abstracted from tapes by Dr. Aaron Silver, Martin Marietta Corporation.

PANEL DISCUSSION

DR. GEORGE W. MORGENTHALER, MODERATOR

In calling the panel to order, Dr. Morgenthaler reviewed the strategies
that had been suggested during the 3-day symposium for reducing the energy
demand, and for increasing energy supply. Demand reductions included
energy conservation in transportation, buildings, industrial processes,
and changed life style. Increasing the supply included consideration of
the impact of new technologies in developing new energy sources and the
incentives which might be given to such development in the economic arena.
The theme of the discussion centered about the question, "what can we do
in the near term and long term to minimize the energy "crunch"?

After appropriate introduction of the respective panel members, Dr.
Morgenthaler presented the format of the panel discussion, namely:

1. Each panelist would make an opening statement concerning the most
    cogent advice he could offer.

2. There would be time for the panelists to engage with one another,
    and to challenge each other's views.

3. There would be a brief period of questions from the audience.

  I. DR. SAMUEL KLAUSNER, UNIVERSITY OF PENNSYLVANIA

Dr. Klausner focused on specific criteria against which the appropriate-
ness of relative strategies for meeting the energy crisis could be assessed.
These were:

1. That energy consumption policies will fail if they take the energy or
    energy source as the object of the policy. The policy should aim
    instead at the development of behavioral factors which influence the
    consumption side of the energy equation. For example, a rationing
    policy for quantities of fuel is a policy for rationing "rights" in
    the fuel and this could affect education if a school fuel ration were
    inadequate. Thus, policies should be directed toward the social

situation in which the need for consumption arises, and should not
only be concerned with the rules of control of materials. Future
energy policies will turn out to be community development policies,
transportation policies, industrial policies, etc., the object being
better behavioral responses. Otherwise, pressures will arise to
break the law, e.g., form a black market.

2. Energy policies in the 1990's will increasingly become a matter of
    political regulation, and decreasingly a matter of economic market
    decisions. This phenomenon is not peculiar to the energy situation,
    but arises from the political polarization of our society. One
    implication is that resource depletion will not be controlled by rising
    prices, but resource allocations will be influenced by the distribution
    of power. Another implication is that the commercial feasibility of
    innovation and introduction of new technologies will be less of a
    factor in the 1990's than it is today. This is because the industries
    themselves, which previously dealt with the consumer market, will
    redirect their efforts to larger community development areas, which
    focus on people's needs.

II. DR. ROLAND SCHMIDT, GENERAL ELECTRIC COMPANY

Dr. Schmidt characterized the supply side of the U.S. energy system as a
complex and intricate network of oil, gas, coal, hydro, and nuclear
technologies. The existence of this network implies that there is no
"pure" strategy for solving the energy problem. Dr. schmidt further
suggested that we should fully explore the R&D options available. He
cautioned against the pitfall that technically intricate proposals for
energy supply tend to appeal to technical people, and that we must be
able to prove the cost-effectiveness of some of these new concepts.
Specifically, he advocated the gradual evolution of currently installed
systems rather than aiming our efforts only at developing revolutionary
new sources.

Dr. Schmidt advocated that we attack the "software" problem as well as
the hardware problems. This includes the social habits of people, i.e.,

conservation efforts, the initiation of energy saving programs by industry, the effective formulation of national energy policies, and the implementation of these by statutes and regulations. He concluded that the two basic questions yet to be answered are:

1. How much can the energy "delta" be reduced?
2. How soon?

III. MR. WALTER SULLIVAN, NEW YORK TIMES

Mr. Sullivan drew several analogies between a biological system (such as the lemming cycle) that overloads its environment, and the corrective measures that must be taken by human counterparts if disaster is to be avoided. Specifically, he developed the theme that man must recognize the consequences of the profligate use of energy and must now hasten to live in hormony with his environment. This implies that ingenuity and technical innovation need be brought to bear upon these problems.

Mr. Sullivan recommended utilization of systems analysis techniques to plan for contingencies and pose the "what if" questions.

IV. DR. JOHN ANDELIN, WASHINGTON, D.C.

Dr. Andelin stated that the energy crisis was definitely real, and called for an assessment of the nation's technology, manpower, and energy resources available to meet the crisis. He advocated a fundamental systems approach to evaluate changes in the supply and demand equation, and an open communications system to disseminate this information to cognizant industrial and government groups.

He advocated the allocation of resources, against some priority scheme in conjunction with the development of an interdependent effort based upon moral, economic, and societal imperatives. Dr. Andelin's scenario consisted of initially a conservation effort, and then the massive exploitation of coal and nucelar energy, coupled with increased exploration of oil and gas. Finally, he would then phase in the development of oil shale, geothermal, solar energy sources, and nuclear breeder reactors.

V.   DR. CHAUNCEY STARR, ELECTRIC POWER RESEARCH INSTITUTE

Dr. Starr agreed with the statements of previous panelists, particularly that of Dr. Schmidt, that the "software" issues are critical in solving the near-term energy crisis. In this respect, he advocated conservation as the first step, which may be achieved by changes in society's life styles and values.

Dr. Starr focused upon technology impacts for both the near (to 1985) and far (1985-2000) term. Specifically for the near term he argued:

1. Expansion of nuclear power is easiest and fastest, and may be accomplished without major economic dislocation.

2. Expansion of coal is not especially simple nor easy because the operation is not fully in place and more reserve is required; the social and environmental costs may be excessive.

3. The exploitation of oil may result in the maximum economic dislocation.

Commenting upon the far-term solution, Dr. Starr declared:

1. Technology lead time is a major issue and we must begin early to solve major problems.

2. Sociological research on the public acceptance of new technologies is needed. For example, the risks and trade-offs between new and old technologies must be assessed.

3. The requirements for adequate resources may determine the direction of the options and alternatives inasmuch as large investments already made act as a deterrent to change, and this precludes already limited alternatives.

VI.   DR. EDWARD TELLER, UNIVERSITY OF CALIFORNIA

Dr. Teller stated that the major impact of the energy crisis falls upon Europe and Japan, and in particular, upon the undeveloped nations. The crisis itself is not only one of energy, but also the danger of starvation due to a shortage of gas for the manufacture of fertilizer. He focused on the main areas:

1. Oil - the tapping of oil by new methods of changing its viscosity.

2. Solid fossil fuel - in situ methods for extracting the oil from shale deposits.

3. Electricity - the development of nuclear reactors to be pushed energetically.